CARBON IN THE GEOBIOSPHERE
— EARTH'S OUTER SHELL —

TOPICS IN GEOBIOLOGY

Series Editors:
Neil H. Landman, American Museum of Natural History, New York, New York, landman@amnh.org
Douglas S. Jones, University of Florida, Gainesville, Florida, dsjones@flmnh.ufl.edu

---

Current volumes in this series

For detailed information on our books and series please visit:

www.springer.com

Carbon in the Geobiosphere
— Earth's Outer Shell —

FRED T. MACKENZIE
Department of Oceanography,
School of Ocean and Earth Science and Technology,
University of Hawaii, Honolulu, U.S.A.

ABRAHAM LERMAN
Department of Geological Sciences,
Northwestern University,
Evanston, Illinois, U.S.A.

 Springer

A C.I.P. Catalogue record for this book is available from the Library of Congress.

ISBN-10 1-4020-4044-X (HB)
ISBN-13 978-1-4020-4044-3 (HB)
ISBN-10 1-4020-4238-8 (e-book)
ISBN-13 978-1-4020-4238-6 (e-book)

Published by Springer,
P. O. Box 17, 3300 AA Dordrecht, The Netherlands.

www.springer.com

Printed on acid-free paper

Aims & Scope Topics in Geobiology Book Series

Topics in Geobiology series treats geobiology — the broad discipline that covers the history of life on Earth. The series aims for high quality, scholarly volumes of original research as well as broad reviews. Recent volumes have showcased a variety of organisms including cephalopods, corals, and rodents. They discuss the biology of these organisms — their ecology, phylogeny, and mode of life — and in addition, their fossil record — their distribution in time and space.

Other volumes are more theme based such as predator-prey relationships, skeletal mineralization, paleobiogeography, and approaches to high resolution stratigraphy, that cover a broad range of organisms. One theme that is at the heart of the series is the interplay between the history of life and the changing environment. This is treated in skeletal mineralization and how such skeletons record environmental signals and animal-sediment relationships in the marine environment.

The series editors also welcome any comments or suggestions for future volumes;

Series Editors:
Douglas S. Jones dsjones@flmnh.ufl.edu
Neil H. Landman landman@amnh.org

Preface

What is the importance of carbon on Earth? Why at this time devote a book to its discussion? Carbon is essential in all biological processes and in a great variety of mineral and inorganic processes near the Earth's surface. Carbon is the chemical element that is the main building block of living organic matter, and no evidence has been found so far of the existence of non-carbon-bearing life forms. Carbon in the atmosphere and waters is critically important to plants that make their organic matter from carbon dioxide and water and produce oxygen in the presence of light and the nutrients nitrogen, phosphorus, and bioessential trace elements, in a process known as photosynthesis. In the absence of light, where life exists below the Earth's surface and in the deep ocean, carbon dioxide is also involved in the production of organic matter. In addition, carbon dioxide in the atmosphere acts as a greenhouse gas that helps maintain temperatures suitable to life and life processes as we know them and an equable climate. Furthermore, carbon dioxide dissolved in rain water makes an acidic solution which, when falling on land, leads to the weathering of minerals and rocks, erosion of land, and transport of dissolved and particulate materials to the ocean.

We chose for the title of this book *Carbon in the Geobiosphere* to convey the view that the naturally occurring inorganic and organic forms of carbon and, in particular, carbon dioxide always played an environmentally important role within that part of the planet Earth that is the home to all living organisms, collectively known as the biosphere. Most of the plants and animals in the biosphere inhabit the region of the Earth's surface that is affected by processes occurring beneath as well as above it. As the book focuses on the integrated, interactive system of the atmosphere, oceans, continental waters, living biota, soils, minerals and organic matter in sediments, and near-surface and deeper interior, we use the term *Outer Shell of the Earth* as extending from the surface to the base of the lithosphere, at a mean depth of -150 km, and to about $+30$ km above the surface, a layer that contains most of the mass of the atmosphere. This shell comprises the geobiosphere with the lower atmosphere, continental, subsurface, and oceanic waters, the biosphere, and the sedimentary and crystalline lithosphere that contain carbon or react chemically with its species in the gaseous phase and natural waters. Among all the organisms, from bacteria to plants and animals, the humans have so far been the strongest and fastest agents of environmental change that affects the geobiosphere. The human species as a geologic agent modifies the Earth's surface environment not only through emissions of gases from the burning of fossil fuels, but also through numerous industrial and agricultural activities that lead to changes in the Earth's vegetation cover, the atmosphere, the quality of soils and groundwaters, concentration of various metals in the geobiosphere, changes in the water regime of rivers and lakes, and a variety of chemical and biological changes that are broadly known as pollution. Such human-induced changes affect the cycles of carbon and other nutrient elements linked to it that are of fundamental importance to life on Earth.

A great deal of attention has been paid in recent years to issues associated with the accumulation of carbon dioxide in the atmosphere, sources of emissions of the gas to the atmosphere and its removal processes, and the future time course of accumulation under different scenarios of growth of the human population and its use of fossil fuels, forest materials, and other natural resources. Human concerns about global warming have become a part of everyday discourse on how increasing carbon dioxide levels might affect the future climate of the world and what are the possible environmental, social, and economic implications of climatic change for human societies worldwide and for different nations of the world. This attention has led to a strong growth and diversification in scientific research and literature on the history and present situation regarding carbon dioxide in the Earth's atmosphere, and a search for reliable indicators in the geological record of past atmospheric carbon dioxide concentrations.

The issues of carbon dioxide in the atmosphere and its potential effects on the global climate represent changes that are gradual but concern all of humankind. Thus it is understandable that responses to the carbon dioxide accumulation in the atmosphere and its future consequences are, and will be, driven to a large extent in the near future by economic, social, and political considerations of societies in different regions of the world. However, for humans to be informed about the basic environmental issue of global warming, it takes considerably more insight into the functioning of the planet Earth than can be gained from reading the popular press, or from other news media, to understand why and how the concentration of carbon dioxide in the atmosphere—that is a relatively very small reservoir of carbon on a global scale—has been changing in the course of Earth's history, and what were and will be the consequences of such changes. The time scale of anthropogenic emissions of carbon dioxide from the increasing consumption of wood and natural biomass, coal, oil, and natural hydrocarbon gases includes less than three centuries of the present industrial age. The effects of these emissions on the chemical composition of the atmosphere and other processes taking place on the Earth's surface can be measured against the longer-term changes that have been occurring in the atmosphere, and the water, soils, rocks, and biota exposed to it.

In considering the geobiosphere and the Earth's outer shell, we provide in Chapter 1 of this book an overview of its structure at scales ranging from the Earth's dimensions to the atomic dimensions of the constituents of the atmosphere, water, and crustal rocks, and also give an outline of the global carbon cycle, its major reservoirs, and the fluxes between them. Because of the importance of fossil fuels to our technological society, the occurrences of these forms of carbon are also reviewed in the chapter. It is emphasized that the global cycles of carbon and many other elements are driven by biological and inorganic geochemical processes. The term biogeochemical cycle that describes the environmental reservoirs and flows of materials between them reflects the joint role of these processes in the environment.

The long period of about 600 million years since the formation of the Earth to the first records of life is addressed in Chapter 2. The primordial atmosphere of the Earth was largely a product of degassing of the Earth's interior, its chemical composition was very different from that of the present-day atmosphere, and its temperature and pressure were likely to have been considerably higher early in Earth's history. We consider in Chapter 2 the early stages of the chemical composition of ocean water that was to a large extent controlled by dissolution of hydrogen chloride, carbon dioxide, and hydrogen sulfide into the hydrosphere and its subsequent reactions with the early lithosphere of the Earth. We also address the question of the occurrence of the gases methane (CH_4), as a chemically reduced form of carbon, and its oxidized form, carbon dioxide (CO_2). The formation of living organic matter by chemical synthesis and, later, by photosynthesis signals the end of the prebiotic stage of Earth's history, and a discussion of the basic chemical reactions that lead to the formation of organic matter concludes this chapter.

In Chapter 3 we address the fundamentals of the thermal energy balance between the incoming solar radiation and the Earth's surface, and the greenhouse effect of the atmosphere that is responsible for a warmer Earth's surface than it would be without the atmosphere. We also point out that the term "greenhouse effect" is something of a misnomer, despite its well-rooted and widely accepted usage everywhere. As a reminder of the historical background, we discuss an early treatment of the greenhouse warming of the Earth's atmosphere by Milutin Milankovitch in 1920–1930, and conclude the chapter with a summary of the recent evidence of the radiation forcing and temperature increase projected for the 21st century, the so-called enhanced greenhouse effect.

The chemical and mineralogical properties of the three common carbonate minerals of calcite, aragonite, and dolomite are discussed in Chapter 4. The chapter describes the mineral characteristics of pure and magnesium-containing calcites, pure and strontium-containing aragonites, and dolomites with slightly variable calcium and magnesium content. The chapter also discusses the essentials of the solid solution theory, and the theoretical and experimental treatment of the dissolution and precipitation rates of these minerals at the conditions approximating those of the Earth's outer shell. Because much of the present mass of limestone in the sedimentary record was formed by marine organisms as skeletal parts and shells, the biogenic and non-biogenic occurrences of these minerals, as well as the relevance of this information to other processes on Earth, are also briefly discussed in this chapter.

The behavior of inorganic carbon as CO_2 and other carbonate ionic species in fresh and ocean waters is treated in Chapter 5. We discuss the implications of the fact that in modern times, part of the CO_2 from industrial and land-use emissions to the atmosphere accumulated in the ocean. We provide calculations for the average thickness of the surface ocean layer that absorbed this CO_2, showing its dependence on how the carbonate equilibria respond to changing environmental conditions and turbulent mixing and water transport in the ocean. We also consider the implications of the fact that the rate of calcite dissolution is related to the degree of carbonate mineral saturation and that although surface seawater is everywhere supersaturated with respect to calcite, calcite dissolves at depth in the oceans because of decreasing saturation state due to increasing internal P_{CO_2} and pressure with depth. Considering the extensive research activity behind the issue of possible storage of industrial CO_2 as liquefied gas in the deep ocean, we discuss in this chapter the properties of the CO_2 liquid and hydrate phases in the environment of ocean water. We also show that because CO_2 forms in the precipitation and storage of mineral $CaCO_3$, it is a possible source of emission of CO_2 gas to the atmosphere from ocean water, where the outward directed flux might be in part counteracted by biological primary production and storage of organic carbon in sediments.

The isotopic fractionation of the two stable isotopes of carbon, ^{13}C and ^{12}C, is treated in Chapter 6. The equilibrium fractionation is discussed for the inorganic system of gaseous CO_2, aqueous carbonate species, and mineral $CaCO_3$ at the environmental conditions of the Earth's surface. The non-equilibrium or kinetic fractionation occurs in the photosynthetic production of organic matter and other biologically mediated reactions. The $^{13}C/^{12}C$ fractionation is considered the first evidence for the occurrence of life on the early Earth, and it is also an indicator of the origin of various organic and mineral substances. In addition, the geologically established long-term record of the isotopically fractionated carbon in limestones and sedimentary organic matter provides bounds on the origin and evolution of the biosphere and the history of oxygen accumulation in the atmosphere.

Chapters 7 through 9 deal with three important parts of the global carbon cycle. First, in Chapter 7 we discuss the storage of carbon in sediments and the recycling that returns it to the atmosphere, biosphere, and ocean, as well as the sedimentary history of carbonate rocks that constitute about 80% of the mass of sedimentary carbon. The carbonate sedimentary record, its mineral composition, and isotopic ratios of $^{13}C/^{12}C$ and $^{18}O/^{16}O$ provide a picture of long-term trends of evolution of ocean water and atmosphere that is supplemented by the isotopic evidence of strontium ($^{87}Sr/^{86}Sr$) in limestones, sulfur ($^{34}S/^{32}S$) in the sulfate minerals of evaporitic rocks, and the chemical composition of fluid inclusions in NaCl crystals in evaporites. This chapter also begins to address the evolving carbon cycle since Hadean time. Next, Chapter 8 focuses on the dissolution of sedimentary and crustal rocks in mineral reactions with acidic, CO_2-containing atmospheric precipitation and groundwaters. This process is the first step in the transport of dissolved and particulate materials from land to the ocean by the release due to weathering of the main constituents of river waters. The weathering rates of the individual rock types that produce the global average river water, relative stabilities of rock-forming minerals, and consumption of CO_2 in weathering are also discussed in this chapter, where we emphasize the importance of the CO_2 flux in weathering in the global carbon cycle. The third part, addressed in Chapter 9, is the processes in the oceanic coastal margin that corresponds roughly to an area equivalent to about 8% of the ocean surface area. In this region there occurs at present a significant delivery of dissolved and particulate materials, an important part of oceanic primary production, storage of land-derived carbon, and calcium carbonate sequestration. The important processes in this relatively small area of the global coastal ocean are the remineralization of organic matter produced *in situ* and transported from land, the production of $CaCO_3$ minerals of different solubilities (calcite, aragonite, and magnesian calcites of variable magnesium content), and the changes in biological production and calcification that are expected due to the rising atmospheric CO_2 concentration that lowers the degree of saturation of ocean water with respect to the different carbonate minerals ("ocean acidification"). Combinations of these factors and physical mixing of the water masses determine the role of the coastal ocean as either a source or sink of atmospheric CO_2.

Building on the preceding chapters, Chapter 10 presents the major aspects of the evolving carbon cycle since geologically distant to more recent time. The chapter starts with the earliest Eon of the Earth's history, the Hadean, and ends in the Pleistocene and Holocene Epochs, and the time of the last glacial to interglacial transition. This chapter draws on the simple picture of the carbon cycle that began in Chapter 1 and the primordial Earth, discussed in Chapter 2, expands it from the material given in the succeeding chapters, and presents our thoughts on the main driving mechanisms of the carbon cycle and the evolution of the ocean-atmosphere-biosphere-sediment system through geologic time.

Chapter 11 discusses the global carbon cycle in the Anthropocene, the period of the industrial age that started about 250 years ago. Because the Anthropocene is very likely to continue into the future, the chapter also considers the last two to three centuries of environmental change in terms of what might lie ahead. The effect of the carbon cycle on climate is only one important aspect of this global change. The human-produced perturbation of the global carbon cycle by the burning of fossil fuels and emissions from land-use has far-reaching effects on the linkages between the carbon cycle and the nutrient nitrogen and phosphorus cycles that are its drivers.

The literature dealing with the past, present, and future of carbon in the geobiosphere and the Earth's outer shell is very extensive. We attempted to cite those publications that bear directly on the ideas, factual material, and data in the diverse topics that we discuss in the book, but even the more than 700 bibliographic entries, from the 1840s to the present, are only an incomplete list. Any errors of omission in the choice and coverage of the subject matter and in the recognition of other authors' contributions in the different fields are entirely ours.

The past and the future are usually studied by means of models of different degrees of complexity. The models are useful to our understanding of the relative magnitude and importance of the different geological, physical, chemical, and biological processes that control the distribution of carbon, and to our ability to describe the past and, at least to some extent, the future. The predictive ability of any model has a great value in forecasting and planning for the future, particularly for the processes that have a human dimension and affect the fabric of human societies. However, these useful features of models encounter serious obstacles because of the often-unforeseen developments in the future that do not conform to the basic assumptions and mechanisms of the models. Therefore our approach in this book is to discuss the fundamental principles of the biogeochemical cycling of carbon and give examples of how they contribute to our understanding of an integrated global picture that has been only recently affected by humans as a geologic agent.

It is our anticipation that this book will serve mainly as a reference text for Earth, ocean, and environmental scientists from various subdisciplines who might be interested in an overview of the carbon cycle and behavior of carbon dioxide, and the effects humans are having on them. Parts of this book have already been used by the authors as material for classroom discussion, so the book might also serve a one-semester course at the upper-undergraduate or graduate level addressing the behavior of the carbon cycle and its human modifications.

April 2006	April 2006
A. L.	F. T. M.
Evanston, Illinois	Honolulu, Hawaii

Acknowledgements

The beginning of this book was a suggestion of Kenneth Howell, editor at Plenum Publishing in New York, to Fred Mackenzie to write a book on the subject of carbon dioxide in Earth's history. After we agreed to do this and a hesitant and slow start, the writing progressed along with a series of editorial transmutations that saw the publisher's name change from Plenum to Kluwer Academic and finally to Springer, in the course of which Ken Howell moved to another position and the editing functions were assumed first by Anna Besse-Lototskaya, and later by Tamara Welschot, and Senior Editorial Assistant Judith Terpos, of the Springer Paleo-Environmental Sciences Department at Dordrecht. We thank Ken, the original editor, his successors in the new organization, and Judith for their encouragement and patience through the time of writing this book and in getting it ready for publication.

We owe a debt of gratitude to many individuals at many institutions for their advice and help in providing us with information, literature references, data, and permissions to cite them. We are especially indebted to Lee R. Kump (The Pennsylvania State University) and Robert Raiswell (University of Leeds) for their review of the initial complete manuscript of this book. Their insights and constructive comments helped us a great deal in improving the presentation and content of the book. Within our individual universities and at other institutions, we thank the following, in alphabetical order.

At Northwestern University: Emile A. Okal, for his ever-readiness to respond helpfully and without regard to the impediment of the distance between himself and A. L. that extended sometimes to seven orders of magnitude (in meters), and other members of the Department of Geological Sciences whose bookshelves and reprint files were always open to us. A. L. is particularly greatful to former and current graduate students Benjamin C. Horner-Johnson, Robert E. Locklair, and Lingling Wu for discussions of the various subjects covered in the book, and to the students in his geochemistry class who were a helpful source of literature references. F. T. M. especially thanks Donna Jurdy, former Chair, and the members of the faculty for giving him the opportunity to be in residence at Northwestern for spring quarters and the unbridled time to write parts of this book.

At the University of Hawaii, Manoa: Michael O. Garcia, Yuan-Hui Telu Li, Christopher Measures, Michael J. Mottl, Brian N. Popp, Francis J. Sansone, Jane S. Schoonmaker, and Richard E. Zeebe. F. T. M. is very grateful to his current and former students, especially Andreas Andersson, Rolf Arvidson, Kathryn Fagan, Michael Guidry, Daniel Hoover, François Paquay, Stephanie Ringuet, Christopher Sabine, Katsumasa Tanaka, and Leah May Ver, for their intellectual interactions, and to the students in his classes who caught misprints in the text. A. L. also thanks the Marine Geology and Geochemistry Division of the Department of Oceanography for providing him the space and support to write unencumbered during the early stages in the development of this book.

At the Centre de Géochimie de la Surface/C.N.R.S., Strasbourg: Norbert Clauer, the former Director of the Center, for his hospitality and extension of the Center's facilities to A. L., while on a sabbatical leave; to Bertrand Fritz, François Gauthier-Lafaye, Philip Meribelle, François Risacher, Claude Roquin, and Peter Stille for discussion of the various subjects dealt with in the book; and to Alain Clement for his continued support in making the computer systems function.

At other institutions: Alberto Borges (Université de Liège), Peter G. Brewer (Monterey Bay Aquarium Research Institute), Ken Caldeira (Department of Global Ecology, Carnegie Institution, Stanford, Calif.), Lei Chou (Université Libre de Bruxelles), Ian D. Clark (University of Ottawa), Tyler B. Coplen (U. S. Geological Survey), William P. Dillon (U. S. Geological Survey), James I. Drever (University of Wyoming), the late Michel Frankignoulle (Université de Liège), Jean-Pierre Gattuso (Laboratoire d'Océanographie, Villefranche-sur-Mer, France), Harold C. Helgeson (University of California, Berkeley), Jochen Hoefs (Universität Göttingen), Dieter M. Imboden (Eidgenössische Technische Hochschule Zürich), Stephan Kempe (Technische Universität Darmstadt), John W. Kimball (Andover, Mass.), Stephen H. Kirby (U. S. Geological Survey), Timothy R. Klett (U. S. Geological Survey), Keith H. Kvenvolden (U. S. Geological Survey), Frank Peeters (Limnologisches Institut, Universität Konstanz), Edward T. Peltzer (Monterey Bay Aquarium Research Institute), Jean-Luc Probst (Ecole Nationale Supérieure Agronomique de Toulouse), Karsten Pruess (Lawrence Berkeley National Laboratory), Christophe Rabouille (Laboratoire des Sciences du Climat et de l'Environnement, Gif-sur-Yvette, France), John A. Ripmeester (National Research Council, Ottawa), Mikhail N. Shimaraev (Limnological Institute, Irkutsk), Stefan Sienell (Director of the Archives, Österreichische Akademie der Wissenschaften, Wien), Francesca A. Smith (Pennsylvania State University), Stephen V. Smith (Centro de Investigación Cientifíca y de Educatión Superior de Ensenada, Mexico), Roland Span (Universität Paderborn), the late Roland Wollast (Université Libre de Bruxelles), and Choong-Shik Yoo (Lawrence Livermore National Laboratory).

For help in the linguistic quandaries of Latin and Greek, we thank Daniel H. Garrison, of the Classics Department at Northwestern, and for additional advice on the use of ancient and modern Greek terms we thank Katerina Petronotis, of the Ocean Drilling Program, Texas A&M University. Our literature searches and finding of publications long out of print or those held by a small number of libraries worldwide were greatly facilitated by the help received from Anna Ren, Geological Sciences librarian, and the dedicated personnel of the Interlibrary Loan Department at Northwestern; at the University of Hawaii, by Kathy Kozuma, secretary, and Jennifer Cole-Conner and Kellie Gushikan, student assistants, and University Library staff; and at the Université Louis Pasteur, Strasbourg, by Betty Kieffer, librarian of the Institut de Géologie, and Janine Fischbach, librarian of the Institut de Physique du Globe, all of whom we thank wholeheartedly for the logistic support and library research.

We are grateful to the following persons and organizations for permissions to cite freely unpublished or copyrighted materials: American Association for the Advancement of Science, Washington, D. C., publishers of *Science* magazine (Fig. 2.6); American Institute of Biological Sciences, Washington, D. C., publishers of *BioScience* magazine (Figs. 6.2 and 6.5); Andreas J. Andersson, University of Hawaii (Figs. 5.6, 9.7, 9.8, 9.12, 9.17, and 9.19); Rolf S. Arvidson, Rice University (Fig. 10.3); Robert A. Berner, Yale University (Figs. 6.9B and 10.2); Nina Buchmann, Eidgenössische Technische Hochschule Zürich (Fig. 8.1); Bruce A. Buffett and David Archer, both of the University of Chicago (Fig. 5.10); Nelia W. Dunbar, New Mexico Bureau of Mining and Geology, Socorro, N. Mex. (photograph of Mount Erebus, front cover); Elsevier, publishers of *Sedimentary Geology* magazine (Fig. 9.18); Kathryn E. Fagan, University of Washington (Fig. 9.7); Richard G. Fairbanks, Lamont-Doherty Earth Observatory, Columbia University, Palisades, N.Y. (Fig. 9.3); Paul Falkowski, Rutgers University (Fig. 10.5); Graham D. Farquhar, Australian National University, Canberra (Fig. 6.3); Volker Hahn, Max Planck Institut für Biochemie, Jena (Fig. 8.1); John M. Hayes, Woods Hole Oceanographic Institution, Woods Hole, Mass. (Fig. 6.9); Ann Henderson-Sellers, Australian Nuclear Science and Technology Organisation, Lucas Heights, NSW, and Peter J. Robinson, University of North Carolina, Chapel Hill (Fig. 3.3A); Intergovernmental Panel on Climate Change, Geneva, Switzerland (Fig. 11.11); J. Richard Kyle, University of Texas, Austin (photograph of the El Capitan Permian reef, front cover, and related information); Kuo-Nan Liou, University of California, Los Angeles (Fig. 3.1); Gary Long, Energy Information Administration, U. S. Department of Energy (Fig. 1.6); C. Nicholas Murray, formerly of the Joint Research Centre, Ispra, Italy (Fig. 5.7); Nature Publishing Group, London, publishers of *Nature* magazine (Fig. 9.3); Emile A. Okal, Northwestern University (Fig. 1.2A); Marion H. O'Leary, California State University, Sacramento (Figs. 6.2 and 6.5); Norman R. Pace, University of Colorado, Boulder (Fig. 2.6); Adina Paytan, Stanford University (data in Fig. 7.14); Pearson Education, Harlow, Essex, U. K. (Fig. 3.3B); Peter W. Sloss, National Geophysical Data Center, Boulder, Colo. (Fig. 9.2); and Ján Veizer, University of Ottawa and Ruhr-Universität Bochum (Figs. 6.9A, 7.11, and 7.13).

    And last, but far from least, thanks to Kare Berg from A. L. and to Judith Mackenzie from F. T. M., to each for her support, encouragement, and help during the writing of this book. This book is dedicated to the memory of our friend and research colleague over the years, Roland Wollast, of the Laboratoire d'Océanographie, Université Libre de Bruxelles, Belgium.

April 2006   April 2006
A. L.   F. T. M.
Evanston, Illinois   Honolulu, Hawaii

Picture Credits on the Front Cover

Top, from left to right:

Mount Erebus, Antarctica, the southernmost volcano on Earth, elevation 3794 m above sea level (77°32′S, 167°17′E), photographed in 1983 by Nelia Dunbar and published in Internet by the National Science Foundation and New Mexico Tech. http://www.ees.nmt.edu/Geop/mevo/mevomm/imagepages/dunbar/index.html
Photograph courtesy of Dr. Nelia W. Dunbar, New Mexico Bureau of Geology and Mineral Resources, Socorro, N. Mex.

Earth photographed on 7 December 1972 by the Apollo 17 crew on the way to the Moon. NASA, http://visibleearth.nasa.gov/view_rec.php?id=1597

Reef limestone of Middle Permian Age, approximately 265 million years ago. El Capitan peak, Guadalupe Mountains, Texas, 2464 m above sea level (31°53′N, 104°51′W). http://www.utexas.edu/features/archive/2003/graphics/texas5.jpg
Photograph and information courtesy of Professor J. Richard Kyle, Department of Geological Sciences, Jackson School of Geosciences, University of Texas, Austin, Tex.

True-color image of a portion of the Great Barrier Reef, off the coast of central Queensland, Australia, north of the Tropic of Capricorn, taken by NASA on 26 August 2000. The width of the reef zone is approximately 200 km. NASA Earth Observatory, http://earthobservatory.nasa.gov/Newsroom/NewImages/images.php3?img_id=4797

Center:

Main features of the global carbon cycle at the Earth's surface. Based on Figures 1.4, 5.11, and 10.7.

Contents

Chapter 1

Brief Overview of Carbon on Earth

In the minds of the broad public, carbon dioxide is associated primarily, if not exclusively, with considerations of global warming. This topic has been the focus of undoubtedly great attention in the last decades of the 20th and in the early 21st century owing to the coverage of the subject of global warming and climate change by the news media drawing their information from the results of scientific studies. The role of carbon dioxide as one of the gases that warm the Earth's atmosphere by absorption of infrared or longwave, outgoing Earth radiation has been known since the work of the French scientist Jean-Baptiste-Joseph Fourier in the early nineteenth century and that of the Irish polymath John Tyndall in the middle part of that century. The similar role of water vapor as a greenhouse gas was also recognized by John Tyndall in 1863. In his studies of the riddle of the causes of the ice ages, the Swedish chemist Svante Arrhenius in the mid-1890s did a series of mathematical calculations and showed that if the amount of atmospheric CO_2 were cut in half, the world would be 4° to 5°C cooler. He also concluded that a doubling of the CO_2 concentration would lead to a 5° to 6°C increase in global mean temperature. Furthermore, he recognized the fact that the burning of coal and oil emits CO_2 to the atmosphere and could lead to warming of the planet because of human activities (Chapter 3). However, it was only an increase in the concentration of atmospheric carbon dioxide, measured systematically by Charles D. Keeling of the Scripps Institution of Oceanography in California since the mid-1950s in the air over the mountain Mauna Loa on the Island of Hawaii, that drew widespread attention to this gas as a product of fossil-fuel burning and land-use changes by the increasingly industrializing world.

In this first chapter of the book we describe the structure of the *Outer Shell of the Earth* on a global scale, the chemical composition of some of its parts on an atomic scale, the essentials of the carbon cycle in modern time, the connections between the inorganic and biological processes within the carbon cycle, and the estimated occurrences of the main types of fossil fuels that are believed to be the major source of the increase in atmospheric carbon dioxide in the industrial age of the last 150 years. This material provides an overview of the global carbon cycle and the framework for discussion of various aspects of the cycle in the chapters that follow.

In the concerns about global warming and the shorter-term increase of atmospheric carbon dioxide, three facts are nearly forgotten: one is the long-term cooling of the Earth's surface in the last 30 million years, since ice cover began to develop in Antarctica; another is the periodic glaciations during the last 1 million years that were accompanied by rises and declines in atmospheric carbon dioxide concentration; and the third is the primary importance of carbon dioxide to plant growth. The long history of carbon on Earth begins with the Earth's accretion 4.55 billion years ago (Fig. 1.1) and it underlies not only the beginning and evolution of organic life on Earth, but also a great variety of the processes that have shaped the geological environment since the early days of the planet.

1   An Unusual Look at Earth's Shells

We, as creatures living on the Earth's surface and capable of seeing things only of certain size that emit or reflect a certain narrow band of electromagnetic radiation, which we call visible light, have a certain perception of the appearance of our global habitat. A clean atmosphere is transparent to us, clear sky is blue, liquid water is of some color varying from blue to greenish or brown that depends on its absorption and reflection of parts of the visible light spectrum and on the presence of other dissolved or suspended materials in it. We see snow flakes and ice crystals as white or transparent, depending on their size and the reflecting quality of their surfaces. In the world of rocks and minerals, we can distinguish with an unaided eye individual mineral grains, but we do not see the atomic structure and different atoms making up the world around us, learning about it by other means. Our knowledge of the Earth's interior (Fig. 1.2) is neither a product of direct observations made on "journeys to the center of the Earth" nor is it based only on studies of rocks recovered from the deepest borehole of about 12 km drilled on land and a borehole of somewhat less than 2 km beneath the ocean floor. The picture of the Earth's interior as well as exterior shells, comprising the atmosphere, hydrosphere, sediments, and living organisms, is a product of researches that synthesize the observational, experimental, and theoretical knowledge gained over a long period of time. Figure 1.2 is a small-scale schematic view of our planet and its surface environment where most of the action involving carbon dioxide and other forms of carbon occurs. Carbon dioxide as the chemical species CO_2 comprises only a very small part of all the carbon in the oceans, fresh waters, and sediments. Although it is a minor component of the atmosphere as a whole, it is the main form of atmospheric

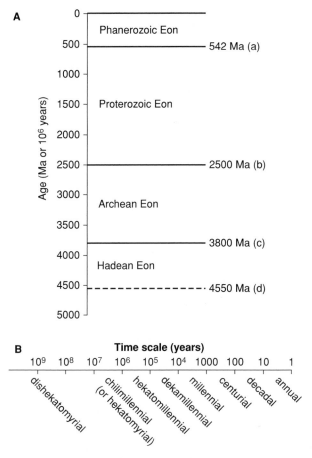

Figure 1.1 Geologic time scale. A. The four Eons. (a) The base of the Phanerozoic Eon is the base of the Cambrian Period; the age shown is that of the International Commission on Stratigraphy 2004 (Gradstein *et al.*, 2004); since 1937, age estimates of this boundary varied from 470 to 610 Ma (Harland *et al.*, 1990). (b) Gradstein *et al.* (2004). (c) The age of the Hadean-Archean boundary varies from 3800 to 4000 Ma because of insufficient age data for the Hadean (Harland *et al.*, 1990). (d) The age of the Earth taken as 4.55 ± 0.07 Ga based on meteorite ages (Patterson, 1956); other meteorite ages between 4.504 and 4.568 Ga indicate a period of 64 million years when parent bodies of meteorites formed (Allègre *et al.*, 1995; Faure and Mensing, 2004). B. Terms used for some of the time segments. Compounds with the Latin-derived millenial are based on Greek terms (Smith, 1964; Woodhouse, 1910) and modern Greek for one billion or 10^9. **Units**: 1 Ga $= 10^9$ yr, 1 Ma $= 10^6$ yr.

carbon and very important to life and regulation of climate on our planet. Because of the chemical reactions that transform carbon dioxide into organic carbon and carbonate minerals in sediments, and the latter two back into carbon dioxide, the processes that affect the CO_2 concentration in the atmosphere and natural waters extend far beyond the

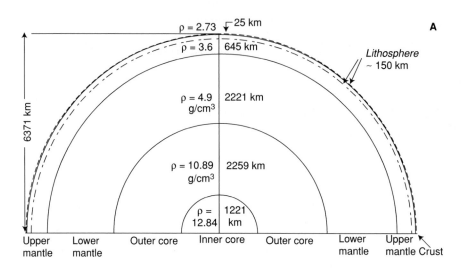

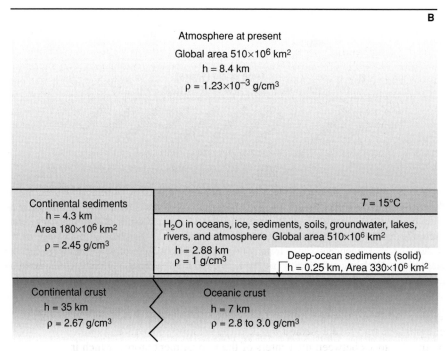

Figure 1.2   A. Earth's interior shells. Shell thickness and densities are from E. A. Okal (*personal communication*, 2005), mean thickness of the lithosphere is from Gung *et al.* (2003). Note the average crust of 25 km thickness and a mean lithosphere of 150 km that includes the crust and part of the upper mantle. B. Atmosphere, water, and sediments above the continental and oceanic crust. Thickness of the atmosphere is its scale height, discussed in Chapter 2. Thickness of deep-ocean sediment is that of the solids in a layer of an average thickness of 500 m (Chester, 2000) and porosity 50%.

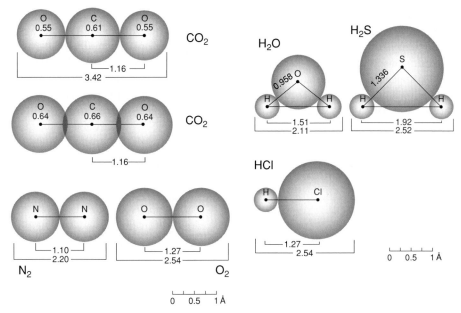

Figure 1.3 Molecules of some of the main atmospheric gases of the present and primordial past: CO_2, N_2, O_2, H_2O, H_2S, and HCl. Units of interatomic distances, atomic radii, and lengths are in ångströms (1 Å $= 10^{-8}$ cm or 10^{-10} m or 100 pm). Interatomic distance, center to center, is the sum of two radii. This convention accounts for two representations of the CO_2 molecule: one, where the oxygen radius is 0.549 Å and C radius is $1.16 - 0.549 = 0.61$ Å; the other, where the C radius has the value for the double bond (0.66 Å, Table 1.1) and O radius is one-half of the interatomic distance in O_2 (0.636 Å). The large sizes of S and Cl, in comparison to O, in single covalent bonding to the hydrogen atoms, may be noted.

last two centuries of the industrial age, as is discussed later in this and other chapters. For a view of our physical environment that is different from the one depicted in Fig. 1.2, we may imagine a fictional observer of our planet from space, who is capable of seeing individual atoms and looking through the thickness of the atmosphere, oceans, and some of the Earth's solid crust. This observer would see mostly oxygen atoms interspersed with other atoms occurring as smaller admixtures or impurities within the mass of oxygen that makes up more than 90% of the volume of the solid crust of the Earth and water in the oceans. The different chemical compounds of atomic and molecular species would be distinguishable to the atom-scale observer by their sizes and their composition that is represented by the different atoms. Volumes of molecules depend on the distances between the centers of their constituent atoms, which in turn depend on the nature of the chemical bonds between them.

The atmosphere would appear to our fictional observer as a mixture of the two most abundant gases, nitrogen and oxygen diatomic molecules of about the same size, as shown in Fig. 1.3, with 78% of the volume being nitrogen and 21% oxygen. The observer would also see the less abundant atoms of argon, about 0.9% of the volume, and small admixtures of carbon dioxide and water molecules, the latter appearing as big oxygen atoms bonded to the much smaller hydrogen atoms in H_2O. Because the volume fraction of water vapor in the atmosphere varies with location and climatic conditions from almost 0% in very dry air to nearly 3% in tropical humid air, and carbon dioxide constitutes currently about 0.038% of the atmosphere, both H_2O and CO_2 would appear to our fictional observer as very small fractions of atmospheric oxygen tied up in a different form. Figure 1.3 shows some of the atmospheric gas molecules, drawn schematically to scale. Their physical dimensions and related parameters of the gases mentioned above are given in more detail in Table 1.1, as are the major gases

Table 1.1   Molecular dimensions of some atmospheric gases of the past and present (the ångström unit is 1 Å $= 10^{-8}$ cm or 10^{-10} m or 100 pm)

Species	Bond length[1] (Å)	Angle[1] (degrees)	Atomic radii in covalent bonds (Å)	Molecule dimension (Å)
CO_2	C–O 1.1600	∠OCO 180°	Double covalent bond: C 0.661, O 0.549[2]	2.32 to 3.42[4]
N_2	N–N 1.0977	—	Triple covalent bond: N 0.545[2]	2.20[5]
O_2	O–O 1.2716	—	Double covalent bond: O 0.549[2]; interatomic distance 1.2074[3]	2.54[5]
H_2O	O–H 0.9575	∠HOH 104.51°	Single covalent bond: O 0.659, H 0.299[2]; O 0.73 and H 0.37[3]	1.51 to 2.11[4]
H_2S	S–H 1.3356	∠HSH 92.12°	Single covalent bond: H 0.299[2]; H 0.37, S 1.02[3]	1.92 to 2.52[4]
SO_2	S–O 1.4308	∠OSO 119.33°	Double covalent bond: O 0.549[2]	2.47 to 3.57[4]
HCl	Cl–H 1.2746		Single covalent bond: Cl 0.99[3]	2.55[5]
Ar			Atomic radius 0.71, radius for single covalent bond 0.97, van der Waals radius 1.88[3]	1.42[5]

NOTE: Atomic radii reported in the literature vary greatly, often without any explanation of what is meant by the atomic radius and its value. Publications and Web Sites by Alcock (1990) and Winter (2003) address such discrepancies.
[1] Lide (1994).
[2] Alcock (1990), with references therein.
[3] Winter (2003), with references therein.
[4] First figure is the distance between the centers of O atoms in CO_2 and SO_2, and between H atoms in H_2O and H_2S (Fig. 1.2). Second figure is the longest dimension of the molecule, as shown in Fig. 1.3.
[5] Double of the interatomic distance for N_2, O_2 (Fig. 1.1), and HCl, and atomic diameter of Ar.

occurring in the present-day and geologically old atmosphere shortly after the time of Earth's formation. Today's atmosphere contains molecular oxygen that was not present in the primordial atmosphere; the latter likely contained much higher concentrations of carbon dioxide and, initially, hydrogen chloride and sulfur in the form of either hydrogen sulfide or sulfur dioxide (Chapter 2).

The continental and oceanic crust, and the weathered products that constitute the inorganic mineral part of soils, are made of aluminosilicate minerals where oxygen is the volumetrically most abundant element owing to its ubiquitous presence as a chemical component and its crystal ionic radius that is large relatively to the smaller radii of other elements bonded to it. Crystal ionic radius depends on the chemical and nuclear characteristics of an element, and its coordination number in the crystal lattice. Although the ionic crystal radii reported by different authors have somewhat different values, they do not affect the general picture of a mineral lattice as made of big oxygens and other smaller elements (Ahrens, 1952; Evans, 1994; Faure, 1998; Moeller, 1952; Shannon, 1976; Whittaker and Muntus, 1970). The crystal ionic radius of O^{2-} in a 6-fold coordination is 1.32 to 1.40 Å, but the ionic radii of the major elements in such common rock-forming minerals as aluminosilicates and carbonates are smaller. Silicon and aluminum are each commonly surrounded by four oxygens (coordination number 4), with the ionic radii of Si^{+4} 0.26 to 0.34 Å, and of Al^{+3} 0.39 to 0.47 Å. For metal cations in a 6-fold coordination, sodium Na^+ has the radius of 1.02 to 1.10 Å, and calcium Ca^{+2} 1.00 to 1.08 Å. Carbon, occurring in the valence state of C^{+4} in carbonate minerals as anion CO_3^{2-}, is a much smaller ion, of radius 0.15 to 0.16 Å. As a whole, the continental granitic crust and the oceanic basaltic crust are 92 to 94% by volume elemental oxygen (Lerman, 1979; Mason, 1958) and they would appear to our fictional observer from space as a layer of oxygen atoms with other, mostly smaller, impurities in it.

2   Global Carbon Cycle

2.1   Cycle Structure

Life on Earth is based on carbon as one of the main components of organic matter. The occurrence of the various forms of carbon in different parts of the Earth's interior and its outer shell, and the processes that are responsible for the transfer of carbon between the different parts of the Earth make a conceptual model known as the geochemical or biogeochemical cycle of carbon. The latter name reflects combinations of geological, biological, chemical, and physical processes that variably control the flows and chemical transformations of carbon in the surface and interior reservoirs of the Earth.

The carbon cycle is usually divided into a deeper part, called the *endogenic cycle*, and the part that includes the surface reservoirs of the sediments, oceanic and continental waters, land and aquatic biomass, soils, and the atmosphere is referred to as the *exogenic cycle*. The main features of the global carbon cycle are shown in Fig. 1.4 and the carbon inventory on Earth is given in Table 1.2. The inventory of the reservoirs represents the geological near-Recent, with atmospheric carbon taken at the

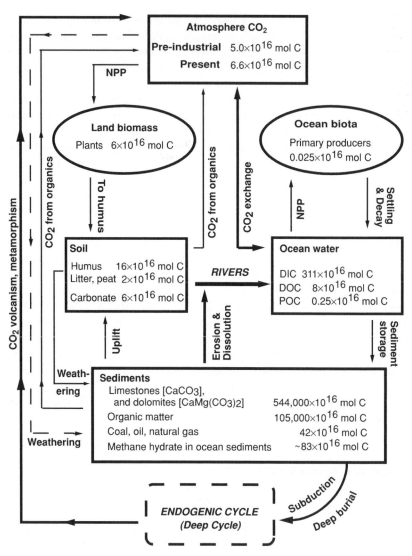

Figure 1.4   Global biogeochemical cycle of carbon. Reservoirs on the Earth surface represent the exogenic cycle. Masses of the major reservoirs are given in Table 1.2. NPP is net primary production, defined in the text. The CO_2 in weathering is shown as taken from the soil (solid line) or from the atmosphere (dashed line). Values of major fluxes are given in Table 1.3. Mass units: 1 mol C = 12.011 g C or 1×10^{16} mol C = 120 Gt C.

pre-industrial level of CO_2 that has probably varied significantly in the geologic past and has risen about 35% during the Industrial Age. Estimates of carbon masses in other reservoirs are subject to variable errors and greater uncertainties. The mass of carbon in land plants has been variably reported in the range of ±15 to 20% of the value given

Table 1.2   Carbon masses in the major environmental reservoirs. From data in Li (2000), Ver *et al.* (1999), and Walker (1977)

Reservoir	Mass of carbon	
	grams C	*mols C*
Atmosphere		
$\quad CO_2$ (at pre-industrial 280 ppmv)	6×10^{17}	5×10^{16}
Ocean		
$\quad$ Dissolved inorganic (DIC)	3.74×10^{19}	3.11×10^{18}
$\quad$ Dissolved organic (DOC)	1×10^{18}	8.33×10^{16}
$\quad$ Particulate organic (POC)	3×10^{16}	2.50×10^{15}
Ocean biota	3×10^{15}	2.50×10^{14}
Land biota		
$\quad$ Phytomass	7×10^{17}	$6 \pm 1 \times 10^{16}$
$\quad$ Bacteria and fungi	3×10^{15}	2.50×10^{14}
$\quad$ Animals	1 to 2×10^{15}	1.25×10^{14}
Land		
$\quad$ Soil humus	1.92×10^{18}	$1.6 \pm 3 \times 10^{17}$
$\quad\quad$ Reactive fraction of humus		2×10^{16}
$\quad$ Dead organic matter, litter, peat	2.5×10^{17}	2.08×10^{16}
$\quad$ Inorganic soil ($CaCO_3$)	7.2×10^{17}	6×10^{16}
Sediments		
$\quad$ Carbonates	6.53×10^{22}	5.44×10^{21}
$\quad$ Organic matter	1.25×10^{22}	1.05×10^{21}
Continental crust	2.576×10^{21}	2.14×10^{20}
Oceanic crust	9.200×10^{20}	7.66×10^{19}
Upper mantle	$(8.9 \text{ to } 16.6) \times 10^{22}$	$\sim 1.1 \times 10^{22}$

in Table 1.2, 700 Gt C (1 gigaton or 1 Gt $= 10^{15}$ g). Large variations in the estimated global mass of peat and soil humus have also been reported in the literature.

Carbon dioxide is emitted from the Earth's mantle through subaerial volcanism, in the spreading zones and ridges on the ocean floor, from magmatic plumes and hot spots, from subduction zones, and through thick piles of sediments. The occurrence of this chemically oxidized form of carbon in magmatic emissions indicates that CO_2 is not an exclusive product of carbon oxidation at the Earth's surface but it has a much longer geological history on our planet. This view is consistent with the chemical thermodynamic properties of carbon and oxygen that tend to combine and form carbon dioxide, as in the reaction

$$C + O_2 = CO_2, \tag{1.1}$$

Over a wide range of temperatures, from those of the Earth's surface to those of the molten silicate melts, chemical equilibrium in reaction (1.1) is strongly shifted to the right, producing very high ratios of CO_2 to O_2. In thermodynamic calculations of reaction equilibria that involve carbon, the standard form of carbon is, by convention, the mineral graphite and oxygen is diatomic gas. The reaction as written may be either a reversible chemical reaction at equilibrium, where carbon dioxide forms from elemental carbon and oxygen, or it may also represent in shorthand notation the chemical balance for the process of one atom of elemental carbon and one molecule of oxygen combining to produce one molecule of carbon dioxide. Examples of the latter type of one-way processes at the Earth's surface are oxidation of carbon in organic matter and burning of carbon in fossil fuels or forest fires. At the relatively low temperature of the Earth's surface, many processes represented by reaction (1.1) do not go spontaneously from the left to the right, as predicted by chemical thermodynamics. A simple daily demonstration of this fact is in the starting of a fire when the temperature of the reactants, carbon and oxygen, has to be raised to the ignition temperature before they begin to combine one with the other and carbon burns. Common examples of such temporal stability or rather metastable existence of elemental carbon in the oxidizing atmosphere of the Earth are graphite and diamond.

As a whole, oxidized carbon is the more abundant form of the element at the Earth's surface where it occurs as carbon dioxide gas, dissolved carbonate ions in waters, and carbonate ions in sedimentary minerals. In addition to the sedimentary carbonates, two types of igneous rocks containing carbon occur in the Earth's crust: kimberlites, the main commercial source of diamonds that are an elemental crystalline form of carbon, and carbonatites, made mostly of calcite, dolomite, and iron-containing dolomite ankerite. Carbonatites are believed to form in the mantle, at high pressures of CO_2, and extruded as volcanic rocks or intruded into the upper lithosphere at the later stages of magmatic differentiation. The occurrences of carbonatites are limited to small areas at about 330 locations worldwide (Barker, 1997), and their total mass is very small in comparison to the mass of sedimentary limestones and dolomites.

On Earth, the biological origin of organic matter is indicated by the abundance ratio of the two stable isotopes of carbon, $^{13}C/^{12}C$: the ratio is lower in biologically produced organic matter than in the source from which CO_2 is taken. This topic is discussed further in Chapter 6. However, organic compounds of non-biological origin are found in a class of meteorites known as chondrites, among which different types of carbonaceous chondrites contain between 0.26 and 3.5 wt % carbon in organic compounds coexisting with aluminosilicate minerals. This abundance of organic carbon in extraterrestrial materials may be compared with its mean abundance of about 0.75 wt % in the sedimentary mass on Earth. Fossil fuels—coal, petroleum, and hydrocarbon gases—are part of the global reservoir of sedimentary organic matter. The known recoverable reserves of fossil fuels (different coals, petroleum, and natural gas, recoverable with the present-day technology) account for only a very small fraction of the total organic carbon in sediments, and these are discussed further in Section 4 of this chapter.

The biggest carbon reservoir on the Earth's surface is sedimentary rocks. The mass of carbon stored over the geologic history of the Earth in sedimentary limestones, dolomites, and undecomposed organic matter exceeds by a very large factor, about 100,000, the mass of carbon in atmospheric CO_2 at the present time (Fig. 1.4, Table 1.2).

Next in size is the carbon reservoir of the oceans or, more generally, of the oceanic and continental surface and ground waters making the hydrosphere. The total mass of carbon in the global oceans, consisting mostly of dissolved inorganic carbon, is about

60 times greater than the mass of atmospheric carbon dioxide (Table 1.2). The main form of carbon in the global water reservoir is dissolved carbon dioxide and its ionic species, and generally less abundant dissolved organic carbon derived from incomplete decomposition of living and dead organic matter. These dissolved forms of carbon are usually denoted as DIC, for dissolved inorganic carbon, and DOC, for dissolved organic carbon (Fig. 1.4). Dissolved inorganic carbon (DIC) includes three major aqueous species—CO_2, HCO_3^-, CO_3^{2-}—and, to a lesser degree, their complexes with metal ions in solution. On the other hand, dissolved organic carbon occurs as a great variety of organic compounds ranging in molecular weight from light, simple organic acids to much heavier and structurally more complex species. Besides the dissolved species of carbon, there are particles containing inorganic and organic forms of carbon in continental and ocean waters. Particulate inorganic carbon or PIC is mostly the grains of carbonate rocks carried by rivers and the calcium-carbonate skeletons of organisms forming in surface ocean waters and sinking to the ocean bottom. Its organic counterpart, particulate organic carbon or POC, consists of undecomposed cells, products of metabolic excretion, soft parts of dead plants and animals sinking through the water column, and organic matter adsorbed on mineral-particle surfaces, such as clays.

In the atmosphere that contains oxygen, carbon dioxide is the only chemically stable form of carbon. More reduced forms of carbon gases, such as carbon monoxide (CO), methane (CH_4), and volatile hydrocarbons, produced by biological or inorganic processes or anthropogenic activities, are eventually oxidized to CO_2. At present, carbon dioxide is the most abundant of the carbon-containing gases and it makes a volume fraction of 3.80×10^{-4} or 380 ppmv (parts per million by volume)[1] of the atmosphere, increasing at a rate of approximately 1.5 ppmv per year. Other gases, such as methane (CH_4) from natural sources or human agricultural and industrial activities, carbon monoxide (CO) and volatile organic compounds mainly from industrial activities, and the most recent emissions of the various synthetic chlorofluorocarbon compounds (CFCs, e.g., Freon 11 and 12) used as refrigerant and spray-propellant gases, are believed to have effects on atmospheric temperature, stratospheric and tropospheric ozone, and climate, despite their occurrences at concentrations much lower than atmospheric CO_2.

In the biosphere, a major part of carbon in living organic matter is represented by land plants. This very large and diversified group includes predominantly photosynthetic plants ranging in size from short-lived unicellular organisms to trees whose lifetimes average several decades. Forests are the main reservoir of biotic carbon on land. In the ocean, most of the photosynthesizing organisms live in the upper layer of the oceans where sufficient light is available. These free-floating organisms are known as the phytoplankton, and the phytoplanktonic group Coccolithophoridae is an important producer of calcium carbonate shells that constitute a major portion of ocean-floor carbonate sediments. Additional aquatic plants include the class of algae and bigger plants rooted to the bottom, known as macrophytes. The oceanic phytoplankton makes only a small fraction, about 0.5% of the carbon mass of global biota. However, amino

---

[1] Volume fraction of a gas in a mixture that is treated as an ideal gas is also its partial pressure. In the atmosphere of total pressure $P = 1$ atm, partial pressure of CO_2 is 3.80×10^{-4} atm.

acid-rich oceanic plants have a much shorter life cycle, about 20 days to 1 month, than the cellulose-rich land plants that turn carbon over approximately every 10 years. These differences account for the fact that fluxes of carbon through the land and oceanic phytomass are not too different.

2.2   Historical Note on the Carbon Cycle

The discovery of carbon dioxide as a gas that forms by fermentation and by burning of charcoal, under the name of *spiritus silvestris*, is attributed to Jan Baptista (or Baptist) van Helmont, a man of medicine, alchemy, and early chemistry in the then Spanish Netherlands, in the first half of the 1600s (e.g., Graham, 1974). The formation of organic matter from carbon dioxide and water under the action of light, the process known as photosynthesis, has been studied since the later part of the 1700s, when molecular oxygen was discovered in the process and carbon dioxide identified as a component of air. Presentation of the first general scheme of the carbon and nitrogen cycles was attributed to the French chemist, Jean Baptiste André Dumas, in 1841 (Rankama and Sahama, 1950). Dumas (1842) described the cycle of CO_2 consumption and production by respiration, pointing to the sources of "carbonic acid" in the air and soil where it forms from decomposition of manure or organic fertilizers. He made estimates of the residence time of atmospheric oxygen with respect to animal respiration and pointed out that it would cause very small changes in the oxygen content of the atmosphere at a centurial time scale. Significantly, he also pointed out (p. 5) that the Earth's primordial atmosphere must have contained all the carbon dioxide and nitrogen that have been taken up by living organisms.

As to geochemical cycles, an early treatment of the subject appeared in 1875, where several chapters on the cycles of chemical elements were included in a book on Earth's history by Friedrich Mohr, a professor at the University of Bonn, with short chapters on the silicon and carbon cycles among them (Mohr, 1875, pp. 397–398). In 1893 Professor Arvid Högbom, a colleague of Svante Arrhenius, presented a lecture at the Physical Society drawing the conclusion that the chief source of atmospheric CO_2 is in the release of the gas during the natural breakdown of limestone. He further theorized concerning the long-term geologic cycle of carbon by addressing the questions of how much of the CO_2 is retained in the atmosphere, how much is absorbed by the ocean, and how changes in the concentration of this gas in the atmosphere might affect climate (Christianson, 1999). By the 1920s, the cycles of the chemical elements that are involved in biological processes—carbon, nitrogen, and phosphorus—and are also transported between soil, crustal rocks, atmosphere, land and ocean waters, and the Earth's interior were well recognized by modern standards. Alfred Lotka's book, *Elements of Physical Biology*, published in 1925, has chapters on the cycles of carbon dioxide, nitrogen, and phosphorus that present a modern treatment of what we call today the biogeochemical cycles (Lotka, 1925). Furthermore, he wrote that his ideas of the nutrient element cycles and mathematical treatment of biogeochemical problems were developed as far back as 1902 and in his publications starting in 1907. The term biogeochemical reflects the fact that biological, physical, and chemical processes play important roles and interact with each other in the element cycles that are mediated by photosynthetic primary production and respiration or mineralization of organic matter.

By 1950, the geochemical cycles of elements in the Earth's interior and on its surface became textbook material (Rankama and Sahama, 1950), with the variable degree of detail in each cycle that reflected knowledge of the igneous and sedimentary reservoir contents and some of the inter-reservoir fluxes at the time. This early, if not first, systematic textbook treatment of the geochemical cycles presented diagrams of the geochemical reservoirs as boxes and fluxes between them, and tabulations of the elemental concentrations or masses in some of the individual-reservoirs. Subsequent decades produced the knowledge we have today of the chemical speciation of the elements in the different compartments of the Earth, their abundances, and mechanisms responsible for their flows. While the earlier models of the global biogeochemical cycles of individual elements were static, describing the cycles without their evolution in time, developments in the mathematical treatment of time-dependent multireservoir systems (e.g., Meadows et al., 1972) found their application in the analysis of geochemical cycles (e.g., Lerman et al., 1975). Since then, there has been a great proliferation of cycle models, and in particular of carbon cycle models (e.g., Sundquist and Visser, 2004), at very different physical and time scales, aimed at interpretation of cycle evolution in the past and its projection into the future for the world as a whole, as well as for such global reservoirs as the atmosphere, land, coastal oceanic zone, and the open ocean.

Considerable attention became focused on the global sedimentary cycle and the cycling of salts in the ocean as a result of Kelvin's (William Thomson, later Lord Kelvin) estimates of the age of the Earth between 24 and 94 Ma, made between 1864 and 1899 (Carslaw and Jaeger, 1959), and the estimates of the age of the ocean from the rate of accumulation of sodium brought in by rivers, as was done, for example, by Joly (1899; Drever et al., 1988) whose age of the ocean was about 90 Ma. Gregor (1988, 1992) summarized and discussed in detail the geological arguments in the second half of the 1800s and the early 1900s for the recycling of oceanic sediments after their deposition and for the existing sinks of dissolved salts in ocean water, such as their removal by adsorption on clays, entrapment in sediment pore water, and formation of evaporites that were contrary to the idea of the ocean continuously filling up with dissolved salts. Garrels and Mackenzie (1971) presented the concepts of the sedimentary cycling of materials, that had laid dormant for some years, in book form, and in 1972 these two authors developed a quantitative model of the complete sedimentary rock cycle. Quantitative estimates of sediment recycling rates, based on mass-age sediment distributions, have been made by Gregor (1970, 1980), Garrels and Mackenzie (1971, 1972), and Dacey and Lerman (1983): the total sedimentary mass has a mass half-age of 600 Ma. The differential weathering rates of different rock types gave the half-age of shales and sandstones of about 600 Ma, longer than the ages of more easily weathered rocks, such as carbonates of half-age 300 Ma and evaporites of about 200 Ma. Later work (Veizer, 1988) showed that the recycling rates of the sedimentary lithosphere and the various rock types within it are mainly a function of the recycling rates of the tectonic realms, such as active margin basins, oceanic crust, and continental basement, in which the sediments were accumulated.

3   Fundamental Equation of a Cycle and Carbon Flows

3.1   The Cycle Equation

In a model of a biogeochemical cycle, such as the one shown in Fig. 1.4, the material balance of each reservoir is the sum of inflows, less the sum of outflows. This difference may be zero, indicating that the reservoir mass is constant and it is in a steady state, or it may be positive if the reservoir grows in size or negative if its mass decreases. A change in the reservoir mass (ΔM) caused by inflow and outflow fluxes (F_{in} and F_{out}, in units of mass/time) in a period of time Δt is:

$$\Delta M = (\Sigma F_{in} - \Sigma F_{out})\,\Delta t \tag{1.2}$$

F_{in} and F_{out} may include not only transfer across the reservoir boundaries, but also production and consumption, respectively, by chemical or biological processes within the reservoir.

The atmosphere receives inputs of CO_2 from the Earth's interior and, more recently, from human industrial and agricultural activities. It loses CO_2 to land vegetation (outflow) and receives back part of the loss by respiration or oxidation of organic matter (inflow); such two-way processes are often called exchange. A similar exchange occurs between the atmosphere and ocean, and there is an additional outflow of atmospheric CO_2 to mineral weathering reactions in sediments and crystalline crust. For the most important inflows and outflows, the atmospheric balance of carbon can be written as:

$$\Delta M_{atm} = \left[\Sigma F_{external} + (F_{in} - F_{out})_{atm\text{-}land} + (F_{in} - F_{out})_{atm\text{-}ocean} - F_{weathering}\right]\Delta t$$

$$\tag{1.3}$$

where $F_{external}$ stands for CO_2 inputs from the Earth's interior and anthropogenic emissions, the term $(F_{in} - F_{out})$ denotes exchange with the land or the ocean, and $F_{weathering}$ is the outflow flux to the mineral weathering reactions that consume CO_2 in the process of dissolution.

3.2   General Pattern of Carbon Flows

The starting point of the carbon cycle is the Earth's mantle, from where it was degassed with other volatile elements in the early stage of the formation of the Earth (Chapter 2). However, the exchange between the mantle and the Earth's surface operates on a much longer time scale (10^8 to 10^9 years), and it is much slower than those among the Earth's surface reservoirs (Fig. 1.4). The magnitudes of the carbon flows through the endogenic and exogenic cycles are summarized in Table 1.3. From the perspective of the Earth's surface, CO_2 in the atmosphere dissolves in rain water, and land- and ocean-surface waters. It is also taken out of the atmosphere and surface waters by photosynthesizing plants. Residues of living plants in part decompose to CO_2 that returns to the atmosphere, and in part they become organic matter of soils and sediments. The solution of CO_2 in fresh water is a mild acid and together with the sulfuric and

Table 1.3   Carbon fluxes between major reservoirs in pre-industrial time (Fig. 1)

	Units	
Flux	10^{15} g C/yr	10^{12} mol C/yr
Net primary production (NPP)	113.5	9450
Land	63.1	5250
Ocean	50.4	4200
Volatilization from soil organic matter	62.5	5200
Weathering consumption of CO_2	0.26	22
Net exchange atmosphere-ocean (dissolution 8×10^{15}, evasion 8.042×10^{15} mol/yr)	0.51	42
River input of dissolved C (DIC + DOC)	0.60	50
DIC	0.38	32
DOC	0.22	18
POC	0.19	16
PIC	0.18	15
Oceanic sediment long-term storage	0.28	23
Carbonates[1]	0.22	18
Organic matter[1]	0.06	5
Volcanism, metamorphism, hydrothermal[2]	0.22	18
Uplift[2]	0.40	33

Note: Data from Ver et al. (1999) unless indicated otherwise. (10^{15} g = 1 Gt).
[1] Phanerozoic averages, from Berner and Canfield (1989), Mackenzie and Morse (1992), Wilkinson and Walker (1989).
[2] Uplift assumed as balanced by subduction of ocean floor over geologically long periods (Mackenzie, 2003).

organic acids in water, it reacts with crystalline rocks of the continental crust, causing mineral dissolution and release of the main dissolved constituents of river waters. Metal ions in rivers (Na^+, K^+, Mg^{2+}, and Ca^{2+}), balanced to a large degree by the negatively charged bicarbonate ions (HCO_3^-), are transported to the ocean. Calcium-carbonate minerals—calcite and aragonite, both of the same chemical composition $CaCO_3$, but different in their crystal structure (Chapter 4), and magnesium-containing calcite— form in the ocean either by inorganic precipitation or as skeletons secreted by marine plants and animals, from microscopic skeletal sizes to large molluscs and corals. This calcium carbonate accumulates over large areas of the ocean floor in the form of settling shells of phytoplankton and zooplankton or as benthic calcifying organisms, such as corals and algae, and their detritus in shallow-water environments, such as reefs. Some of the geologically old carbonate sediments became the limestones in the sedimentary cover of the present-day continents, and some of the calcium carbonate was transported into the Earth's mantle by the process of the spreading of the ocean floor plates and subduction of their margins in the zones of oceanic trenches. This subduction process is already part of the endogenic cycle (Fig. 1.4) that breaks down $CaCO_3$ at the high temperatures in the mantle and returns CO_2 to the surface, mostly in gases emitted by continental and oceanic volcanoes. The isotopic composition of carbon in CO_2 in some of the volcanic emissions indicates that it is a mixture of carbon from the mantle and carbon from the Earth's surface.

3.3   Production and Remineralization of Organic Matter

On Earth with life and free oxygen in the atmosphere, carbon continuously undergoes chemical reduction and oxidation—in a nutshell, these two processes represent the cycle of life. Carbon dioxide, where the valence state of carbon is +4, is reduced when organic matter is being photosynthesized to a general composition of CH_2O. It is further reduced to the valence of -4 when methane, CH_4, is formed by bacterial or inorganic processes. As mentioned previously, the formation of organic matter from carbon dioxide and water under the action of light, the process known as photosynthesis, has been studied since the later part of the 1700's, when molecular oxygen was discovered in the process and carbon dioxide was identified as a component of air. Photosynthesis—uptake of carbon dioxide by plants and release of molecular oxygen in light, and the reverse process of respiration—consumption of oxygen and release of carbon dioxide, have been studied for about 150 years before it was realized in the 1930's by Cornelis van Niel (Bassham, 1974; Gaffron, 1964; Meyer, 1964; Whitmarsh and Govindjee, 1995) that organic matter is produced by a combination of carbon dioxide and hydrogen from water, whereas free molecular oxygen is formed from H_2O. This and later studies using stable isotopic tracers found that free molecular oxygen is produced from water molecules rather than from carbon dioxide:

$$CO_2 + 2H_2O \overset{\text{Photosynthesis}}{\underset{\text{Respiration}}{\rightleftarrows}} CH_2O + H_2O + O_2 \qquad (1.4)$$

The cycle of photosynthesis and respiration is the most important process of the formation and decomposition of living matter on our planet. Accumulation of free molecular oxygen in the atmosphere over geologic time is a result of the imbalance between photosynthesis and respiration: photosynthesis produces more organic matter than is respired, which results in some amount of free oxygen left over and available for accumulation in the atmosphere or for other oxidation reactions, and the remaining organic matter is buried in sediments. The rate of biological production of organic carbon, as in reaction (1.4), is also known as gross primary production or P, and the rate of respiration is denoted R. For free oxygen to form and some organic matter to be stored in sediments, a long-term geological condition must have been $P > R$. If the combined rate of oxidation of organic carbon by natural processes and human activities exceeds the rate of primary production then more CO_2 is returned to the environment than removed from it, and this condition of $P < R$ may lead to CO_2 accumulation in the atmosphere. Approximately 60% of photosynthesis on Earth occurs in terrestrial plant ecosystems and the remainder in aquatic systems (Falkowski and Raven, 1997).

The organic matter of terrestrial plants becomes incorporated in soils, where it is known as humus (a mixture of complex organic humic and fulvic acids), and the residues of fresh-water and oceanic plants are buried in sediments. Some of this organic matter decomposes more slowly, producing CO_2 and organic acids and returning the nutrient elements nitrogen and phosphorus to soil and sediment interstitial water, where they again become available to new plant growth. However, some fraction of this organic matter is refractory to decomposition and it may become a practically permanent addition to the sediment. Because the land surface has a higher elevation than the ocean surface, there is a continuous transport of running water to the ocean carrying dissolved material and solid particles eroded from rocks, soils, and organic matter on land. Particles containing organic carbon (POC) are brought to the ocean where some of it decomposes. CO_2 produced by decomposition of land organic matter is added to CO_2 that is present in ocean water.

On the Earth's surface with a climate and vegetation cover similar to those of pre-industrial time, the biggest carbon fluxes are those of net primary production (NPP $= P - R$) on land and in the ocean (Table 1.3). In near-Recent time, NPP on land recycles atmospheric CO_2 every 9 years:

$$\text{Residence time} = (CO_2 \text{ mass})/\text{NPP}$$
$$= (4.95 \times 10^{16} \, \text{mol C})/(5.25 \times 10^{15} \, \text{mol C/yr}) = 9.4 \, \text{yr}$$

The residence times of carbon in the individual reservoirs are a measure of the carbon cycle dynamics. Changes in the reservoir sizes and fluxes in the geologic past must have been also been reflected in great changes in the residence times of carbon in the different reservoirs. In particular, changes in atmospheric carbon dioxide and the density of vegetation cover of land, discussed in the next section, were likely to result in changes in the residence times. The near-Recent mean residence time of carbon in land vegetation, from the data in Tables 1.2 and 1.3, is about 11 years. A greater land coverage by forests or by grasses would have resulted in a considerably longer or shorter, respectively, residence time of carbon in land phytomass. Recycling time by oceanic phytomass is about 0.06 year or 20 days.

Volatilization of carbon from soil humus is due to oxidation or mineralization of the reactive fraction of humus that is 15 to 20% of the total. This implies a faster recycling of soil organic matter than may be suggested by the size of the whole reservoir.

In the ocean, the photosynthetically produced organic matter has its own subcycle. As organic matter settles into the deeper ocean, it undergoes oxidation that returns CO_2 to ocean water. This process is also known as the biological pump. Some of the respired CO_2 is transported back to the surface layer by water mixing, but some of it is used in dissolution of $CaCO_3$ that rains down from the surface layer. An increase in the concentration of CO_2 in solution makes it more acidic and may bring it to a level of undersaturation with respect to calcite and aragonite. Both of these minerals dissolve in the deep ocean and their rates of dissolution are sufficiently fast to return most of the $CaCO_3$ to ocean water. Preservation of $CaCO_3$ in ocean-floor sediments is limited in the present-day ocean to depths less than approximately 3700-4000 m, which is due to a combined effect of higher concentrations of dissolved carbon dioxide and an increase in the solubility of calcite with increasing pressure.

3.4 Carbonate System and Air-Sea Exchange

Carbon dioxide is soluble in water where it reacts to produce bicarbonate and carbonate ions:

$$CO_2 + H_2O \rightleftarrows H^+ + HCO_3^- \rightleftarrows 2H^+ + CO_3^{2-} \qquad (1.5)$$

Increase in concentration or partial pressure of CO_2 in the atmosphere results in its higher concentration in solution at an equilibrium and this further causes the reaction to go to the right, making a higher concentration of the hydrogen-ion, H^+, that accounts for the acidic properties of a carbon-dioxide solution.

Dissolved inorganic carbon in the seawater can be either a source or a sink of atmospheric carbon dioxide depending on environmental conditions. For example, greater biological production of organic matter and its burial in sediments removes carbon dioxide from the atmosphere-ocean system; the formation of calcium carbonate and its deposition in sediments removes carbon from the atmosphere-ocean system, yet it also changes the relative abundances of the three dissolved species in seawater (CO_2, HCO_3^-, and CO_3^{2-}) and affects the exchange of CO_2 with the atmosphere. Transfer of CO_2 between ocean water and the atmosphere depends on such factors as the mass of CO_2 in the atmosphere, temperature, salinity and chemical composition of ocean water, the production and respiration of organic matter, and the production of carbonate minerals.

The bicarbonate and carbonate ions react with the calcium-ion in water, making calcium-carbonate minerals calcite and aragonite, of composition $CaCO_3$:

$$Ca^{2+} + 2HCO_3^- \rightleftarrows CaCO_3 + H_2O + CO_2 \qquad (1.6)$$

The reaction of CO_2 with water that produces the bicarbonate and carbonate ions, and the reactions of the latter with calcium are behind the abundance of limestones and dolomites that once formed as oceanic sediments and became part of the continental crust and land surface. Reaction (1.6) proceeding from the left to the right ($\rightarrow$) represents the formation of calcium carbonate minerals and concomitant production of CO_2 that may result in a CO_2 transfer from seawater to the atmosphere. As the bicarbonate ion, HCO_3^-, is the main form of DIC at the environmental conditions of ocean water and continental freshwaters, its reaction with Ca^{2+} removes 1 mol of carbon into sediments and makes 1 mol of CO_2 that tends to restore the balance of the carbonate species, according to reaction (1.5). Dissolution of calcium carbonate by reaction with dissolved CO_2 is reaction (1.6) going from the right to the left ($\leftarrow$). This is essentially the process of chemical weathering of limestones and silicate rocks on land where they are exposed to atmospheric precipitation and soil and ground waters containing elevated concentrations of CO_2 from the remineralization of organic matter in soils. It is also a process of dissolution of carbonate minerals by an elevated atmospheric CO_2 partial pressure, as is believed to be taking place under the present conditions of the rising atmospheric CO_2.

4   Carbon in Fossil Fuels

After the extensive use of wood and charcoal, energy sources for the industrialization of the world became progressively coal, oil, and natural gas. Energy released as heat in the oxidation of carbon, as shown in reaction (1.1), has been used for the generation of steam

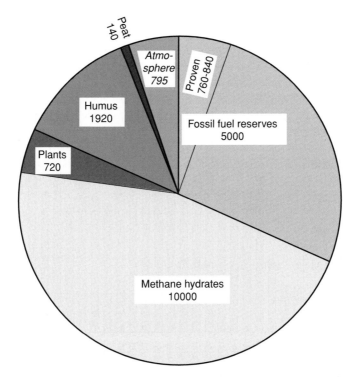

Figure 1.5   Carbon content (gigatons or Gt C) of the major organic carbon reservoirs and as CO_2 in the present-day atmosphere. Fossil fuel reserves are from Kvenvolden (1988, with references to earlier sources). Proven reserves of coal, oil, and natural gas from BP (2004), allowing for variation in the carbon content of different coals. Methane hydrates in oceanic sediments are from Kvenvolden and Laronson (2001). Peat resources are from WEC (2001), and the carbon content of peat is from Shimada *et al.* (2001). Other estimates are from Table 1.3.

and, later, electricity that are further converted to mechanical and other forms of work. In addition to carbon, hydrogen that is contained in liquid and gaseous hydrocarbons (oil and gas) is also oxidized in burning to water, releasing more heat. Water vapor (H_2O) and carbon dioxide (CO_2) are the main greenhouse gases in the atmosphere, and the mass of carbon in the present-day atmosphere is much smaller than in the fossil fuel reserves, as shown in Fig. 1.5. However, the carbon mass in proven recoverable reserves at the end of 2003 is comparable to that in the atmosphere, indicating that the atmospheric CO_2 concentration would double if these reserves were consumed and no CO_2 removed from the atmosphere to the land and oceanic carbon reservoirs. The occurrences of methane hydrates in the tundra and mostly in oceanic sediments represent a potential source of a strong greenhouse gas that may be emitted to the atmosphere in a warming climate, as well as a potentially usable form of fuel. The estimated mass of carbon, about 10,000 Gt C, in methane hydrates is a mean of a range from 1,000 to 100,000 Gt C, and it exceeds the conventional fossil fuel reserves by

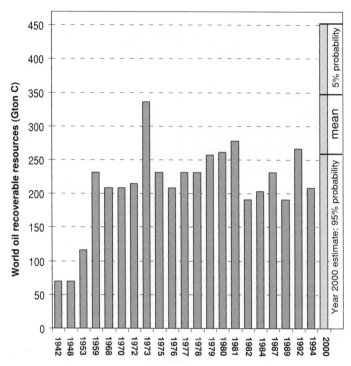

Figure 1.6   Historical development of estimates of world oil resources by various authors since 1942 (from Wood and Long, 2000; G. Long, *personal communication*, 2005). The year 2000 estimates of the U. S. Geological Survey distinguish between reserves estimated at a level of 95% probability, mean amount, and a higher estimate at a 5% probability level. Conversion: 1 barrel oil = 0.1364 ton oil (BP, 2004), carbon content 85 wt %.

a factor of about 2. Both of these reservoirs store much more carbon, about 20 times more, than the present-day atmosphere.

Recoverable reserves of fossil fuels are considered recoverable at the current state of technology. Estimates of proven recoverable reserves depend at least to some extent on the accuracy of the data from different countries, but as a whole the proven reserves represent only a fraction of the total. In 2003 (BP, 2004), the most abundant fossil fuel continued to be coal (high-grade coal, 374 to 446 Gt C, and low-grade coal, 163 Gt C), followed by oil (133 Gt C), and natural gas (91 Gt C). At the 2003 production rates, the coal reserves would be depleted in 192 years, oil in 41 years, and natural gas in 67 years. The greater amount of the estimated reserves of 5,000 Gt C (Fig. 1.5) suggests that the actually usable reserves will continue to increase with time. World oil reserves show such irregular trends of increasing estimates in 1942–81 and 1982–2000 (Fig. 1.6). The year 2000 mean estimate is about 2.5 times greater than the proven recoverable oil reserves in 2003, making it about 100 years to depletion at the 2003 oil production rate. Time to depletion would be longer at lower production rates, such as are anticipated for some countries in the future.

5 Feedbacks in the Carbon Cycle

In the ocean, atmospheric CO_2 dissolves in the surface water layer that is up to a few hundred meters thick and well mixed by winds, where it is being used in photosynthetic production of organic matter and in production of calcium carbonate. Net removal of carbon from ocean water by sedimentation of undecomposed organic matter and calcium carbonate must be balanced by inputs of carbon from other sources. A large and important source is oxidation of organic matter exposed on land and in older sediments and dissolution of limestones on the continents.

Mean fluxes of long-term storage or burial in sediments of carbonate and organic carbon are in the ratio carbonate/organic = 3.6 (Table 1.3). However, their reservoir sizes are in the ratio carbonate/organic = 5.2 (Table 1.2). This indicates that the rate of recycling of organic matter in sediments is faster than that of the carbonates on a geologically long time scale.

As atmospheric CO_2 is the source both of organic matter stored on land and in oceanic sediments, and the atmosphere and ocean water also lose carbon by the formation of $CaCO_3$ and its storage in sediments, this source would have been exhausted rapidly if it were not replenished. Replenishment comes from oxidation of old organic matter occurring in sediments in disseminated form and in more concentrated formations, such as black shales. Tectonic processes, responsible for the uplift of continental masses and mountain building, expose older sediments to the atmosphere. Weathering of organic matter provides CO_2 input to the atmosphere, weathering reactions of minerals consume CO_2, and volcanic emissions also add or, in part, return subducted carbon as CO_2 to the atmosphere.

In the course of the geologic history of the Earth during its lifetime of approximately 4.6 billion years, the partitioning of carbon dioxide among the atmosphere and other biogeochemical reservoirs has varied greatly. The rise of atmospheric carbon dioxide by approximately 35% above its pre-industrial concentration of some 300 years ago has led to the current environmental issue related to CO_2, that of global warming and climate change. However, this change is small but its rate is rapid when compared to many greater changes in the geologic past. The latter changes have variably affected the planet's climate, making it sometimes warmer or colder than now, with many consequent effects, not always well understood, to the evolution of life over the geologically long time span. At the shorter time scale of the Pleistocene Epoch that covers the last 1.8 million years of Earth's history, major environmental changes have followed the periods of advancing and retreating glaciations, and the role of humans as a geologic agent became pronounced during the most recent Holocene Epoch of 11,000 years and, most strongly, in the time period called the Anthropocene, the name that was originally proposed for the industrial age since the late 1700s (Crutzen, 2002; Crutzen and Stoermer, 2000). The likely course of the early geologic history of carbon dioxide and its role in shaping the Earth's surface was controlled by the sources and fate of volatiles that degassed from the Earth's interior or were added late in its accretionary history and formed the primordial, abiotic atmosphere that is discussed in the next chapter.

Chapter 2

Earth's Volatile Beginnings

In this chapter we discuss Earth's early prebiotic history and the conditions under which life emerged. The chapter focuses in particular on the sources and fate of the volatiles that degassed from the Earth's interior and formed its atmosphere, hydrosphere, and sedimentary rocks, giving rise to the global carbon cycle. The chapter also discusses the cooling history and compositional changes of the primordial atmosphere and the early oceans up to the time when life appears. The environmental conditions and the material and energy sources of the early organismal groups are primarily inferred from those of the chemosynthesizing and photosynthesizing organisms of the present.

1 The Major Volatiles

1.1 Sources of Volatiles and Degassing

Earth's history in its early stages went from the formation of the planet by accretion of a cloud of solid particles and gases to its partial melting after accretion that resulted in the structural separation of its major subdivisions—the inner and outer core, the mantle, and the lithosphere and crust. After the formation of the Earth about 4.55 billion years ago, a long period of some 600 to 800 million years preceded the origin of life, the first fossil records of which are dated at 3.75 to 3.9 billion years before present (Holland, 1997; Mojzsis *et al.*, 1996; Rosing, 1999). However, it is possible that life emerged even earlier if the findings from the Jack Hills Formation of Western Australia hold up to scientific scrutiny. Zircon ($ZrSiO_4$) grains in this formation have uranium-lead ages ranging from 4.2 to 4.4 billion years. In addition, rounded detrital zircon grains are found in gravelly deposits of this formation and have $\delta^{18}O$ signatures suggesting formation in a cold, wet environment (Valley *et al.*, 2002). Thus a primitive, cool hydrosphere may have been present only 150 to 350 million years after Earth's formation; if so, life on Earth may have had a longer history than previously thought.

The time of a scant or no rock record on Earth, known in the geological time scale as the Hadean (Hidden Life) Eon (Fig. 1.1), was a long period of environmental conditions very different from those of later time, when photosynthetic organisms had emerged and greatly modified the environment of the Earth's surface. In this early, prebiotic period of Earth's history, when the temperatures near the Earth's surface are believed to have been considerably higher than at present, near 1000°C, the lighter chemical constituents had separated or been degassed from the solids and melt, forming the fluid layer of the primordial atmosphere. These volatile constituents were to become the major components of the outer shell of the Earth—its atmosphere, hydrosphere, biosphere, and sediments.

The chemical elements and their compounds that escaped from the molten or solid Earth's interior by degassing made the primordial atmosphere. These are known as the *volatiles* (Rubey, 1951, 1955) or *hyperfusibles* (Poldervaart, 1955), the major ones being water (H_2O), carbon (C), nitrogen (N), sulfur (S), and chlorine (Cl). Rubey (1951) and later investigators, who studied the balance of the volatiles at the Earth's surface, demonstrated that their complete inventory includes some of the same constituents that had been added to the Earth's surface by the weathering of igneous rocks and, in general, by transfer of materials from the crust and upper lithosphere to the oceans and sediments. The difference between the total masses of volatiles and the masses contributed by the weathering of igneous rocks is known as *excess volatiles*. The masses of H_2O, C, N, S, and Cl in the present-day Earth's surface environment are approximately the masses of the volatiles on the primordial Earth that became incorporated in the differentiated outer shell of the cooler, later Earth. Estimates of the volatile masses from several sources are summarized in Table 2.1.

Water is the most abundant volatile and its mass exceeds the masses of other volatiles by a factor of 30 to 300. The individual estimates of H_2O and N, shown in Table 2.1, agree within 10 to 15%, but there is much greater variation among the estimates of C, Cl, and S, attributable probably to the differences in estimates of the masses of sedimentary rock salt, mineral sulfates, sulfides, and the organic and inorganic carbon reservoirs. Our estimates, as given in Table 2.2, represent the total masses of the volatiles in the Earth's surface reservoirs: the atmosphere, oceans, and sediments. For these reservoirs, as well as for the oceanic crust, continental crust, and upper mantle, additional data on the volatile element masses are shown in Box 2.1.

It should be noted that the continental and oceanic crust and the upper mantle are large and significant reservoirs of some of the volatiles. However, the mass of H_2O on the Earth's surface is greater than the mass of water in the upper mantle. Similarly, the mass of chlorine is much greater in ocean water and sediments than in the mantle.

EARTH'S VOLATILE BEGINNINGS   25

Table 2.1   Volatiles in primordial Earth's surface environment (masses in grams of chemical species shown)

Major species	Rubey, 1951 Excess volatiles	Poldervaart, 1955 Hyperfusibles	Walker, 1977			Table 2.2 Total volatiles on Earth's surface
			Total volatiles	From igneous rocks	Excess volatiles	
H_2O	1.66×10^{24}	1.66×10^{24}	1.590×10^{24}	0.031×10^{24}	1.559×10^{24}	1.430×10^{24}
C	2.48×10^{22}	6.22×10^{22}	7.609×10^{22}	0.804×10^{22}	6.805×10^{22}	7.784×10^{22}
N	4.20×10^{21}	4.50×10^{21}	4.892×10^{21}	0.04×10^{21}	4.852×10^{21}	4.890×10^{21}
Cl	3.00×10^{22}	3.40×10^{22}	3.120×10^{22}	0.1×10^{22}	3.020×10^{22}	4.311×10^{22}
S	2.20×10^{21}	3.00×10^{21}	5.220×10^{21}	0.8×10^{21}	4.420×10^{21}	1.245×10^{22}
Total volatiles	1.721×10^{24}	1.768×10^{24}	1.707×10^{24}	0.041×10^{24}	1.667×10^{24}	1.568×10^{24}

Table 2.2 Masses (in grams) of the five major volatiles in the upper mantle, crust, and surface reservoirs

Species	Upper mantle[1]	Oceanic crust[2]	Continental crust[3] (8)	Earth surface[4]	Sediments[5]	Ocean[6]	Atmosphere[7]
H_2O	$(2 \text{ to } 5.5) \times 10^{23}$		6.2×10^{23}	1.430×10^{24}	7.15×10^{22}	1.358×10^{24}	1.173×10^{19}
C	$(8.9 \text{ to } 16.6) \times 10^{22}$	9.200×10^{20}	2.576×10^{21}	7.784×10^{22}	7.780×10^{22}	3.850×10^{19}	7.850×10^{17}
N	1.11×10^{22}	?	2.240×10^{20}	4.890×10^{21}	1.000×10^{21}	2.263×10^{19}	3.867×10^{21}
Cl	5.55×10^{20}	?	1.680×10^{21}	4.311×10^{22}	1.735×10^{22}	2.576×10^{22}	
S	2.00×10^{23}	5.256×10^{21}	5.936×10^{21}	$(1.07 \text{ to } 1.42) \times 10^{22}$	$(9.5 \text{ to } 13) \times 10^{21}$	1.230×10^{21}	$0 \text{ to } 5.66 \times 10^{14}$

[1] Upper mantle data from compilation by Li (2000), Wood (1996), and references in Box 2.1.

[2] Mass of oceanic crust: 6.5 km × 3.61 × 10^8 km^2 × 2.7 g/cm^3 = 6.57 × 10^24 g. C and S concentrations from Li (2000).

[3] Mass of continental crust: 30 km × 1.49 × 10^8 km^2 × 2.5 g/cm^3 = 11.2 × 10^24 g. Species concentrations from Li (2000).

[4] Sum of the masses in sediments, ocean, and atmosphere.

[5] Total sediment mass (continental and oceanic) 2.09 × 10^24 g (Li, 2000). Other estimates: 1.7 × 10^24 g (Poldervaart, 1955), 1.8 × 10^24 g (Gregor, 1968), 3.2 × 10^24 g (Garrels and Mackenzie, 1971), 2.7 × 10^24 g (Ronov, 1980). Sulfur content of sediments 0.45 to 0.62 wt % (compilations of Li, 2000, and Lerman, 1979).

[6] From species concentrations in gram/liter (Li, 2000) and ocean volume 1.37 × 10^21 liter.

[7] From Table 2.3, Recent atmosphere.

[8] Water content of crystalline crustal rocks 3.5 wt %, range 2.5 to 4.5 wt % (Poldervaart, 1955).

Box 2.1   Water in some of the Earth's reservoirs

Water content of some hydrated minerals

Mineral	Idealized chemical composition (Deer et al., 1962a b)		Formula H_2O (weight %)
	Stoichiometric	As oxides	
Brucite	$Mg(OH)_2$	$MgO \cdot H_2O$	30.9
Gibbsite	$Al(OH)_3$	$Al_2O_3 \cdot 3H_2O$	34.6
Kaolinite	$Al_2Si_2O_5(OH)_4$	$Al_2O_3 \cdot 2SiO_2 \cdot 2H_2O$	14.0
Illite	$KAl_5Si_7O_{20}(OH)_4$	$0.5K_2O \cdot 2.5Al_2O_3 \cdot$ $7SiO_2 \cdot 2H_2O$	4.7
Nontronite	$(0.5Ca, Na)_{0.66}Fe^{3+}_4$ $(Si_{7.34}Al_{0.66})O_{20}(OH)_4$	$0.33Na_2O \cdot 2Fe_2O_3 \cdot 0.33$ $Al_2O_3 \cdot 7.34SiO_2 \cdot 2H_2O$	4.2
Talc	$Mg_3Si_4O_{10}(OH)_2$	$3MgO \cdot 4SiO_2 \cdot H_2O$	4.8
Serpentine	$Mg_3Si_2O_5(OH)_4$	$3MgO \cdot 2SiO_2 \cdot 2H_2O$	13.0
Chlorite	$Mg_5Al_2Si_3O_{10}(OH)_8$	$5MgO \cdot Al_2O_3 \cdot 3SiO_2 \cdot 4H_2O$	13.0
Biotite & phlogopite	$K(Mg,Fe)_3AlSi_3O_{10}(OH)_2$	$0.5K_2O \cdot 3(Mg,Fe)O$ $\cdot 0.5Al_2O_3 \cdot 3SiO_2 \cdot H_2O$	3.9

Water in the upper mantle: $(2 \text{ to } 5.5) \times 10^{20}$ kg

Upper mantle: thickness from (10 to 50) to 700 km, mass
   18.5% of Earth mass (Li, 2000) or 1.11×10^{24} kg.
Water content: 200 ppm (Wood, 1996), 400 ppm (Dixon
   and Clague, 2001), 100 ppm MORB mantle, 525 ppm
   basalts (Dixon *et al.*, 1997), 450 ppm source Kilauea
   basalts (Wallace, 1998)

Rate of formation of new oceanic crust (Mottl, 2003): $(6.0 \pm 0.8) \times 10^{13}$ kg/yr

Water taken up in serpentinization of oceanic crust: 2×10^{12} kg/yr

Serpentinization taken as 25% of the crust; water content
   of serpentinized rock 13 wt %

Water flow from mantle to the surface: $(4.5 \text{ to } 4.8) \times 10^{11}$ kg/yr

Dehydration of subducting crust $\sim 3.6 \times 10^{11}$ kg/yr;
   release in subaerial volcanism 0.9 to 1.2×10^{11} kg/yr
   (Kerrick and Connolly, 1998)

Water photodissociation in the atmosphere: 4.8×10^8 kg/yr
   (Walker, 1977)

The opposite holds for sulfur, the abundance of which is greater in the upper mantle. The
occurrence of large amounts of the volatiles in the Earth's mantle shows that degassing
removed only some fraction of the volatiles that made the primordial atmosphere and

ocean and that constitute the present-day hydrosphere, part of the atmosphere, and part of the sediments accumulated since the Earth's early days.

As there was very little free oxygen in the primordial atmosphere before photosynthetic organisms emerged, what were the chemical species of the five major volatiles during the several hundred million years of the prebiotic Earth's history? One view holds that the primordial anoxic atmosphere contained water (H_2O), molecular nitrogen (N_2), ammonia (NH_3), methane (CH_4), and molecular hydrogen, H_2 (Poldervaart, 1955; Schopf, 1980; Urey, 1952). In particular, Gold (1999) advanced a view of methane as the main carbon gas that formed with other hydrocarbons at the time of the Earth's accretion and became a major component of the primordial atmosphere and precursor of biological chemosynthesis (Section 5). Another view, based on a thermodynamic analysis of the mineral-fluid systems approximating the primordial Earth and its external fluid shell, argues for an early, prebiotic atmosphere as composed of H_2O, carbon dioxide (CO_2), molecular nitrogen, reduced and/or partly oxidized sulfur in the form of H_2S and/or SO_2, and hydrogen chloride (HCl), with possibly small amounts of carbon monoxide, CO, and molecular hydrogen, H_2 (Rubey, 1951; 1955; Walker, 1977). Table 2.3 compares estimates of volatiles in the primordial and present-day atmospheres.

A question of whether all the volatiles were more or less simultaneously degassed or whether some were added at different stages in the cooling history of the planet bears on the composition of the primordial atmosphere. The Earth's mantle contains large masses of CO_2, H_2O, and sulfur, and early degassing of CO_2 and other gases together with H_2O would have made a primordial atmosphere very different from the case if a large fraction of H_2O were absent from the atmosphere but remained dissolved in the primordial melt. Holland (1984) argued that if little melting occurred during the accretion process of the Earth, then most of the CO_2 and H_2O would have been held in the gaseous phase. However, significant melting of the accreted material would have taken up a large fraction of H_2O in the melt, with an estimate that a mass of water equivalent to 50% of the present-day hydrosphere might have been dissolved in a 50 km thick layer of melt. At present, the mass of H_2O in the upper mantle, a layer 700-kilometer thick, is one order of magnitude smaller than the mass of water in the oceans (Table 2.2).

The view that most of the volatiles on the Earth's surface had been released from the mantle by degassing rather than remaining on the surface from an early stage of planetary accretion was advocated by several authors (Holland, 1984; Li, 2000; Rubey, 1951; Walker, 1977). Degassing was probably not uniform and for such volatiles as carbon dioxide, water, and perhaps sulfur, large fractions remain in Earth's mantle and possibly deeper in the planetary interior. The total CO_2 transfer from the mantle includes the flux at mid-ocean ridges and spreading zones, flux at volcanic arcs near the areas of subduction of the oceanic crust, and the flux from mantle plumes and hot spots and, possibly, from very deep sedimentary basins. The main pathways of transfer in the opposite direction, from the oceans to the mantle, are chemical reactions between ocean water and basalts in the oceanic crust, and subduction of calcium carbonate and organic carbon in sediments accumulating on top of the basaltic ocean floor. The global magmatic flux of carbon to the hydrosphere and atmosphere has been

Table 2.3   Volatiles in primordial and present-day atmosphere. M_i is molecular mass

Major species	M_i (g/mol)	Primordial			Recent[2]		
		Mass (g)[1]	Mass (mols)	Vol or mol %	Mass (g)	Mass (mols)	Vol or mol %
H_2O	18.0153	1.430×10^{24}	7.938×10^{22}	90.58	1.17×10^{19} [3]	6.51×10^{17}	0 to 3%
CO_2	44.0098	2.852×10^{23}	6.481×10^{21}	7.40	2.878×10^{18} [4]	6.539×10^{16}	0.0370
N_2	28.0135	4.890×10^{21}	1.746×10^{20}	0.20	3.867×10^{21}	1.380×10^{20}	78.082
HCl	36.4606	4.434×10^{22}	1.216×10^{21}	1.39			
H_2S	34.0819	1.323×10^{22}	3.882×10^{20}	0.44			
Other Species							
H_2	2.0159			<0.1?	1.889×10^{14}	9.370×10^{13}	0.000053
CH_4	16.0428			≥0.01?	4.822×10^{15}	3.006×10^{14}	0.00017
CO	28.0104			?	1.238×10^{15}	4.420×10^{13}	0 to 0.000025
SO_2	64.0648			*see* H_2S above	1.133×10^{15}	1.768×10^{13}	0 to 0.00001
NH_3	17.0306			0.001?	9.033×10^{12}	5.304×10^{11}	0 to 0.0000003
Ar	39.9480				6.590×10^{19}	1.650×10^{18}	0.933
O_2	31.9988				1.185×10^{21}	3.703×10^{19}	20.947
Primordial total	20.28	1.778×10^{24}	8.764×10^{22}	100.00			
Present-day dry[5] atm.	28.97				5.121×10^{21}	1.768×10^{20}	100.00

[1] From masses at the Earth's surface in Table 2.2.
[2] From Walker (1977) and Mackenzie (2003).
[3] From mean water content of atmosphere (mass/area), 2.3 g H_2O/cm^2 (from data in Lerman, 1979).
[4] Corresponds to 370 ppmv CO_2 in year 2001.
[5] As water vapor is a variable component of the atmosphere, the chemical composition and many of the atmospheric properties are often given for a dry atmosphere containing no H_2O. The term dry atmosphere also indicates that the atmosphere has no condensable components within a certain range of temperature and pressure (Box 2.2).

estimated at 6×10^{12} mol C/yr, with a possible range from 4×10^{12} to 10×10^{12} mol C/yr. This total flux is comprised of the flux at mid-oceanic ridges, $(2.2 \pm 0.9) \times 10^{12}$ mol C/yr; flux from volcanic arcs, about 2.5×10^{12} mol C/yr, of which 80% is derived from the subducting plate; and flux from mantle plumes, $\leq 3 \times 10^{12}$ mol C/yr (Marty and Tolstikhin, 1998). In another estimate, mid-ocean ridges emit CO_2 at the rate of 1×10^{12} to 3×10^{12} mol C/yr and consume about 3.5×10^{12} mol C/yr in calcium carbonate formation in hydrothermal veins in submarine basalts. Emissions of CO_2 from subaerial volcanism amount to 2×10^{12} to 2.5×10^{12} mol C/yr (Kerrick, 2001a, b). An estimate of total transfer of carbon from the oceans to the mantle at 5.4×10^{12} mol C/yr, consisting of 1.2×10^{12} mol C/yr of sedimentary carbonate, 0.8×10^{12} mol C/yr in organic carbon, and 3.4×10^{12} mol C/yr in metabasalts (Kerrick and Connolly, 1998), is in reasonable agreement with the total flux from the mantle to the surface at 6×10^{12} mol C/yr, as mentioned above.

1.2   Reduced or Oxidized Carbon

Elemental carbon, such as it occurs in the minerals graphite and diamond (C), or chemically reduced carbon in organic matter is thermodynamically unstable in the presence of even very low concentrations of oxygen gas over a wide range of temperatures. The products of carbon oxidation are carbon monoxide (CO) and carbon dioxide (CO_2), with the latter predominating under the conditions of the cooling Earth and the very low, prebiotic concentrations of free oxygen gas in the primordial atmosphere (Eugster, 1966). The arguments in favor of the occurrence of carbon, nitrogen, and sulfur as gases—CO_2, N_2, and H_2S—rather than methane CH_4, carbon monoxide CO, ammonia NH_3, or sulfur dioxide SO_2, in the primordial atmosphere are based on the oxidation state of the upper mantle, chemical composition of basaltic magmas, and analysis of chemical equilibria in the systems containing these gaseous species and minerals and melts in the presence of low partial pressures or low fugacity of molecular oxygen, O_2, and hydrogen, H_2. However, the presence of methane at concentration levels of 100 to 300 ppmv in the Earth's atmosphere between the ages 3500 and 750 Ma, in the Archean and Proterozoic Eons, would have been sufficient to overcome the effects of the lower solar luminosity in the Early Archean, known as the faint young Sun (Chapter 3), and provide the necessary warming to maintain liquid water on Earth (Pavlov et al., 2000, 2003). Furthermore, the formation of organic molecules by photolysis of methane in the stratosphere, the so-called organic smog, was proposed as a shield against the ultraviolet radiation that penetrated the primordial atmosphere (Sagan and Chyba, 1997).

If the fugacities or pressures of oxygen were controlled by mineral equilibria in the mineral-melt system approximating the composition of the early mantle, taken as the system $FeO-MgO-SiO_2$ (as summarized by Holland, 1984), and the low abundance of hydrogen gas in the Earth's atmosphere was due to its low molecular weight and consequent escape into interplanetary space (Goody, 1976; Walker, 1977), then the prevalence of CO_2 over CH_4, N_2 over NH_3, and H_2S over SO_2 may be taken as characteristic of the primordial atmosphere, particularly after the outer part of the Earth cooled significantly below $1000°C$. However, the abundance of methane relative to CO_2 in the primordial atmosphere might have been relatively high in the presence of H_2O at the fugacity of 5 atm (Holland, 1984): the fugacity ratio of $CH_4/CO_2 = 1$ in a system with elemental iron and low oxygen fugacity of about 10^{-13} at temperatures above 130°C. But much lower ratios of methane to carbon dioxide, $CH_4/CO_2 \approx 10^{-12}$ to 10^{-4}, near 1150°C, were given for a system controlled by magnetite and silicate mineral assemblages, with the higher value of the ratio corresponding to the oxygen fugacity of 10^{-8}. In the present-day atmosphere, the partial pressure ratio is $CH_4/CO_2 = 4.6 \times 10^{-3}$, where methane emissions come from the biosphere and human agricultural and other activities (Table 2.3). In an inorganic system containing H_2O and O_2, as in the present-day atmosphere at 25°C, there would be virtually no methane and the equilibrium partial pressure ratio CH_4/CO_2 would have a vanishingly low value of 10^{-142}. The reaction between CO_2 and H_2O that produces methane and molecular oxygen is:

$$CO_2 + 2H_2O = CH_4 + 2O_2 \tag{2.1}$$

In the modern, oxygen-containing atmosphere, oxidation of methane is primarily accomplished by reactions with the hydroxyl radical and to a lesser extent in soils and sediments by methanotrophic bacetria. The equilibrium constant of reaction (2.1)

$$K = \frac{f_{CH_4} f_{O_2}^2}{f_{CO_2} f_{H_2O}^2} \tag{2.2}$$

increases with increasing temperature (Fig. 2.1B) that favors the formation of methane. In (2.2), f is the fugacity, defined as the product of the gas partial pressure (P_i) and the fugacity coefficient (χ_i) that is a function of temperature and total pressure:

$$f_i = \chi_i P_i \tag{2.3}$$

In Section 2, we discuss the Earth's prebiotic atmosphere that might have contained all the volatiles at a temperature close to the critical temperature of water, 374°C. To establish whether methane might have been a significant component of such an atmosphere, we need to address the fugacity ratio f_{CH_4}/f_{CO_2} at the temperatures and pressures of the H_2O liquid-gas phase boundary (Fig. 2.1A) in a range from below the critical point (350°C, 165 bar) down to 100°C and $P_{H_2O} = 1$ bar. Within this temperature and pressure range, the pressure effect on the equilibrium in (2.2) is insignificant and the differences between the fugacities and partial pressures of methane and carbon dioxide are also negligible (Duan et al., 1992b; Lemmon et al., 2003). Thus the partial pressure ratio CH_4/CO_2 can be written as:

$$\frac{P_{CH_4}}{P_{CO_2}} \approx \frac{f_{CH_4}}{f_{CO_2}} = \frac{K f_{H_2O}^2}{f_{O_2}^2} \tag{2.4}$$

The results, shown in Fig. 2.1C, indicate that the CH_4/CO_2 ratios in the Earth's atmosphere would be very low in the temperature range from 100 to 350°C and 1 to 165 bar pressure even at the very low oxygen fugacities, characteristic of the magmatic iron-silicate reactions at higher temperatures, that were used in this calculation. Thus it is unlikely that abiogenic methane forming in reaction (2.1) could have been a significant component of the early atmosphere. Much lower oxygen fugacities (10^{-57} to $10^{-29.5}$)

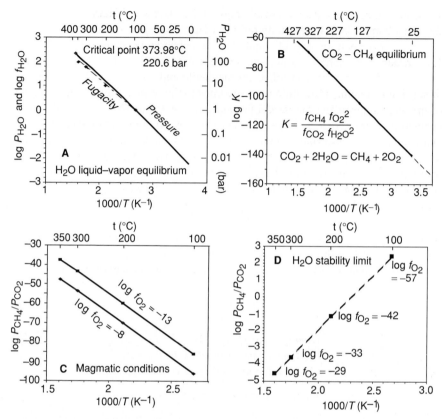

Figure 2.1   A. Vapor pressure and fugacity of H_2O from 0°C to the critical point (Lide, 2004; fugacity coefficients from Duan *et al.*, 1992b). B. Equilibrium constant for CO_2, H_2O, methane, and oxygen reaction as a function of temperature (from Robie *et al.*, 1978). C. Calculated CH_4/CO_2 partial pressure ratios along the H_2O *T-P* curve (A) from equation (2.4) and H_2O fugacity values characteristic of magmatic reaction temperatures (Holland, 1984). D. CH_4/CO_2 partial pressure ratios corresponding to the O_2 fugacity values at the stability limit of H_2O, equation (2.5), calculated from K_{H_2O} (Robie *et al.*, 1978), water vapor fugacity from A, and $f_{H_2} = 0.001 f_{H_2O}$ (Table 2.3). Note the difference between the CH_4/CO_2 partial pressure ratios in C and D.

characterize the stability limit of H_2O in the same temperature range, according to the reaction:

$$H_2O = H_2 + \tfrac{1}{2}O_2$$

$$f_{O_2} = \left(\frac{K_{H_2O} f_{H_2O}}{f_{H_2}} \right)^2 \tag{2.5}$$

At such hypothetically low oxygen fugacities, the CH_4/CO_2 ratios might have been significantly higher (Fig. 2.1D, with the hydrogen fugacity assumed as 0.1% of f_{H2O},

Table 2.3), but the occurrence of such low oxygen values is unlikely if they were determined by magmatic reactions that maintained oxygen in the atmosphere.

1.3 Water

Water, H_2O, is by far the most abundant volatile on Earth, being the main constituent of the hydrosphere. The hydrosphere contains all forms of H_2O: water vapor and its condensed forms in the atmosphere; liquid water in the oceans, on the land surface in lakes and rivers; groundwater in soils and deeper below the land surface; water filling the pore space of oceanic sediments; and continental and oceanic ice. Water that is part of the biosphere is contained in tissues of living organisms and cycles rapidly between living organic matter and the hydrosphere, but this transient water in the biosphere is only a very small fraction of the total.

Generally not included in the water balance of the hydrosphere is H_2O bound to minerals that constitutes part of their crystal structure and chemical composition, such as water in the commonly occurring sedimentary mineral gypsum, $CaSO_4 \cdot 2H_2O$, other hydrated minerals, and the hydroxyl groups, OH^-, that are part of the chemical stoichiometric formulas of clays and other aluminosilicate minerals (Box 2.1). Also not included in the hydrosphere and, therefore, in the volatile inventory is water contained in the Earth's mantle. The mass of water in the hydrosphere is about 1.4×10^{21} kg (Table 2.2). The exchange of water between the mantle and Earth's surface is a two-way process that has likely taken place almost continuously, but probably at variable rates during Earth's history. Although this process is not completely understood, water may be added to the surface environment from the mantle at mid-ocean spreading zones of rifts and rises and at the sites of submarine and continental volcanism, and through thick piles of sediments, and it may be removed by subduction of ocean floor and overlying sediments.

The circulation of ocean water through the mid-ocean ridge systems also leads to reactions with the hotter oceanic crust and some removal of water into hydrated minerals forming as alteration products of the oceanic crust and subducted sediments. The main mineral groups that form as alteration products of the iron and magnesium silicates in the oceanic lithosphere and crust are serpentines and chlorites that are sheet-structure silicates containing water as part of their crystal chemical composition, usually in the form of the hydroxyl ion (OH^-) bound in some coordination to Mg^{+2}, Al^{+3} and other metal ions in the tetrahedral and octahedral sites of the lattice. Serpentines and chlorites contain 13 wt % H_2O as part of their idealized chemical composition (Box 2.1). The amount of water consumed in hydrothermal alteration reactions of the newly formed oceanic crust is about 2×10^{12} kg/yr, with an upper limit of 8×10^{12} kg/yr if all the newly formed crust were serpentinized. In fact, only some fraction of between 5 to 25 vol % of the oceanic crust is believed to be serpentinized (Kerrick and Connolly, 1998; Mottl, 2003; Peacock, 1993) and subduction of the serpentinized crust and its deeper metamorphism release some of the water at a rate of about 3 to 4×10^{11} kg/yr. This rate has been estimated to be three to four times greater than the emission of water from subaerial volcanism, such that the total H_2O flux from the Earth's interior to the surface is from about 4 to 5×10^{11} kg/yr. In comparison to water flow through the oceanic lithosphere, the rate of water removal from the upper atmosphere by photochemical destruction and escape of hydrogen is much smaller, 4.8×10^8 kg/yr (Box 2.1).

Thus it follows from the preceding estimates that subduction of H_2O at present likely exceeds the mass returned to the surface. Water fluxes on a mol/year basis are by a factor of about 10 smaller than the CO_2 fluxes. Relative to the mass of water in the present-day hydrosphere, about 1.4×10^{21} kg H_2O, the crustal subduction and return fluxes correspond to long residence times of the order of 10^8 to 10^9 years. Faster rates of formation and subduction of the oceanic crust on the prebiotic Earth might have resulted in shorter water recycling times.

Comets contain H_2O and other volatiles and it has been hypothesized by many investigators that cometary impacts might have added significant amounts of water to the Earth, in particular from the smaller, water-rich comets that may vaporize in the atmosphere. From the data on the lunar impact record, one estimate of water from cometary inputs gave 3 to 39% of the water mass in the present oceans delivered over a relatively short period of 700 million years, between 4.5 and 3.8 Ga (Chyba, 1987). From astronomical observations and their later revisions (Frank and Sigwarth, 1993), an earlier estimate of the cometary water flux to the Earth's atmosphere of 30×10^{10} molecules H_2O cm^{-2} s^{-1} was revised down to 6×10^{10} molecules H_2O cm^{-2} s^{-1}, closer to the value of 1×10^{10} molecules H_2O cm^{-2} s^{-1} given by Reid and Solomon (1986). For the whole Earth, the latter two fluxes translate into water input rates of 2.9×10^{14} and 4.8×10^{13} g H_2O/yr, that would supply in 4.5 billion years 91 to 15% of all the water on the Earth's surface. The higher percentage is unlikely because it would create a major problem in the balance of the water budget of the Earth.

1.4   Other Volatiles

Among the primordial volatiles, the least controversial volatile history is probably that of hydrogen chloride, HCl. There are no major sources of chlorine in igneous rocks, but there is much of it in ocean water occurring as the aqueous chloride ion (Cl$^-$) and in the sediments, where it occurs in chloride minerals, mostly the mineral halite (NaCl), common rock salt, and other minerals forming under evaporitic conditions. This evidence suggests that HCl in the primordial atmosphere likely dissolved in water, either when H_2O condensed to liquid or when it accumulated in a significant volume on the Earth's surface, and subsequently reacted with crustal rocks, releasing metal cations in the reaction between acid and metal-silicate minerals.

At present, most of the CO_2 and H_2O have been removed from the atmosphere, free oxygen accumulated through excess of its production by plant photosynthesis over consumption in oxidation reactions, and the noble gas argon is mostly the isotope ^{40}Ar that is produced by radioactive decay of ^{40}K in the lithosphere. Perhaps the best known argument in favor of the occurrence of ammonia, NH_3, in the primordial atmosphere is based to a large extent on the Miller-Urey experimental amino-acid synthesis in a mixture of gases subjected to electrical discharge, as an analogue of the primordial atmosphere (Miller, 1953; Miller and Orgel, 1974). As for the occurrence of molecular nitrogen, N_2, Lovelock (1988, 1995) argued that because this form of nitrogen is more oxidized than nitrogen in ammonia, N_2 is a product of long-term biological photosynthetic activity that modified the composition of the atmosphere through nitrogen oxidation and reduction reactions (nitrification and denitrification) rather than one of its primordial constituents. Atmospheric concentrations of ammonia, NH_3, at a level of about 10 ppmv (Table 2.3), absorbing part of the infrared radiation, were proposed as a greenhouse gaseous constituent of the primordial atmosphere that could have overcome the effect of the faint young Sun in an atmosphere not particularly thick in CO_2 (Sagan and Chyba, 1997). The abundance of hydrogen, H_2, in the prebiotic atmosphere might have also been relatively high, near 0.1 mol %, if the volcanic outgassing rates were several times higher than at present, while the rate of hydrogen escape from the atmosphere was a relatively slow diffusion-controlled process (Pavlov et al., 2001).

2   Primordial Atmosphere-Ocean System

For the five primordial volatiles—H_2O, CO_2, N_2, HCl, and H_2S (Table 2.3)—to exist in a gaseous state, the temperature of the atmosphere should have been sufficiently high to prevent their condensation. The critical temperature is an upper limit for this state, above which the substance exists as a supercritical fluid that can not be condensed to a liquid. Critical temperatures, pressures, molar volumes, and densities of the five major volatiles, some additional gases of potential significance on the prebiotic Earth, and two other atmospheric gases, oxygen and argon, are given in Table 2.4. The highest critical temperature is that of H_2O, 374°C. This may be taken as a characteristic temperature of that stage of the early Earth before the volatiles began to condense to liquid water and form an aqueous solution of reactive and nonreactive gases. This hypothetical starting stage of the atmosphere is useful to consider for a clearer understanding of its early evolution, bearing in mind that most of the volatiles that were released from the mantle to the atmosphere might have been released over some period of geological time or when the Earth's surface cooled already to below the critical temperature of H_2O

Table 2.4   Critical constants of some volatiles and atmospheric gases (from Ambrose, 1994): temperature (T), pressure (P), molar volume (V), and density (ρ)

Gas	T_c (°C)	T_c (K)	P_c (bar)	V_c (cm³/mol)	ρ (g/L)
H_2O	373.99	647.14	220.6	56	321.7
CO_2	30.99	304.14	73.75	94	468.2
CO	−140.24	132.91	34.9	93	301.2
CH_4	−82.99	190.16	45.99	98.6	162.7
N_2	−146.94	126.21	33.9	90	311.3
NH_3	132.35	405.5	113.5	72	236.5
H_2	−240.18	32.97	12.93	65	31.0
HCl	51.55	324.7	83.1	91	400.7
H_2S	100.05	373.2	89.4	99	344.3
SO_2	157.65	430.8	78.84	122	525.1
Ar	−122.28	150.87	48.98	75	532.6
O_2	−118.56	154.59	50.43	73	438.3

or even below its boiling point. Because the temporal sequence and rates of Earth's cooling, the formation of the early oceans, and the mineral-water reactions that resulted in the accumulation of dissolved solids in ocean water are known very incompletely, we resort to an analysis of the prebiotic atmosphere and ocean in three representative stages:

(1) A hot atmosphere where the five volatiles—water, carbon, nitrogen, sulfur, and chlorine—could occur as gaseous species H_2O, CO_2, N_2, H_2S, and HCl;

(2) A cooler atmosphere, after the water had condensed and accumulated as a liquid on the Earth's surface, the hydrogen chloride was removed from the atmosphere to the ocean, and the hydrogen sulfide possibly reacted with the crust in making sulfide minerals;

(3) An atmosphere where carbon dioxide and nitrogen remained the two main constituents, and CO_2 dissolved in the primordial hydrosphere and reacted with crustal rocks.

For purposes of discussion, each of these three stages is considered as a discreet slice of Earth's history, corresponding to a certain set of environmental conditions. In reality, each stage was not likely to be a discreet, stepwise jump, but all of them probably evolved more or less concomitantly, perhaps over some hundreds of millions of years between the formation of the Earth and the emergence of life.

2.1   Hot Atmosphere

At the critical temperature of H_2O, $374°C$ or $647 K$ (Table 2.4), assuming no interactions among the gases in the primordial mixture, the atmosphere would be made of the mass of all the five volatiles occurring within a spherical layer of a certain thickness and exerting a certain pressure on the Earth's surface. The pressure at the base of such an atmosphere is determined by its mass (Table 2.3), the Earth's surface area A, and gravitational acceleration g that is taken here as a constant for the entire thickness of the atmosphere:

$$P = \frac{mg}{A} = \frac{1.778 \times 10^{21}\,\text{kg} \times 9.8\,\text{m s}^{-2}}{5.10 \times 10^{14}\,\text{m}^2} \times 1 \times 10^{-3} \frac{\text{kPa}}{\text{kg m}^{-1}\,\text{s}^{-2}}$$

$$= 341.6 \times 10^2\,\text{kPa} \quad \text{or} \quad 341.6\,\text{bar} \qquad (2.6)$$

As the Earth's surface cooled to a temperature where water could exist as a liquid, other gases could dissolve in it and react with the crustal rocks, releasing dissolved solids to the primordial hydrosphere, including the ocean. The course of geochemical evolution of the cooling Earth, downward from a high temperature point of $600°C$ and water vapor pressure of some 300 atm, was considered by Garrels and Mackenzie (1971), who concluded that the early hydrosphere containing dissolved HCl and CO_2 must have been a very acidic solution reacting vigorously with the newly formed crust. In analyses of the primordial atmosphere near 4500 Ma before present, CO_2 pressures in

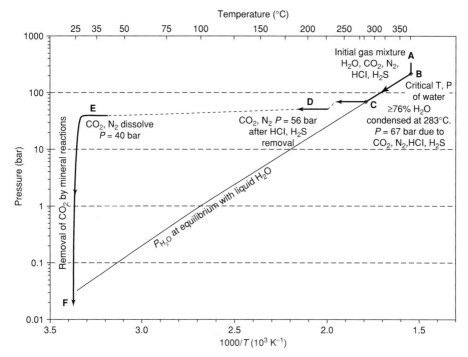

Figure 2.2   Schematic cooling pathway of the Earth's prebiotic atmosphere and surface. Details in the text.

the range from slightly above 0.1 to 10 bar were considered, the high value taken for the water-covered Earth (Kasting, 1987; Walker, 1986). Among the reactive gases in the early atmosphere, it is convenient to separate HCl and H_2S from CO_2 because of their different chemical behavior in solution and in mineral-water reactions, and particularly because of the role of CO_2 in the formation of sedimentary carbonate rocks and, later, in biological production of organic matter.

The assumed initial temperature and pressure of the primordial atmosphere are shown by point A in Fig. 2.2. The figure shows that early cooling lowers the pressure to the critical point of water and continues along the H_2O liquid-vapor phase boundary, from B to C. At point C, the temperature is 283°C and pressure 67.1 bar for pure H_2O, where 76% of the water mass has condensed to liquid. The remaining four gases—CO_2, N_2, HCl, and H_2S—make a pressure of 67 bar (Table 2.3, equation (2.6)). Removal of HCl and H_2S from the atmosphere by dissolution and reactions with crustal rocks leaves behind CO_2 and N_2 that amount to 7.6 mol % of the initial atmosphere, making the residual mass 0.29×10^{24} g and the pressure $P = 56$ bar (point D, Fig. 2.2). If the Earth's surface cooled to 25 to 35°C, then dissolution of CO_2 and N_2 in the primordial ocean would lower the atmospheric pressure further to about 40 bar (point E, Fig. 2.2), as is explained in a later section. A possible liquefaction of CO_2 below the critical temperature of $31°C$ at elevated pressures is not considered here (CO_2 liquid-gas phase transition at the environmental conditions in the ocean and sediments is discussed in Chapter 5). Additional drawdown of atmospheric CO_2, as in the path E–F, could be accomplished by mineral-CO_2 reactions.

It should be reiterated that the path from A to F, shown in Fig. 2.2 as a sequence of discreet steps, was not necessarily the evolutionary path of the Earth's surface during the cooling period, but it was a combination of processes occurring simultaneously. The final stage of a prebiotic atmosphere with a lower CO_2 concentration is likely to be a product of continuous reactions of HCl, H_2S, and CO_2 with crustal minerals and the path from D to F might have been more direct than shown in the figure. The individual segments of the path B–F in Fig. 2.2 are discussed in more detail in the following sections.

2.2   Early Hydrosphere

Hydrogen chloride, dissolving in the early hydrosphere, would have produced the chloride-ion concentration higher than that found in present-day seawater:

$$[Cl^-] = \frac{1.216 \times 10^{21} \text{ mol HCl}}{1.43 \times 10^{21} \text{ kg H}_2\text{O} + 4.43 \times 10^{19} \text{ kg HCl}} = 0.825 \text{ mol/kg} \quad (2.7)$$

What was the chemical composition of the primordial ocean at this stage? The chloride-ion in solution was likely to be balanced by metal-cations dissolved from crustal minerals that reacted with the HCl-containing waters. A concentration of 0.8 mol Cl^-/kg of solution could be balanced by concentrations of between 0.8 mols of univalent cations (Na^+, K^+) and 0.4 mols of divalent cations (Ca^{2+}, Mg^{2+}) or any combination of the two. For a simple estimate of ocean water composition at that stage, taken in units of grams of dissolved solids in 1 kg of solution, the cation concentrations making up 0.8 equivalents/kg might be any combination of the major metal ions in ocean water—Na^+, K^+, Mg^{2+}, and Ca^{2+}. The sum of their masses should be equivalent to the Cl^--ion mass of 1.22×10^{21} mol or 4.31×10^{19} kg Cl^-. For the individual metal ions, the masses equivalent to 1.22×10^{21} mol Cl^- are 2.80×10^{19} kg Na^+, 4.75×10^{19} kg K^+, 1.48×10^{19} kg Mg^{2+}, or 2.44×10^{19} kg Ca^{2+}.

The relative abundances, by weight, of the four metal ions in the source igneous rocks vary from a sequence of K > Na > Ca > Mg in granites to Ca > Mg > Na > K in mafic rocks and basalts. These differ from their sequence in the present-day ocean—$Na^+ > Mg^{2+} > Ca^{2+} > K^+$—that is a product of long-term geochemical reactions that modify the chemical composition of ocean water by processes involving, for example, storage of calcium carbonate in sediments and removal of potassium into clay minerals.

From the preceding estimates of the chloride-ion mass and major metalcation equivalent masses in the early ocean water, a mixture of the cations in different proportions might have been in the range from about 1.5 to 3×10^{19} kg. Then the salinity of ocean water would have been higher than the present-day salt content of 35 g/kg:

$$\text{Salinity} \approx \frac{4.31 \times 10^{22}\,\text{g Cl}^- + (1.5 \text{ to } 3) \times 10^{22}\,\text{g cations}}{1.43 \times 10^{21}\,\text{kg H}_2\text{O} + (5.81 \text{ to } 7.31) \times 10^{19}\,\text{kg solutes}} = 39 \text{ to } 49\,\text{g/kg}$$

$$(2.8)$$

Salinity is a measure of the total concentration of dissolved ionic solutes or dissolved solids in ocean water. The unit that was in use for a long time, grams of solutes in 1 kg of ocean water solution (that is, 1 kg of H_2O and solutes), was superseded by a dimensionless "practical salinity unit" (psu) on the UNESCO salinity scale, based on an electrical conductivity standard that has a value of 35 for an average ocean water salinity, close to the measure of 35 g/kg. Salinity does not include dissolved non-ionized gases or organic compounds. At present, the average global salinity of the ocean is close to 35 g/kg or 35 psu or simply expressed as 35. It follows from the preceding that the primordial ocean was a "chloride ocean": that is, Cl^- was the major anion that balanced the metal cations in solution and resulted in a slightly higher salinity than that of the present-day.

The role of carbon dioxide in the primordial atmosphere-ocean system is discussed in the next section, but beforehand the possible behavior of hydrogen sulfide, H_2S, is briefly addressed. Of the five excess volatiles, H_2S is the least abundant by mass and it represents only 0.44 mol % of the primordial atmosphere (Tables 2.1, 2.3). As the Earth cooled, hydrogen sulfide was probably more likely to react with iron occurring in crustal silicate minerals rather than remain as a gas dissolved in ocean water. However, if sulfur in H_2S were oxidized to the sulfate ion, SO_4^{2-}, in ocean water and SO_4^{2-} were balanced by the major metal cations, then the mass of solutes in the early ocean would have been greater by about 3.73×10^{22} g SO_4^{2-} (3.882×10^{20} mol $H_2S \times 96.06$ g SO_4^{2-}/mol) plus an equivalent mass of balancing cations, such as either 1.78×10^{22} g Na^+ or 0.94×10^{22} g Mg^{2+} or equivalent amounts of K^+ or Ca^{2+}. With these additions, the early salinity would have been:

$$\text{Salinity} \approx \frac{4.31 \times 10^{22}\,\text{g Cl}^- + 3.73 \times 10^{22}\,\text{g SO}_4^{2-} + (3.5 \pm 1.1) \times 10^{22}\,\text{g cations}}{1.43 \times 10^{21}\,\text{kg H}_2\text{O} + (11.54 \pm 1.1) \times 10^{19}\,\text{kg solutes}}$$
$$= 68 \text{ to } 81\,\text{g/kg} \qquad\qquad (2.9)$$

The evidence from the occurrences of sedimentary iron sulfides and the isotopic composition of sulfur in Archean sediments, older than 2500 million years, does not indicate that oxidation of sulfur was already occurring at that time (Hayes *et al.*, 1992). As this chapter deals with the prebiotic Earth at an even earlier time, 4500 to 3900 million years, the inclusion of sulfate in the preceding estimate of the early ocean salinity gives a range of salinity values for an early ocean, despite the fact that it might have contained chloride but no sulfate among its major dissolved constituents.

3 Carbon Dioxide

3.1   Carbon Dioxide Before Dissolution

For illustrative purposes, it is more convenient to treat carbon dioxide in the early atmosphere-ocean system as if it were a gas that was originally present in the primordial atmosphere at the higher temperature and total pressure, and then attained an equilibrium distribution between the atmosphere and ocean water after the Earth's surface cooled to a temperature where liquid water could exist. This stage of development of the Earth's surface is a hypothetical reference state, but as such it has also been used in an analysis of the primordial atmosphere by Rubey (1955). The starting point of this section is a model atmosphere containing CO_2 and N_2 only, as shown by point D in Fig. 2.2, after H_2O, HCl, and H_2S had been removed by condensation to liquid, dissolution, and chemical reactions with crustal minerals. Such a CO_2-N_2 atmosphere was made predominantly of carbon dioxide and hence the mean molecular weight of the atmosphere was considerably higher than at present:

Component	Molecular weight (g/mol)	Mass (mol)	Mass (gram)	Mol fraction
CO_2	44.010	6.481×10^{21}	2.852×10^{23}	0.974
N_2	28.013	0.175×10^{21}	4.890×10^{21}	0.026
Total	43.59	6.656×10^{21}	2.901×10^{23}	1.000

A simple estimate of the atmosphere thickness can be made from the atmosphere volume and the molar volumes of the component gases at a specified temperature and pressure. The atmosphere volume is a spherical shell of thickness h enveloping the Earth of radius r_0, and the atmosphere thickness is much smaller than the Earth radius ($h \ll r_0$):

$$V_a = \frac{4}{3}\pi(r_0 + h)^3 - \frac{4}{3}\pi r_0^3 = 4\pi r_0^2 h \left(1 + \frac{h}{r_0} + \frac{h^2}{3r_0^2}\right) \approx 4\pi r_0^2 h \qquad (2.10)$$

The ideal gas law for a CO_2-N_2 atmosphere, as given above, is:

$$P = \frac{(n_{CO_2} + n_{N_2})RT}{V_a} \qquad (2.11)$$

where n is the number of mols of each gas, R is the gas constant, and T is temperature in kelvin (K). From the preceding equation for an ideal gas, in combination with (2.6) and (2.10), the atmosphere thickness h can be obtained for a range of temperatures from about 282 to 35°C that is shown in Fig. 2.2 as:

$$\begin{aligned}
h &= \frac{(n_{CO_2} + n_{N_2})RT}{mg} \\
&\approx \frac{(6.481 + 0.175) \times 10^{21} \times 8.315 \times 10^7 \times (555 \text{ to } 308)}{2.9 \times 10^{23} \times 980} \times 10^{-5} \, \text{km cm}^{-1} \\
&= 10.8 \text{ to } 6.0 \, \text{km} \qquad (2.12)
\end{aligned}$$

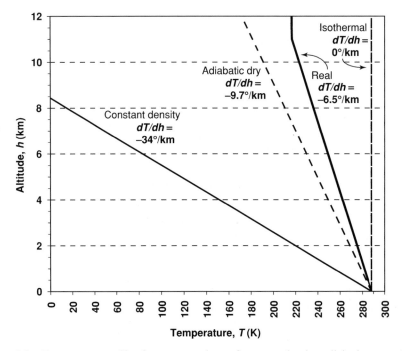

Figure 2.3   Temperature profiles for an atmosphere of constant density, adiabatic atmosphere, isothermal atmosphere, and the "real atmosphere," represented by the U.S. Standard Atmosphere 1976. Earth's surface temperature is at 15°C or 288 K.

In the *linear* or a *constant-density atmosphere* model that is described in more detail in Box 2.2 at the end of this chapter and Fig. 2.3, both the pressure and temperature decrease from the Earth's surface upward, maintaining the density constant and temperature decreasing linearly from some value T_0 at the Earth's surface to $T = 0$ at the top of the atmosphere, at height H. This height H is the atmosphere thickness:

$$H = \frac{RT_0}{gM} \quad (\text{km}) \tag{2.13}$$

The thickness of the atmosphere depends on two parameters: temperature at the Earth's surface, T_0 in kelvin, and the mean molecular weight of the atmosphere, M gram/mol. The higher the temperature at the base of the atmosphere, the thicker it would be. Heavier gases in the atmosphere would make its thickness smaller. This constant-density model assumes that none of the gaseous components of the atmosphere condense at lower temperatures. The present-day atmosphere of constant density is 8.4 km thick (Box 2.2, at the end of the chapter), at the temperature of 15°C or 288.15 K and $P = 1$ bar at the Earth's surface. The thickness of a CO_2-N_2 atmosphere, at the Earth's surface temperatures used previously, 282°C and 35°C, would be from equation (2.13):

$$H \approx \frac{8.315 \times 10^7 \, \mathrm{erg \, cm^3 \, K^{-1} \, mol^{-1}} \times (555 \text{ to } 308 \, \mathrm{K}) \times 10^{-5} \, \mathrm{km \, cm^{-1}}}{980 \, \mathrm{cm \, s^{-2}} \times 43.59 \, \mathrm{g \, mol^{-1}}}$$

$$= 10.8 \text{ to } 6.0 \, \mathrm{km} \tag{2.14}$$

The latter result is the same as that obtained from equation (2.12). At elevated pressures and temperatures, CO_2 and other gases depart from the ideal-gas behavior and their properties in the pressure-temperature field are estimated from other equations of state, such as the van der Waals equation of state or many others that give more accurate results. At the two P-T points of the primordial atmosphere considered above, CO_2 as a van der Waals gas is denser, or its molar volume is smaller, than an ideal gas by 4% at 282°C, 56 bar (D in Fig. 2.2), and it is about 18% denser than an ideal gas at 35°C, 40 bar. In an atmosphere consisting of CO_2 as the main component, the higher density would translate into a smaller thickness of the model atmosphere.

3.2   Partitioning Between the Atmosphere and Water

As CO_2 must have been dissolving in the hydrosphere as the Earth cooled and liquid water formed on its surface, two relevant questions need to be addressed:

First, how did the volatile mass of CO_2 partition itself between the atmosphere and ocean?

Second, what were the effects of chemical reactions of CO_2 with crustal rocks on the atmospheric CO_2 concentration and composition of ocean water?

The volatile mass of CO_2 initially present in the primordial atmosphere (6.481×10^{21} mol CO_2) would partition itself between the atmosphere and the early ocean containing aqueous Cl^--anion and metal cations. This stage of an early "chloride ocean" was discussed in Section 2.2. From an atmosphere containing initially n_a^o mols CO_2, n_w mols dissolved in the hydrosphere, leaving $n_a^o - n_w$ mols in the atmosphere. Then the partial pressure of CO_2 remaining in the atmosphere, assuming ideal gas behavior, is:

$$P_{CO_2} = \frac{n_a^o - n_w}{V_a} RT \quad \text{(bar)} \tag{2.15}$$

where V_a is the atmosphere volume, as given in equation (2.10). To express the partial pressure of CO_2 in terms of the atmosphere thickness at constant density, substitution for RT from (2.13) and for V_a from (2.10) gives the relationship:

$$P_{CO_2} = \frac{(n_a^o - n_w)g M_a}{4\pi r_0^2} \quad \text{(bar)} \tag{2.16}$$

The dissolved CO_2 concentration, from the number of mols of gas in the early ocean, n_w, and mass of ocean water, M_w (that is, the mass of H_2O and dissolved ionic solids), is:

$$[CO_2] = \frac{n_w}{M_w} \quad (\mathrm{mol \, kg^{-1}}) \tag{2.17}$$

Dissolution of an ideal gas in water at equilibrium obeying Henry's law is a quotient of the dissolved gas concentration and its partial pressure in the atmosphere (Chapter 5):

$$K_0' = \frac{[CO_2]}{P_{CO_2}} \quad (mol\,kg^{-1}\,bar^{-1}) \tag{2.18}$$

From the preceding relationships, (2.15) through (2.18), the quotient of the masses of dissolved CO_2 to atmospheric CO_2 is:

$$\frac{n_w}{n_a^o - n_w} = \frac{gK_0'M_wM_a}{4\pi r_0^2} \tag{2.19}$$

The CO_2 solubility coefficient, K_0', depends on temperature, total pressure, and solution composition, as discussed in Chapter 5, but other parameters in the equation are constants. The terms on the right-hand side of equation (2.19) are effectively a partition coefficient of CO_2 between the atmosphere and ocean water, denoted here q:

$$q = \frac{gK_0'M_wM_a}{4\pi r_0^2}$$
$$= \frac{K_0'M_wRT}{V_a} \tag{2.20}$$

With this notation, the relationships for the masses of dissolved and atmospheric CO_2 at equilibrium become simpler:

$$n_w = \frac{n_a^o q}{1 + q} \tag{2.21}$$

$$n_a = \frac{n_a^o}{1 + q} \tag{2.22}$$

It was shown earlier in this chapter that after H_2O had condensed to liquid and HCl and H_2S had been removed from the atmosphere, the main volatiles remaining in the atmosphere were CO_2 and N_2, in proportions of 96.8 and 3.2 mol %, respectively. As carbon dioxide dissolved in the hydrosphere and further reacted with crustal rocks, the proportions of CO_2 and N_2 would change, with CO_2 decreasing and N_2 increasing, which would also result in some decrease in the mean molecular weight of such a dry CO_2-N_2 atmosphere. The masses of atmospheric and dissolved CO_2 can be estimated first for a dry atmosphere of unchanged proportions of the two gases and a mean molecular weight $M_a = 43.59$ g/mol.

The CO_2 partition coefficient q, at 25 to 35°C, has the following values:

$$q = \frac{980\,cm\,s^{-2} \times (0.030 \pm 0.005\,mol\,kg^{-1}\,bar^{-1}) \times 1.50 \times 10^{21} kg \times 44.01\,g\,mol^{-1} \times 10^{-6}\,bar\,g^{-1}cm\,s^2}{5.1 \times 10^{18}\,cm^2}$$
$$= 0.44\,to\,0.31 \tag{2.23}$$

where the Henry's law solubility coefficient K_0' is taken as represented by a range from 0.025 to 0.035 mol kg^{-1} bar^{-1} (Chapter 5; fresh to ocean water; the effect of pressure on K_0 at these temperatures is small and it is disregarded), and the mass of the hydrosphere is taken as $M_w = 1.5 \times 10^{21}$ g (water and dissolved solids, equation (2.8)). The partitioned masses in the ocean and atmosphere, with an initial CO_2 mass in the primordial atmosphere $n_a^o = 6.481 \times 10^{21}$ mol (Table 2.3), are:

in the ocean,

$$n_w = \frac{6.481 \times 10^{21} \text{ mol} \times (0.44 \text{ to } 0.31)}{1.44 \text{ to } 1.31} = 1.98 \times 10^{21} \text{ to } 1.53 \times 10^{21} \text{ mol } CO_2$$
(2.24)

and in the atmosphere,

$$n_a = \frac{6.481 \times 10^{21} \text{ mol}}{1.44 \text{ to } 1.31} = 4.50 \times 10^{21} \text{ to } 4.95 \times 10^{21} \text{ mol } CO_2$$
(2.25)

Thus 24 to 31% of the mass of volatile CO_2 could dissolve in an early hydrosphere, cooled from some higher temperature to 25 to 35°C, without any reactions between CO_2 and crustal rocks that would release metal cations and increase alkalinity in ocean water. At this stage, the concentration of dissolved CO_2 in the hydrosphere would have been very high, 1.02 to 1.32 mol CO_2/kg solution, and its partial pressure in the atmosphere of thickness 6 km that follows from equation (2.14) would have been:

$$P_{CO_2} = \frac{(4.50 \text{ to } 4.95) \times 10^{21} \text{ mol} \times 83.15 \text{ bar cm}^3 \text{ K}^{-1} \text{ mol}^{-1} \times (298 \text{ to } 308 \text{ K})}{5.1 \times 10^{18} \text{ cm}^2 \times 6 \times 10^5 \text{ cm}}$$

$$= 36 \text{ to } 41 \text{ bar}$$
(2.26)

The second constituent of this atmosphere is nitrogen, N_2, that is about 50 times less soluble in water than CO_2. The consequence of the lower solubility is that only a small fraction of N_2, about 0.8% of its original mass of 0.175×10^{21} mol, dissolves in the hydrosphere, leaving most of the gas in the atmosphere at the partial pressure of

$$P_{N_2} = 1.4 \text{ to } 1.5 \text{ bar}$$
(2.27)

Thus the total pressure of the CO_2-N_2 atmosphere at equilibrium with the oceans at the temperature of 25 to 35°C is

$$P_{Total} = P_{CO_2} + P_{N_2} = 38 \text{ to } 43 \text{ bar}$$
(2.28)

Its composition was about 96.6 mol % CO_2 and 3.4 mol % N_2. This composition is represented by point E in Fig. 2.2.

The preceding estimates of total pressure and density are for a dry atmosphere that does not include water vapor. Because of the dissociation of CO_2 in water and input of metal ions from mineral dissolution, an additional drawdown of atmospheric CO_2 was likely to have occurred early in Earth's history. Subsequent reduction of the primordial atmospheric CO_2 pressure could have been caused by its storage in sedimentary $CaCO_3$. A decline in the CO_2 pressure from about 38 to 10 bar, the lower value as cited in Section 4 (p. 46), would have required storage of a $CaCO_3$ mass equivalent to about 65% of the carbonate rocks preserved in the geologic record (Table 1.2).

4 Summary and Speculations

The occurrence and distribution of CO_2 in the early, prebiotic atmosphere and ocean can be summarized as follows:

- After the Earth's surface cooled, and chlorine and sulfur were removed from the atmosphere by dissolution in the early hydrosphere and chemical reactions with crustal rocks, the remaining main two constituents of the atmosphere were CO_2 and N_2 at concentrations of 96.6 vol % CO_2 and 3.4 vol % N_2.
- If the initial mass of CO_2 were partitioned between the atmosphere and early hydrosphere, approximately 76 mol % or $4.86 \times 10^{21} \pm 0.09 \times 10^{21}$ mol remained in the atmosphere and 24 mol % or $1.53 \times 10^{21} \mp 0.09 \times 10^{21}$ mol dissolved in the hydrosphere.
- The total pressure of the CO_2-N_2 atmosphere was $P \approx 40$ bar, at an Earth's surface temperature of 25 to 35°C that gives an atmosphere thickness of 6 km.

The atmospheric masses for these three stages of the prebiotic Earth that were considered in this chapter are shown in Fig. 2.4: all the primordial volatiles in the atmosphere; water condensed, HCl and H_2S removed from the atmosphere, and all of CO_2 and N_2 remaining in the atmosphere; and, finally, the mass of CO_2 and N_2

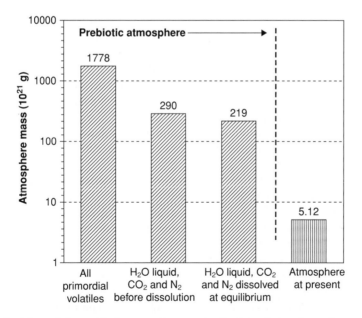

Figure 2.4 Mass of the prebiotic atmosphere as the Earth cooled and reactive gases were progressively removed by reactions and dissolution in the hydrosphere. Details are given in the text. Masses are shown for each model stage of the atmosphere, in units of 10^{21} gram.

remaining in the atmosphere after their equilibration with surface waters at 25 to 35°C. The much smaller mass of the present-day atmosphere is also shown in the figure. If most of the primordial volatiles were degassed soon after the formation of the Earth, then their condensation was likely to be completed during the prebiotic period in about 600 million years, from about 4500 to 3900 million years ago, or perhaps during an even shorter time period. The balance of primordial CO_2 in the atmosphere and hydrosphere leads to a number of such important questions as, for example: Could most of the H_2O exist as liquid water on the early Earth's surface if the CO_2 atmospheric pressures were high and possibly causing a stronger greenhouse warming (Kasting and Toon, 1988)? Could such high concentrations of atmospheric CO_2 have been maintained on the prebiotic Earth for any length of geological time, without the consumption of CO_2 in weathering reactions? Could dissolved CO_2 and its aqueous ionic species, HCO_3^- and CO_3^{2-}, occur at high concentrations in the early hydrosphere without calcium carbonate precipitation and their removal into this sedimentary sink?

It is unlikely that in the presence of liquid water on Earth, CO_2 could exist as a gas in the atmosphere and in solution, without reacting with crustal silicate rocks. Such reactions would release metal cations into the water that would electrically balance the bicarbonate and carbonate anions in solution. The concentrations of dissolved carbonate species—$CO_{2(aq)}$ or $H_2CO_3^o$, HCO_3^-, and CO_3^{2-}, collectively known as dissolved inorganic carbon or DIC—in the early ocean might have been considerably higher than in geologically later times and at present, if they were balanced in solution by sodium or potassium ions, Na^+ or K^+, rather than by Ca^{2+} and Mg^{2+}, insofar as carbonate minerals calcite, aragonite, and dolomite are much less soluble than sodium bicarbonate or carbonate. Stephan Kempe and coworkers postulated the existence of the "soda ocean" of low Ca^{2+} concentration and the carbonate ions as the main anionic species, during the period from 4500 to 1000 million years (Kempe and Degens, 1985; Kempe and Kaźmierczak, 1994; Kempe et al., 1989). The soda ocean, compositionally analogous to saline soda lakes, was dominated by sodium-carbonate rather than sodium-chloride, and the prevalence of the chloride ion developed as late as 1000 million years ago. It was, however, shown earlier in this section that chloride was most likely to be the major constituent of the primordial ocean, unless the degassing of HCl were delayed.

As the weathering of the continental and/or oceanic crust was supplying calcium along with other metal ions to the ocean, there remains an issue of the chemical behavior of the Ca^{2+}-ion in the presence of a high concentration of dissolved CO_2 in the prebiotic ocean. In a model study of the chemical composition of the prebiotic ocean during the Hadean Eon, Morse and Mackenzie (1998) concluded that in the primordial ocean, in the temperature range from about 100 to 70°C, and partial pressure of CO_2 near 10 bar, the pH of ocean water would have been significantly lower, <6.0, and dissolved inorganic carbon and alkalinity (Chapter 5) much higher than at present. Such conditions would have made it at least theoretically possible to maintain supersaturation with respect to $CaCO_3$ minerals at a level of up to 20, as compared to a supersaturation range of 3.5 to 5.5 for near-Recent seawater, and for seawater to contain much less calcium that it contains at present. Figure 2.5 shows some of the calculated results for a model Hadean seawater, at 85°C and a range of CO_2 partial pressures from 3×10^{-2} to 10 atm. At the

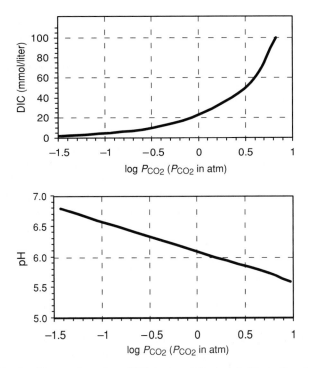

Figure 2.5   Dissolved inorganic carbon (DIC, in mmol/liter) and pH as a function of P_{CO_2} between 0.03 and 10 atm, at 85°C, in a model Hadean ocean of the following chemical composition: Na^+ concentration twice that of the modern seawater of salinity 35 g/kg; K^+, Mg^{2+}, and Ca^{2+} concentrations as in modern seawater; Cl^- balances total charge, at a concentration similar to that of Na^+ (from data in Morse and Mackenzie, 1998).

higher end of the CO_2 pressure, DIC of 100 mmol/liter is much higher than the modern value of approximately 2.5 mmol/liter, and the pH of ocean water is slightly acidic, as compared to the modern surface water value of about 8.2. However, even at the higher levels of supersaturation of ocean water with respect to $CaCO_3$ minerals, removal of most of CO_2 from the early atmosphere and ocean down to the levels comparable to those of the present would have required deposition and storage of a mass of $CaCO_3$ similar to the total mass of preserved limestone and dolomite. Although there is no evidence in the geologic record that either supports or negates such a process of calcium carbonate removal during the first several hundred million years of Earth's history that fall in the Hadean and Early Archean Eons, the patterns of weathering and dissolution of the continental and oceanic crust and the chemical behavior of CO_2 in water suggest that the deposition of $CaCO_3$ is a geologically old phenomenon on our planet.

Calcium is brought to the ocean from the weathering of Ca-silicates in the continental and oceanic igneous crust and by dissolution of old limestones that are now part of the sedimentary cover of the continents. Before limestones formed, the source of calcium in the primordial ocean must have been the weathering of the crystalline crust.

This concept is represented by the reaction:

$$CaSiO_3 + CO_2 = CaCO_3 + SiO_2 \tag{2.29}$$

that is a shorthand notation for the weathering of silicates by reactions with CO_2 and its transfer to the sedimentary carbonate reservoir. The reaction is generally known as the Urey reaction, following Urey (1952). According to Berner and Maasch (1996), the reaction and interpretation of its role in the global carbon cycle were introduced by J. J. Ebelmen much earlier, in 1845.

As the metal cations from the weathering of the igneous crust began to accumulate in the ocean, calcium was the first likely constituent to react with dissolved CO_2 and form sedimentary $CaCO_3$. On the prebiotic Earth, removal of essentially all CO_2 from the atmosphere and ocean by carbonate deposition would have required a mass of Ca equivalent to the mass of CO_2. This mass, about 6.48×10^{21} moles (Table 2.2), had to come from a primary crustal source. The present-day mass of carbonate sediments and sedimentary organic matter, all taken as $CaCO_3$, corresponds to a layer about 0.5 km thick over the entire Earth's surface. The mass fraction of Ca in $CaCO_3$ is 40%, but it is considerably smaller, 10% or less, in basalts. Thus a layer of igneous crust of an order of magnitude of 2 km or more was needed to supply calcium to bind all of the CO_2 into $CaCO_3$. At the weathering rates of silicate rocks, comparable to those in the tropical environment of the South American Shield and in the Andes at about 15 ± 10 m per 10^6 yr (Mortatti and Probst, 2003), the process might have taken 100 to 200 million years, a period of time of the same order of magnitude as the length of the prebiotic stage of the Earth, about 600 million years.

Reaction (2.29) can proceed from the right to the left only at higher temperatures and pressures, as in metamorphic reactions occurring during the subduction of oceanic carbonate sediments or the burial of carbonates in deep sedimentary basins, leading to the breakdown of $CaCO_3$ and eventual return of CO_2 to the Earth's surface. These are the processes of the endogenic carbon cycle, shown in Fig. 1.4. Geological interest in reaction (2.29) as representing the mineral assemblage of wollastonite ($CaSiO_3$), quartz (α- or β-SiO_2), calcite ($CaCO_3$), and gaseous CO_2 dates to earlier times because it also describes metamorphism of a common type of sediment, sandy or siliceous limestones. The calculated equilibrium temperature at 1 bar total pressure for the reaction is 280°C, and it rises to about 620°C at a lithostatic overburden equal to about 1000 bar (Kern and Weisbrod, 1967). In the continental crust, where the average geobaric-geothermal gradient is about 10 bar/°C, 1000 bar would be attained at a depth of about 3.5 km where the temperature would be about 130°C, suggesting that limestone would not react with silica at shallower depths and no CO_2 would be released under those conditions.

It is tempting to speculate about the Earth's surface environment at a stage when most of the CO_2 had been removed from the atmosphere but before oxygen formed. If the prebiotic atmosphere that was made primarily of CO_2 and a small fraction of N_2 lost its CO_2 to the sediments, the remaining nitrogen at the pressure of about 1 bar might have been the main atmospheric gas for a long time. The greenhouse functions of the Earth's atmosphere were likely to be maintained by water vapor, as they are now, and by CO_2 at a concentration much reduced relative to the earlier time.

5   An Early Biosphere

The biosphere, as the world of living organisms is known, emerged when the Earth was some 600 to 800 million years old, presumably as cells meeting their nutrient and energy needs from inorganic chemical compounds in the environment. The diversity of the biosphere as it evolved in the subsequent eons is represented by the five taxonomic kingdoms or three main genetic groups that are shown in Fig. 2.6 and Table 2.5. The cladogram or branching diagram in Fig. 2.6 shows the diversification of the living world that started from an unknown root organism. Such an organism was

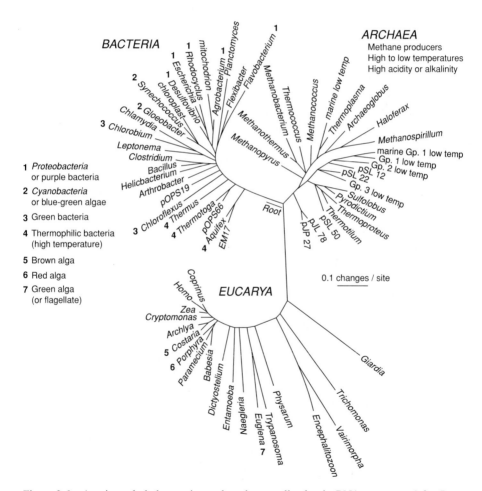

Figure 2.6   A universal phylogenetic tree based on small-subunit rRNA sequences (after Pace, 1997; cladogram reproduced by permission of the author and American Association for the Advancement of Science, Washington, D. C., publishers of *Science*).

Table 2.5   Genetic and taxonomic classification of the major groups of the biosphere (from Margulis and Schwartz, 1998; Pace, 1997, 2001; Woese, 2000).

Genetic classification	Taxonomic kingdom	Some examples of geologically prominent groups
Eucarya	Animalia	Phyla of marine skeleton builders: *Porifera* (sponges), *Cnidaria* (including corals), *Arthropoda, Mollusca, Brachiopoda, Bryozoa, Echinodermata*
	Plantae	Mosses, grasses, ferns, conifers, flowering plants
	Fungi	?
	Protoctista	Phyla of *Granuloreticulosa* (foraminifera), *Haptophyta* (coccolithophorids), *Bacillariophyta* (diatoms), *Actinopoda* (radiolarians), *Chrysophyta* (golden algae), *Xanthophyta* (yellow algae), *Phaeophyta* (brown algae), *Rhodophyta* (red algae), *Chlorophyta* (green algae)
Bacteria	*Monera* (*Prokaryota*)	Subkingdom *Eubacteria*, phyla of *Proteobacteria* (purple bacteria, aerobic and anaerobic, use oxidized and reduced nitrogen and sulfur compounds), *Cyanobacteria* (blue-green bacteria, chloroxybacteria, and oxygenic and anoxygenic photosynthesizers), *Chloroflexa* (green non-sulfur phototrophic), *Chlorobia* (anoxygenic green sulfur bacteria), *Thermotogae* (thermophilic fermenters)
Archaea		Subkingdom *Archaea* (methanogenic, halophilic, and thermoacidophilic bacteria)

Note: Additional information on calcium-carbonate secreting taxa is given in Table 4.2.

probably a product of a long series of natural experiments in biopoesis[1] that resulted in a mechanism synthesizing organic matter from the chemical compounds occurring in the environment. It should be noted that some of the thermophilic bacteria and archaea that live at temperatures near 95°C are relatively close to the common root, as shown in Fig. 2.6. Perhaps they indicate that the primordial environment of biopoesis was hot. The temperature range of bacterial survival is between about $-15°$ to $+115°C$, the currently known high temperature limit of 113°C is that of the proteobacterium *Pyrolobus* living inside the hot deep-ocean vents. Bacteria have been reported to live in the deep ocean, at pressures greater than 700 bar or 7,000-meter-deep ocean water, in deep sediments, and in the upper part of the continental and oceanic crust. As a scientific curiosity, two species of bacteria have survived pressures up to 16 kbar (1600 MPa) in a laboratory experiment and returned to normal physiological activity after depressurization to atmospheric conditions (Darling, 2005; Kerr, 2002; Pennisi, 2002; Sharma *et al.*, 2002). Characteristic linear dimensions of bacterial cells are from $<1\mu m$ to about 10 μm, and the dry mass of *Escherichia* spp. cells is about 2.5×10^{-13} g/cell, containing 50% carbon, which translates into 1×10^{-14} mol C/cell.

---

[1] The word biopoesis is used here in its literal meaning of "the making of organic life", from the Greek roots as given in the *Oxford English Dictionary*.

Table 2.6   Biological synthesis of organic matter (glucose, $C_6H_{12}O_6$, in the notation CH_2O) by fermentation and photosynthesis, in order of increasing energy requirement. ΔG_R^o is the standard Gibbs free energy of the reaction at $25°C$, in kJ per 1 mol carbon of CH_2O.[1]

Process	Reaction[2]	$\Delta G_R{}^o$ (kJ/mol C)
Chemosynthesis	1. $CO_2 + 2H_2 = CH_2O + H_2O$	4.17
	2. $2CO_2 + CH_4 + 2H_2 = 3CH_2O + H_2O$	47.7
	3. $CO_2 + 2H_2S = CH_2O + H_2O + S$	59.6
	4. $CO_2 + CH_4 = 2CH_2O$	69.4
	5. $CO_2 + 4Fe(OH)_2 + 3H_2O = CH_2O + 4Fe(OH)_3$	130.3
	6. $2CH_4 + 3H_2O = CH_2O + CO_2 + 6H_2$	132.6
	7. $CH_4 + H_2O = CH_2O + 2H_2$	134.7
Sulfur oxidation and non-O_2 photosynthesis	8. $2CO_2 + H_2S + 2H_2O = 2CH_2O + 2H^+ + SO_4^{2-}$	120.0
	9. $3CO_2 + 2S + 5H_2O = 3CH_2O + 4H^+ + 2SO_4^{2-}$	140.1
	10. $CO_2 + NH_3 + H_2O = 2CH_2O + NO_3^- + H^+$	318.7
O_2 photosynthesis	11. $CO_2 + H_2O = CH_2O + O_2$	478.4
	12. $CO_2 + NO_3^- + H^+ + 2H_2O = CH_2O + NH_3 + 3O_2$	797.7

[1] Standard Gibbs free energy of formation values from Lide (1994) and Drever (1997)
[2] Standard free energy of formation of glucose ($C_6H_{12}O_6$), $\Delta G_f^o = -918.78$ kcal/mol, computed from the free energy change of the photosynthesis reaction, taken as $\Delta G_R^o = 2870$ kJ/mol or 478.4 kJ/mol C (Kimball, 2004). Gaseous species: CO_2, CH_4, H_2, and O_2. Liquid: H_2O. Aqueous species: H_2S, NH_3, H^+, NO_3^-, SO_4^{2-}, and glucose $C_6H_{12}O_6$ (assumed). Solids: rhombic sulfur S, and Fe^{2+} and Fe^{3+} hydroxides.

Concentrations of live and dead bacterial mass in sediments vary greatly, but in at least one study of a shallow pond above a layer of permafrost in Northern Alaska, the living mass of bacteria in the upper 5 cm layer of the sediment was reported as 2.5×10^{-6} mol C/cm^3, where it accounted for a small fraction, less than 0.1%, of total organic matter that was mostly of plant origin (Hobbie, 1980). Two different modes of production of living organic matter are recognized in the upper 5 cm layer of the sediment: autotrophic, where the organisms derive their building material or food from inorganic chemical compounds, and heterotrophic or organotrophic, where organisms utilize organic compounds. Those autotrophic organisms that require no light for the synthesis of organic matter are known as chemotrophs and those that use light are phototrophs.

There exist bacteria that alternate between the autotrophic and heterotrophic modes. A number of chemosynthetic and photosynthetic reactions producing organic matter are shown in Table 2.6. The term *primary production* designates the autotrophic production of organic matter by unicellular organisms on land and in water, as well as by the higher plants. Hence the name primary producers that is often used for these organisms. The variety of primary producers of organic matter in the present-day world—the chemosynthesizing and photosynthesizing autotrophs—inhabit diverse environments that include such extremes as the hot acidic plumes on the deep ocean floor, high-temperature springs, hypersaline lakes, and cold habitats. In such environments, some of the bacteria live within narrowly defined niches of temperature and salt content of water, as can be observed by the zones of different color that occur around the outflow of hot springs or in saline lakes and lagoons with horizontal salt concentration and temperature gradients in the surface water. The synthesis of organic matter by chemical reactions that does not make free oxygen is also called anoxygenic, to distinguish it from the widespread production of oxygen in photosynthesis that is as an oxygenic process. These two processes are shown as reactions 8–12 in Table 2.6. Chemosynthetic reactions producing organic matter, written as CH_2O (for glucose, $C_6H_{12}O_6$), between CO_2, CH_4, H_2, H_2O, and ferrous iron (Fe^{2+}) are shown as reactions 1–7 in Table 2.6, in order of increasing energy need. Because the process of organic matter formation results in a reduction of entropy as discrete and disorganized inorganic and organic materials are integrated into the ordered units of an organism, energy must be supplied from an external source. The early autotrophs might have synthesized organic matter from carbon dioxide and hydrogen or methane (Table 2.6, reactions 1, 2, 4, and 7), whereas the present-day methanogenic bacteria are heterotrophs.

Organic matter can form from carbon dioxide and methane, and it can be decomposed or disproportionated as acetate (CH_3COOH) into these two compounds (Fenchel et al., 1998):

$$CO_2 + CH_4 \rightleftarrows 2CH_2O \qquad (2.30)$$

Sulfur oxidation, as done by some of the sulfur bacteria that appeared later in the geological record, is given by reactions 8 and 9. Reaction 11 is photosynthesis that requires a relatively large energy input and even more energy is needed to produce organic matter with the reduction of nitrate, reaction 12, as done by nitrate-reducing bacteria. Primary producers use either oxidized nitrogen (nitrate, NO_3^-) or its reduced form (ammonia, NH_3, or the ammonium ion in solution, NH_4^+), sometimes interchangeably. Less energy is required in the biological oxidation of ammonia, reaction 10, than in the reduction of nitrate, reaction 12. The energy-change values for the production of organic matter in Table 2.6 are positive, whereas the decomposition of organic matter by heterotrophic bacteria (Chapter 9) is characterized by negative values indicating energy release. Among the latter reactions is the production of methane, the process of methanogenesis, that is at present practiced by heterotrophic bacteria. Although methane is thermodynamically unstable in the presence of even very low concentrations of molecular oxygen, as discussed in Section 1.2, reactions between CO_2 and molecular hydrogen, H_2, that produce CH_4 and water or methane and organic matter are characterized by negative values of the Gibbs free energy change. Methane, as a product of such reactions in sediments or soils, is further either consumed by methanotrophic bacteria or eventually released to the atmosphere.

Apart from the chemical reduction and oxidation processes that underlie the carbon cycle in the living organic matter, the processes of autotrophy and organotrophy are mutually complementary: primary production of organic matter is followed by its remineralization that returns to the environment at least some of the carbon dioxide and other chemical species consumed. The process of organic matter decomposition or remineralization is a negative feedback to the withdrawal of CO_2 from the surroundings by primary production and this functional complementarity might have emerged close to one another in the early days of Earth's history. In today's world, there is a great variety of both types of trophic organisms, the primary producers and bacteria that remineralize organic matter, and the functional diversity in these processes is great, as evidenced by a variety of taxa performing similar functions of primary production and decomposition on land and in water. However, the functional complementarity within the biosphere extends beyond the fixation and return of carbon dioxide: methane-producing (methanogens) and methane-consuming (methanotrophs) bacteria occur in soils, at a not too great spatial separation one from another; nitrogen fixation (nitrification) is counteracted by denitrification, with different bacteria performing each function on land, in ocean water, and sediments; and bacterial reduction of sulfate in oceanic sediments and oxidation of reduced sulfur are additional examples of the two-way processes that are performed by different taxonomic groups.

In photosynthesis, light energy is used to oxidize hydrogen in the half-cell oxidation reaction:

$$2H_2O + light \rightarrow 4H^+ + 4e^- + O_2 \tag{2.31}$$

The complementary half-cell reduction reaction is:

$$CO_2 + 4H^+ + 4e^- \rightarrow CH_2O + H_2O \tag{2.32}$$

and the sum of (2.31) and (2.32) represents the overall reaction for oxygenic photosynthesis, not including other nutrient elements, such as nitrogen, phosphorus, and sulfur, that are needed in the formation of living organic matter:

$$CO_2 + 2H_2O + light \rightarrow CH_2O + H_2O + O_2 \tag{2.33}$$
$$\Delta G_R^o = 2485.3 \, kJ/mol \, C - 2006.9 \, kJ/mol \, C = 478.4 \, kJ/mol \, C$$

More on the mechanisms of photosynthesis can be found in Chapter 6. It may be reiterated that free molecular oxygen is produced from H_2O rather than from CO_2. Light is a substrate in reaction (2.33) and a portion of the energy absorbed is stored in the products of the reaction, 470 to 490 kJ per 1 mol of carbon for oxygenic photosynthesis. Except for cyanobacteria and chloroxybacteria, all photosynthetic bacteria are anaerobes and do not evolve oxygen in their metabolic processes. The pigments involved in their photosynthetic apparatus are bacterial chlorophylls. All other photosynthetic organisms, including cyanobacteria, chloroxybacteria, eukaryotic algae, and the higher plants release oxygen in the process of photosynthesis. The net free energy change in reaction (2.33), 478.4 kJ/mol C, is the algebraic sum of two terms (Kimball, 2004): the energy per 1 mol of carbon that is needed to break the C–O and H–O bonds and that comes from sunlight, 2485 kJ/mol C, and the energy of the reaction that makes glucose (1/6 of $C_6H_{12}O_6$) and molecular oxygen, -2006.9 kJ/mol C.

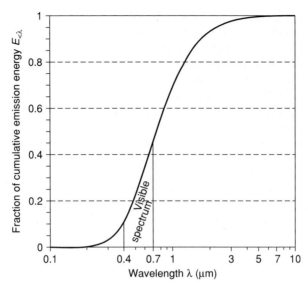

Figure 2.7   Fraction of cumulative energy of wavelengths $< \lambda$ emitted by the Sun as an ideal black body of surface temperature T = 5780 K. More than 99% of energy is delivered by wavelengths shorter than 10 μm. The visible light band is 35% of total energy E = σT^4 (Chapter 3). A faint young Sun that radiated 0.75 of the present-day energy would have a surface temperature T = 5379 K. Calculation given in Chapter 3, Box 3.3.

These numbers shed light on the question of what fraction of the incoming sunlight is utilized by photosynthesizing land and marine organisms. In the present-day world, plants and algae absorb the visible part of the solar energy spectrum in photosynthesis, in the range of wavelengths from 0.4 to 0.7 μm or 400 to 700 nm. This wavelength range accounts for 35% of solar radiation energy outside the Earth's atmosphere and it was about 33% in the early days of the Earth and faint young Sun when solar luminosity was about 25 to 28% lower than now (Fig. 2.7). The fraction of the sunlight energy that is converted photosynthetically to organic matter is known in the botanical and agricultural literature as quantum yield, and it is measured as a ratio of the rate of O_2 production or CO_2 consumption (e.g., mol min^{-1}) to the solar energy flux on the plant (e.g., J mol^{-1} min^{-1}). As there are 8 electrons involved in the reduction of H_2O and CO_2 in photosynthesis, a theoretical maximum quantum yield is $1/8 = 0.125$ when 1 mol CO_2 is fixed by a plant that absorbs 8 moles of photons in the absence of photorespiration. This applies only to C_3 plants because in C_4 plants additional energy is required for intracellular CO_2 storage (Dutton, 1997; Singsaas *et al.*, 2001). Quantum yields in species of cyanobacteria, algae, diatoms, and vascular plants, from laboratory and field measurements, are commonly between 0.03 and 0.10, although lower yields have also been reported in land plants grown under natural conditions (Bray, 1961; Brody and Emerson, 1959). On the global scale at present, the fraction of solar energy received by the Earth that is used by photosynthesis on land and in the ocean can be estimated from the following data:

Gross primary production, about twice net primary production (Table 1.3)	1.9×10^{16} mol C/yr
Land (149×10^6 km^2 less 25×10^6 km^2 of ice and extreme deserts) and ocean surface	4.8×10^{14} m^2
Solar constant outside Earth's atmosphere (Table 3.1)	1367 W/m^2
Fraction of sunlight reaching Earth's surface (Fig. 3.1)	0.44
Fraction of incoming energy due to visible light (Figs. 2.7, 3.3a)	0.35 to 0.50
Energy uptake in photosynthesis, equation (2.33), net	478.4 kJ/mol C

Bearing in mind that the incoming solar radiation is spread over the spherical surface area of the Earth that is four times larger than the intercepting cross-section, an average photosynthetic efficiency, f, is:

$$f = \frac{(1.9 \times 10^{16} \text{ mol C/yr}) \times (0.48 \times 10^6 \text{ J/mol C})}{(4.8 \times 10^{14} \text{ m}^2) \times (0.44 \times [0.35 \text{ to } 0.50] \times 1367/4 \text{ W/m}^2) \times 3.16 \times 10^7 \text{ s/yr}}$$

$$= 0.010 \pm 0.002$$

which is about 1% if only the net energy change of the photosynthesis reaction is counted. The result of the preceding calculation is lower than direct measurements of quantum yield made in relatively small vegetated areas over short periods of time: the reported light energy that is utilized in photosynthesis may be smaller by a factor of two or three than that of the global average value of 53 to 75 W m^{-2} used in the calculation; also, most of the global gross primary production occurs in an area of land and ocean surface that is smaller than 94% of the Earth's surface, a value used in the calculation. A smaller surface area and a lower light energy input to plants would make the efficiency factor f greater than 1%. It is clear, however, that the availability of the visible sunlight was not a limiting factor in the distant past to anoxygenic as well as oxygenic photosynthesis, the energy requirements of which are shown in Table 2.6.

The first result of the emergence of oxygenic photosynthesis was the production of molecular oxygen that, as a substance newly added to the atmosphere, was its early pollutant. The bacterial oxidation of reduced iron and sulfur along with small quantities of reduced atmospheric gases using the photosynthetically produced O_2 were likely the main oxygen sinks until about 2200 Ma ago when the geologic record shows that oxidizing conditions began to develop on the Earth's surface.

The diverse material of this chapter presents some necessary background information for discussions later in this book that deal with the heat balance and temperature of the atmosphere (Chapter 3), the main carbon-containing minerals of inorganic and biogenic origin (Chapter 4), the isotopic record of carbon in sediments (Chapter 6), and the long-term history of the carbon cycle (Chapter 10).

Box 2.2   Three models of the atmosphere

I.   Atmosphere of constant density

An atmosphere of constant density is an often used and useful model because it presents the atmosphere as a homogeneous layer or a spherical shell of finite thickness enveloping the Earth. The equation of state of an atmosphere obeying the ideal gas law is:

$$P = \frac{nRT}{V} \quad \text{or} \quad P = \frac{\rho RT}{M} \quad \text{(bar)} \tag{2.2.1}$$

where n is the number of mols of gas, V is the atmosphere volume (cm^3), T is temperature (kelvin), and R is the gas constant. The quotient n/V mol/cm^3 is the molecular density of the atmosphere. For an atmosphere of mean molecular weight M (g/mol), the molecular and mass densities are interrelated through

$$n/V = \rho/M \quad \text{(mol cm}^{-3}) \tag{2.2.2}$$

where ρ is the mass density (g/cm^3). With increasing altitude, the atmospheric pressure decreases as

$$dP = -\rho g \, dh \tag{2.2.3}$$

where g is the gravitational acceleration and h is the height above ground level. If the atmosphere density ρ is constant and g does not change significantly with altitude within the atmosphere, then equations (2.2.1) and (2.2.3) can be combined to give:

$$-\rho g dh = \frac{\rho R}{M} dT \quad \text{(bar)} \tag{2.2.4}$$

and, upon rearrangement of the terms, the temperature gradient is obtained as:

$$\frac{dT}{dh} = -\frac{gM}{R} \quad \text{(deg km}^{-1}) \tag{2.2.5}$$

The temperature gradient is negative and the temperature decreases from $T = T_0$ at the Earth's surface taken as zero altitude, $h = 0$, to $T = 0$ at some altitude $h = H$. Because the temperature gradient is linear, the *constant-density model* is also known as the *linear atmosphere model*. In this model, both the temperature and pressure decline to zero at some altitude H that depends only on the mean molecular weight of the atmosphere, M, and the temperature at the Earth's surface, T_0. Integration of dT/dh, between the limits of $T = T_0$ at $h = 0$ and $T = 0$ at $h = H$, gives the height of the atmosphere:

$$H = \frac{RT_0}{gM} \quad \text{(km)} \tag{2.2.6}$$

Higher temperature at the Earth's surface would increase the atmosphere thickness, whereas heavier gases would make H smaller. The present-day atmosphere is of thickness:

$$H = \frac{8.3145 \times 10^7 \text{erg K}^{-1} \text{mol}^{-1} \times 288.15 \text{ K}}{980 \text{ cm s}^{-2} \times 28.97 \text{ g mol}^{-1} \times 10^5 \text{cm km}^{-1}} = 8.4 \text{ km} \quad (2.2.7)$$

The temperature gradient in this atmosphere is:

$$\frac{dT}{dh} = -\frac{980 \text{ cm s}^{-2} \times 28.97 \text{ g mol}^{-1} \times 10^5 \text{cm km}^{-1}}{8.3145 \times 10^7 \text{ erg K}^{-1} \text{mol}^{-1}} = -34.1 \text{ K km}^{-1}$$
$$(2.2.8)$$

The density of this atmosphere is:

$$\rho_0 = \frac{P_0 M}{R T_0} = \frac{1.013 \text{ bar} \times 0.1 \text{ J cm}^{-3} \text{ bar}^{-1} \times 28.97 \text{ g mol}^{-1}}{8.315 \text{ J mol}^{-1} \text{ K}^{-1} \times 288.15 \text{ K}} \quad (2.2.9)$$
$$= 1.23 \times 10^{-3} \text{g cm}^{-3}$$

A small correction can be introduced to account for the decrease in the gravitational acceleration g in the free air above the Earth's surface. Acceleration due to the force of gravity, g, decreases inversely to the square of the distance: g decreases as $1/r^2$ with an increase in r. The value at the surface of a spherical Earth, g_0, would decrease with an increasing altitude as $1/(r_0 + h)^2$, where r_0 is the radius of the Earth and h is altitude above the surface. Thus at the higher altitudes in the atmosphere,

$$\frac{g}{g_0} = \frac{r_0^2}{(r_0 + h)^2} \quad \text{or} \quad g = \frac{g_0}{(1 + h/r_0)^2} \quad (2.2.10)$$

Substitution of g from (2.2.10) in (2.2.5) and integration give:

$$H = \frac{R T_0}{g_0 M} \cdot \frac{1}{1 - R T_0/(g_0 M r_0)} \quad (2.2.11)$$

The decrease in g with altitude above ground level is, from (2.2.10):

$$\frac{d \ln g}{dh} = -\frac{2}{r_0(1 + h/r_0)} \approx -\frac{2}{r_0} = -0.0003 \text{ km}^{-1} \quad (2.2.12)$$

This results in a very small increase in the atmosphere thickness: for a 10-km-thick atmosphere, a correction in thickness would be only 0.016 km or 16 m.

II. Isothermal atmosphere

The atmosphere thickness H, derived in the preceding section, is numerically identical to the atmospheric scale height that is the altitude where the atmospheric pressure is $1/e$ of its ground-level value, in an isothermal atmosphere. The decrease in pressure with an increasing altitude, h, is given by the relationship:

$$P = P_0 e^{-\frac{gM}{RT}h} \quad (\text{bar}) \quad (2.2.13)$$

that is also known as the hydrostatic equilibrium or barometric equation, obtained from (2.2.3). At constant temperature, T, the pressure decreases to $1/e$ or about 0.37 of P_0 at the altitude $h = H$:

$$H = \frac{RT}{gM} \quad \text{(km)} \tag{2.2.14}$$

H is called the scale height of the atmosphere. An isothermal atmosphere at $T = 288$ K has the same scale height, 8.4 km, as the thickness of the constant-density atmosphere with the ground temperature $T = 288$ K. Mean density of the atmosphere within the layer of thickness $H = 8.4$ km is obtainable by integration of (2.2.13) written in terms of density, ρ, in the following form:

$$\rho = \rho_0 e^{-\frac{gM}{RT}h} \tag{2.2.15}$$

The mean density of the 8.4-kilometer-thick layer is smaller than that in the constant-density model:

$$\frac{1}{H}\int_0^H \rho \, dh = \frac{\rho_0}{H}\int_0^H e^{-\frac{gM}{RT}h} \, dh = \rho_0 \left(1 - \frac{1}{e}\right) \approx 0.63\rho_0 = 0.77 \times 10^{-3} \text{ g cm}^{-3} \tag{2.2.16}$$

III. Adiabatic atmosphere

In a system that is called adiabatic, no heat is added or removed and no heat flows through the system. In an adiabatic process, pressure-volume work is done by the system or on the system, and a change in temperature occurs because of a change in internal energy. This contrasts with an isothermal system where heat must be added or removed to maintain a constant temperature. The equation for the temperature gradient in an adiabatic atmosphere, also called the lapse rate, is commonly derived in textbooks of atmospheric sciences from two equations. One equation defining an adiabatic system is:

$$C_p \, dT = v \, dP \quad \text{(J mol}^{-1}\text{or bar cm}^3 \text{ mol}^{-1}) \tag{2.2.17}$$

where C_p is the heat capacity of the gas at constant pressure (J mol^{-1} K^{-1}), and the molar volume of the gas, v from (2.2.1), is any one of the following quotients:

$$v = \frac{M}{\rho} = \frac{V}{n} = \frac{RT}{P} \quad \text{(cm}^3 \text{ mol}^{-1}) \tag{2.2.18}$$

The second equation is for the hydrostatic equilibrium in the atmosphere column, defined in the preceding section as (2.2.3):

$$dP = -\rho g \, dh$$

Substitution of (2.2.3) for dP in (2.2.17), of M/ρ for v, and elimination of ρ gives the temperature gradient in a dry adiabatic atmosphere:

$$\frac{dT}{dh} = -\frac{gM}{C_p} \quad \text{or} \quad \frac{dT}{dh} = -\frac{g}{c_p} \ (\mathrm{K\,km^{-1}}) \tag{2.2.19}$$

where the units of heat capacity c_p are per unit of mass ($\mathrm{J\ g^{-1}\ K^{-1}}$) rather than per mol as in C_p. The adiabatic temperature gradient is negative and temperature decreases with increasing altitude. The value of the gradient in a dry atmosphere is:

$$\frac{dT}{dh} = -\frac{980\ \mathrm{cm\ s^{-2}} \times 10^5\ \mathrm{cm\ km^{-1}}}{1.007\ \mathrm{J\ g^{-1}K^{-1}} \times 10^7\ \mathrm{g\ cm^2\ s^{-2}\ J^{-1}}} = -9.73\ \mathrm{K\ km^{-1}} \tag{2.2.20}$$

Temperature decrease with altitude is, from equations (2.2.19) and (2.2.20):

$$T = T_0 - \frac{gM}{C_p}h = T_0 - 9.7h \quad \text{(K)} \tag{2.2.21}$$

where h is in kilometers.

In an atmosphere containing water vapor, the vapor condenses at some altitude where the temperature is lower, and condensation releases heat to the atmosphere. The result is that in a water-containing atmosphere, the temperature decrease with altitude is not as strong as in dry air and the gradient is smaller in absolute terms, varying from -5 to -7 deg/km. The adiabatic temperature gradient determines the vertical stability of the atmosphere: air warmer than the adiabatic would rise and air cooler than the adiabatic would sink. Adiabatic conditions are usually maintained in the troposphere over several kilometers in altitude.

Adiabatic temperature gradients in the atmospheres of other planets have been calculated as -10.7 on Venus, -4.5 on Mars, and $-20.2\ \mathrm{K\ km^{-1}}$ on Jupiter (Wayne, 1991).

The standard relationship between the temperature and pressure in an adiabatic atmosphere is derivable by substitution of RT/P for v in (2.2.17), and integration:

$$\frac{T}{T_0} = \left(\frac{P}{P_0}\right)^{R/c_p} \tag{2.2.22}$$

where T_0 and P_0 are reference temperature and pressure, such as their values at ground level. The practice of writing the power exponent in (2.2.22) varies, and here we adopt the use of the gas constant $R = 8.315\ \mathrm{J\ mol^{-1}\ K^{-1}}$ and heat capacity at constant pressure $C_p = 29.17\ \mathrm{J\ mol^{-1}\ deg^{-1}}$, giving the power exponent $R/C_p = 0.285$.

Mean density of the adiabatic atmosphere, within the layer extending from the ground level to altitude $H = 8.4$ km, can be computed as for the isothermal atmosphere, using relationships (2.2.22), (2.2.21), and (2.2.1). The density-temperature relationship, from (2.2.1) and (2.2.22), is:

$$\frac{\rho}{\rho_0} = \left(\frac{T}{T_0}\right)^{\frac{c_p}{R}-1} \tag{2.2.23}$$

The mean density of an atmospheric column of thickness H follows from the preceding, by substitution of (2.2.21) for T in (2.2.23), and integration:

$$\frac{1}{H}\int_0^H \rho\,dh = \frac{\rho_0}{H}\int_0^H \left(1 - \frac{gM}{c_p T_0}h\right)^{\frac{c_p}{R}-1} dh$$

$$= \rho_0\left[1 - \left(1 - \frac{gM}{c_p T_0}H\right)^{\frac{c_p}{R}}\right] \approx 0.69\rho_0 = 0.85\times 10^{-3}\ \text{g cm}^{-3} \tag{2.2.24}$$

The latter density value is between those of the linear and isothermal atmospheres.

Atmospheric composition affects the value of the adiabatic temperature gradient through heat capacity. For present-day air, c_p varies from 1.007 J g^{-1} K^{-1} near 300 K and 1 bar pressure to 1.15 J g^{-1} K^{-1} near 1000 K and 100 bar (Lide, 1994; C_p from 29.17 to 33.32 J mol^{-1} K^{-1}). For some of the other primordial volatiles, listed in Table 2.1, the ranges of the C_p values between 298.15 and 700 K, at 1 atm, are (Gurvich et al., 1994):

$$H_2O,\quad 33.598 \text{ to } 37.557 \text{ J mol}^{-1}\ \text{K}^{-1}$$
$$CO_2,\quad 37.137 \text{ to } 49.569 \text{ J mol}^{-1}\ \text{K}^{-1}$$
$$N_2,\quad\ \ 29.124 \text{ to } 30.754 \text{ J mol}^{-1}\ \text{K}^{-1}$$

As nitrogen is the main component of the present-day atmosphere, the heat capacities of dry air and N_2 are similar.

Figure 2.3 shows temperature profiles in the model of the constant-density atmosphere, dry adiabatic atmosphere, isothermal atmosphere, and a generalized profile for the troposphere. Note the change in the gradient near 11 km altitude where the lower boundary of the tropopause lies and the temperature gradient changes from the negative in the troposphere to the positive in the stratosphere.

Chapter 3

Heat Balance of the Atmosphere and Carbon Dioxide

It is well known that the temperature of the Earth's surface without the atmosphere would have been much colder and probably not suitable for life in the form as we know it. As the conditions on the surface of the two nearest neighboring planets, Venus and Mars, are very different from those on Earth and no life has yet been detected outside our planet, it is difficult to dispute the view that life on Earth is tied to the presence of liquid water and a gaseous atmosphere within the range of temperatures and pressures that characterize the terrestrial near-surface environment. The Earth, at its distance of 1 AU (astronomical unit) from the Sun, lies within the "habitable zone" of our solar system that extends from 0.95 to 1.15 AU (Kasting *et al.*, 1993). Venus, at 0.7 AU from the Sun, and Mars at 1.5 AU are outside the habitable zone. On Venus, that is closer to the Sun than the Earth, the atmosphere is much hotter and denser (450°C and about 90 bar total pressure), whereas on Mars, farther away from the Sun than the Earth, the surface temperature is about −33°C, the atmosphere is very thin, about 0.007 bar, and the occurrence of liquid water and ice is suspected but not proven definitively. This chapter addresses the role of the atmosphere and its constituents, in particular carbon dioxide, in controlling the temperature of the Earth's surface. First, the chapter outlines in a simplified form the main aspects of the heat balance at the Earth's surface and, second, it describes the role of the atmosphere in maintaining the Earth's surface within the range of temperatures that were sometimes colder than at present and sometimes warmer, but as a whole evidently suitable for the existence and evolution of life during the past 4 billion years, a period that represents about 85% of time since the origin of the planet. Furthermore, this chapter discusses some simple relationships of the thermal radiation balance of the Earth's atmosphere and the temperature and greenhouse effect based on them. It also addresses the temperatures of the primordial atmosphere, of the Phanerozoic Earth of the last 550 million years, and of the near future under the effects of the human perturbration (see also Chapter 11).

1   Heat Sources at the Earth's Surface

The Earth's surface receives heat from the Sun and from the interior, the Sun emitting radiation that reaches the Earth and is converted to thermal energy or heat, and the interior supplying heat to the surface in the process known as the geothermal flux. Although both the solar radiation and geothermal flux warm the land, the atmosphere, and surface waters, the magnitudes of these two heat sources are hugely different: at present, solar radiation supplies about 2900 times more heat than the geothermal flux from the Earth's interior.

The solar energy flux[1] from outside the Earth's atmosphere averages over the whole Earth's surface about 342 W m^{-2}, of which 30% is reflected (Fig. 3.1), allowing 239 W m^{-2} to be absorbed, reemitted, and used in geochemical and biological processes. This value is much greater than the next biggest source of energy, the geothermal flux of 0.082 W m^{-2} (Stacey, 1992). For a comparison with the magnitudes of other energy uses, the global geothermal energy flux is 42×10^{12} W, global photosynthesis of organic matter by land and marine plants uses energy at the rate of about 140×10^{12} W, global combustion of fossil fuels in the year 2002 produced energy at the rate of 11×10^{12} W, and food consumption by the six billion people of the world population is equivalent to about 0.6×10^{12} W, at an average food energy intake of 2000 kcal per person per day or about 100 watt per person. The very large amount of solar radiation energy received by the Earth is often compared with energy consumption in various natural and human-generated processes, some of which are mentioned above. The comparisons invariably show that human-generated activities consume only very small fractions of the equivalent of the solar energy available on Earth, and most of them use energy from fossil fuels and sources other than solar. It is a puzzling fact in our historical and technological development that the most plentiful energy source, sunlight, did not become more widely used. Humans as a whole have been, and still are, almost exclusively dependent on fossil fuels and biomass burning as their main sources of energy other than food.

Geothermal flux from the interior to the surface is a combination of two processes: the cooling of the planet since its formation and heat produced by the decay of radioactive isotopes of potassium, thorium, and uranium in the Earth's mantle and crust. The amount of heat from the mantle and crust declines with planetary age: the cooling of the Earth since 4.6 billion ago has removed some heat from the interior and at that time the masses of the main heat-producing radioactive isotopes ^{40}K, ^{232}Th, ^{235}U, and ^{238}U were greater (Sclater *et al.*, 1980; Sleep, 1979; Stacey, 1992). At present, the amount of heat produced by radioactive decay in the mantle and crust is 28×10^{12} W, which represents 67% of the global geothermal flux of 42×10^{12} W or a global average of 0.082 W m^{-2}. Near the origin of the Earth 4.5 billion years ago, the heat flux from radioactive decay was about four times greater, 117×10^{12} W or 0.23 W m^{-2}, but even this value is much lower than the solar energy flux.

---

[1] Heat flow is usually expressed in units of energy supplied to a unit of area in a unit of time, such as in watt per square meter (W m^{-2} = J s^{-1} m^{-2}) or in the older units of langley for the solar energy flux outside the atmosphere (1 langley = 2 cal cm^{-2} minute^{-1} $\approx$ 1395 W m^{-2}) and Heat Flux Units for the geothermal flux (1 HFU = 1 μcal s^{-1} cm^{-2} = 1 × 10^{-6} cal s^{-1} cm^{-2} = 4.184 × 10^{-2} W m^{-2}).

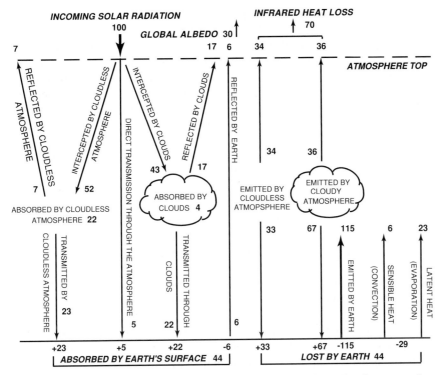

Figure 3.1   Radiation balance of the Earth's surface and atmosphere, showing energy fluxes in percent of incoming radiation. The solar radiation flux at the top of the atmosphere is taken as 100 units (after Liou, 1992, by permission of the author).

2   Solar Heating and Radiation Balance

The energy balance of the incoming solar radiation in the Earth-atmosphere system is shown in Fig. 3.1. About 30% of the incoming radiation, taken as 100 units, is reflected back into space by the atmosphere, clouds, and the Earth's surface. This reflection is called the albedo of the Earth. The incoming radiation is usually referred to as shortwave radiation and the outgoing radiation is longwave radiation. Of the remaining 70% of the incoming radiation, a fraction $44/70 = 0.63$ is transmitted through the atmosphere and absorbed by the Earth's surface and a fraction $26/70 = 0.37$ is absorbed in the atmosphere. Energy absorbed by the Earth-atmosphere system, 70 units of the incoming radiation, is reradiated back to space in the form of longwave infrared radiation. However, because of the absorptive properties of the different atmospheric gases some of the outgoing radiation is absorbed by the atmosphere and then reradiated upward and downward, causing the warming of the atmosphere and the Earth's surface. In the literature, energy balance diagrams vary greatly in their representation of the fluxes within the atmosphere and at the Earth's surface, but all show the conservation of the incoming energy that is equal to the outgoing, 70% of incoming solar radiation or 239 W m^{-2}. More details on the absorption or trapping of the infrared radiation in the atmosphere are given in Section 3.3.

The surface temperatures of the planets orbiting the Sun are usually derived in various textbooks from the balance of the solar illumination of the planet and the planet's reradiation of the energy absorbed, when both the Sun and the planet are assumed to be *ideal black bodies*. In short, an ideal black body absorbs all the incoming radiation energy and reemits it, the energy of emission obeying the Planck law of radiation and the Stefan-Boltzmann law. The latter is the integral of the Planck radiation distribution over all wavelengths (Box 3.3) and it relates energy emitted per unit of area, E, to the fourth power of the temperature of the radiating body:

$$E = \sigma T^4 \quad (\text{W m}^{-2}) \tag{3.1}$$

where the Stefan-Boltzmann constant $\sigma = 5.670 \times 10^{-8}$ W m^{-2} K^{-4} and T is temperature in kelvin (K). For example, the Sun's surface at the temperature of 5780 K as an ideal black body emits into space 63.3×10^6 W m^{-2} or about 63 megawatt per square meter of its surface.

Solar constant is the solar energy flux on a unit of area perpendicular to the Sun rays, at the distance from the Sun equal to 1 Astronomical Unit (1 AU), which is the mean distance of the Earth from the Sun. Inherent in the definition of the solar constant is that the energy flux radiated by the Sun is the same in all directions, the energy is not consumed along the outward paths, and all the radiation received by a planet is converted to heat. More generally, for planets other than the Earth in nearly circular orbits, the solar constant is also defined as the energy flux from the Sun surface that is distributed over the area of a sphere of radius equal to the distance from the Sun to the planet:

$$\text{Solar constant} = \sigma T_s^4 \frac{R_s^2}{D_p^2} \quad (\text{W m}^{-2}) \tag{3.2}$$

where T_s is the temperature of the Sun's surface, R_s is the radius of the Sun, and D_p is the mean distance from the Sun to the planet. The surface temperature of the Sun is 5780 K, its radius is between 695,000 and 696,000 km, and the mean distance to Earth is 149,579,890 km. These values give the solar constant for the Earth (S_0) between 1366 and 1370 W m^{-2} and some other values are given in Table 3.1. The currently accepted value is 1367 W m^{-2} that corresponds to the color temperature of 5780 K for the emitting shell of the Sun (Hoyt and Schatten, 1997).

The variation in the solar constant values shown in Table 3.1 undoubtedly reflects variations in the instrumental techniques that have been evolving during the 19th and 20th centuries and the accuracy of calibration of the satellite measurements since 1978 (Willson, 1997). Apart from the variation in the value of the solar constant due to instrumental and observational factors, the solar constant varies by about 0.1% during the 11-year sunspot cycle (more radiation when the sunspot number is larger) and it also varies periodically on the time scale of one year because the Earth's orbit around the Sun is not a circle but an ellipse where the Sun is located in one of the foci. During one revolution of the Earth around the Sun, the amount

Table 3.1   Solar constant and total solar
irradiance from satellite measurements (W m^{-2})

Year	Solar constant
1902–1904	$1485 \pm \sim 100$[1]
1964	$1353 \pm \sim 27$[2]
1904–1971	$1366 \, (1322-1465)$[3]
1980–1986	$1368.8-1364.6$[4]
1990–1996	$1366.2-1365.0$[4]
1994	1373[5]
1997	1367[6]
1996–2002	$1366.0-1367.8$[7]

[1] Abbe (1911).
[2] Willett (1964).
[3] Mean and range of 30 values in the 20th century
    compiled by Hoyt and Schatten (1997).
[4] Willson (1997), satellite data from maxima and
    minima of two sunspot cycles.
[5] Lide (1994).
[6] Hoyt and Schatten (1997).
[7] Crommelynk (2002).

of solar radiation received by the cross-sectional plane of the Earth that is perpendicular to the Sun's rays (S_e) varies from slightly more to slightly less than the solar constant (S_0), by about $\pm 3.4\%$, as given in equation (3.3). Between the Earth's position closest to the Sun, at the perihelion, and farthest from it, at the aphelion, the solar constant varies about the mean value as (Milankovitch,[2] 1920, p. 6; also Box 3.1):

$$S_e = \frac{S_0}{(1 \pm e)^2} = \frac{S_0}{(1 \pm 0.0167)^2} = (1.034 \text{ to } 0.967) \times S_0 \quad (\text{W m}^{-2}) \qquad (3.3)$$

where S_0 is the mean value of the solar constant and $e = 0.0167$ is the eccentricity of the Earth's solar orbit, with the major semi-axis a and minor semi-axis b $(a > b)$, as defined in Box 3.1 along with some other essential geometric relationships for the ellipse. Mean distance from the Earth to the Sun places the Sun at the center of the circle of radius 1 AU. In calculation of the mean solar constant and the radiation energy balance of the Earth, the periodic intra-annual variations due to the changing distance are usually disregarded. The actual distance from the Earth's surface receiving solar radiation to the Sun's surface emitting it is also not considered: the Earth to Sun mean distance of 149,579,890 km should be the distance between the centers of mass of the two bodies; relative to this distance, the Earth's mean radius of 6,371 km is negligible and the Sun's radius of approximately 696,000 km amounts to 0.5% of the mean distance.

---

[2] Records in the Archive of the Austrian Academy of Sciences (Sienell, 2003) suggest that Milankovitch must have adopted this spelling of his name while living and working in Vienna before World War I. The Serbo-Croatian spelling, in the Latin letters of the Croatian branch of the language, was used by him earlier as Milanković. He continued to use Milankovitch before and after World War I, after his return to Belgrade in future Yugoslavia.

Box 3.1   Ellipse and the Earth's solar orbit

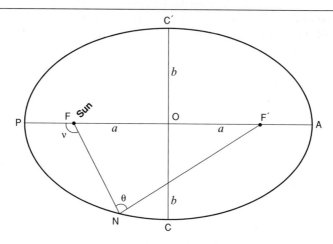

Geometric relationships in an ellipse. As a representation of the Earth's orbit around the Sun, the shape is greatly exaggerated: the distance between the foci, segment FF′, in the Earth's orbit is only about 1.7% of the major axis of the ellipse, PA, which makes the orbit nearly circular. Sun at the focus of point F. P is perihelion, A is aphelion, FN and F′N are radii vectors.

The ellipse, as shown above, is a closed curve with the major (PA) and minor (CC′) axes intersecting at the center O (Sanger, 1964). The major semi-axis OP = OA = a, and the minor semi-axis OC = OC′ = b. The points F and F′ are the foci of the ellipse and they are important in defining many of its geometric properties.

Eccentricity of the ellipse: $$e = \sqrt{1 - \frac{b^2}{a^2}} \tag{3.1.1}$$

Eccentricity of Earth's solar orbit: $e = 0.01673$

Eccentricity approaches 1 when $b \ll a$, and 0 when $b = a$, making the ellipse a circle. For the Earth's orbit as a circle, the radius is 1 AU (Astronomical Unit) = 149,597,890 km.
Focal distances FO and OF′ are:

$$\text{FO} = \text{OF}' = ae = \sqrt{a^2 - b^2} \tag{3.1.2}$$

$$\text{FO} + \text{OF}' = 2ae = 2\sqrt{a^2 - b^2} \tag{3.1.3}$$

Distances:

$\text{FP} = a(1 - e)$, closest approach to the Sun at F, P is perihelion   (3.1.4)

$\text{FA} = a(1 + e)$, greatest distance from the Sun at F, A is aphelion   (3.1.5)

$\text{CF} = \sqrt{b^2 + \text{FO}^2} = \sqrt{b^2 + a^2 e^2} = a$ (3.1.6)

Distance equalities: FP = F'A and FA = F'P

Radius vector. Straight line segment drawn from a focus to a point on the ellipse, such as FN or F'N, is called the radius vector. The sum of two radii vectors drawn to the same point is a constant:

$$FN + F'N = 2a \qquad (3.1.7)$$

The length of the radius vector (FN or F'N) changes as point N moves along the circumference of the ellipse, while the sum of their lengths is constant, (3.1.7). The length of the radius vector can be defined in terms of either the angle between the two radii vectors to the same point (angle FNF' or θ) or in terms of the angle between the major axis and the radius vector (angle PFN or v) or its complementary angle NFO or $\pi - v$. The equation for the radius vector, $r_V = FN$, is:

$$r_V = a \pm \sqrt{a^2 - \frac{2b^2}{1 + \cos\theta}} \qquad (3.1.8)$$

There are two values of r_V in (3.1.8), both positive, because the same angle θ can be formed by the two radii vectors, once in the eastern half of the ellipse and once in the western half.

Two other relationships define the radius vector in the Earth's solar orbit, one in terms of angle v, related to Sun's longitude, and the other in terms of time of the annual revolution (Milankovitch, 1920, pp. 11, 13):

$$r_V = \frac{b^2}{a\sqrt{1 + e\cos v}} \qquad (3.1.9)$$

$$r_V = a\left[1 - e\cos\frac{2\pi}{T}(t - t_0)\right] \qquad (3.1.10)$$

In (3.1.10), e is the eccentricity, as given in (3.1.1), T is the period of rotation (1 sidereal year $= 3.1558 \times 10^7$ s), t_0 is the time of the position closest to the Sun (for Earth at present, the passage through the perihelion is on January 2), and t is time elapsed since t_0. The quotient $(t - t_0)/T$ is a fraction of one complete revolution around the Sun or a fraction of one sidereal year. Equation (3.1.10), with the two terms shown in brackets, is valid for orbits of small eccentricity, such as the Earth's.

The balance between the energy of the incoming solar radiation received by the Earth and the energy reradiated by the Earth determines its equilibrium temperature, T. The Earth is a sphere spinning around its axis of rotation such that one half of its surface is always illuminated by the Sun and the other half is dark. The boundary between the illuminated and dark hemispheres is a great circle on the spherical Earth's surface, also called the terminator circle. The diameter of the Earth is so much smaller than the diameter of the Sun that the incoming Sun's rays are considered parallel. The rays pass through a cross-section of the Earth outside its surface, the area of which is πR_e^2, where R_e is the mean radius of the Earth. Thus the amount of energy falling on the Earth is:

$$S_{in} = \pi R_e^2 S_0 \quad (W) \tag{3.4}$$

and this amount is reradiated by the entire surface of the Earth, $4\pi R_e^2$, because the Earth rotates on its axis faster than it revolves around the Sun. The amount per unit of the Earth's surface area that is reradiated back is therefore

$$S_{out} = \frac{\pi R_e^2 S_0}{4\pi R_e^2} = \frac{S_0}{4} \quad (W\,m^{-2}) \tag{3.5}$$

and the balance between the incoming and outgoing radiation is from equations (3.1) and (3.5):

$$\frac{S_0}{4} = \sigma T^4 \quad (W\,m^{-2}) \tag{3.6}$$

For bodies that are not ideally black, a coefficient of emissivity, $\varepsilon \lesssim 1$, is introduced in (3.6), to allow for incomplete reradiation of the absorbed energy:

$$\frac{S_0}{4} = \varepsilon \sigma T^4 \quad (W\,m^{-2}) \tag{3.7}$$

The emissivity coefficient values for the Earth's surface are close to unity. Values smaller than 1 would make the equilibrium temperature T higher. For the Earth's surface, $\varepsilon = 0.92$ was used by Milankovitch (1920, p. 192), and a range from 0.94 to 0.99 for water, cultivated land, rocks, and vegetation leaf cover, with a mean value of $\varepsilon = 0.95$ was cited by Tardy (1986). Emissivity coefficients for similar types of surfaces and deserts, salt pans, tropical forests, snow-covered vegetation, sea ice, and snow-covered ice are in the range from 0.88 to 0.99 (Henderson-Sellers and Robinson, 1986). An average emissivity coefficient $\varepsilon \approx 0.95$ for the Earth's surface would add about 3.5 K to an equilibrium temperature. Annual variations in the Earth's distance to the Sun, mentioned in equation (3.3), would translate into temperature variations by a factor of $(1 \pm 0.0167)^{1/2}$ or about ± 2 K.

For the Earth radiating as an ideal black body, its temperature is from equation (3.6):

$$T = \sqrt[4]{\frac{S_0}{4\sigma}}$$

$$= \left(\frac{1367\,W\,m^{-2}}{4 \times 5.67 \times 10^{-8}\,W\,m^{-2}\,K^{-4}} \right)^{\frac{1}{4}} = 278.6\,K \quad \text{or} \quad 5.5°C \tag{3.8}$$

The result is a temperature that is about 10°C lower than the global average of the Earth's surface. The discrepancy becomes even greater when the reflected fraction of the incoming solar radiation, the Earth's albedo, is taken into account. The clouds in the Earth's atmosphere, snow and ice on the surface, and other geomorphic elements reflect different fractions of the radiation falling on them. The average albedo value for the Earth is about 30% or $A = 0.30$ (Fig. 3.1). The remaining fraction of 70%, $1 - A$, is absorbed and reradiated by the Earth. Accordingly, the incoming radiation flux outside the Earth's atmosphere, S_0, is reduced by the fraction A, and equation (3.8) then gives a lower temperature:

$$T = \sqrt[4]{\frac{S_0(1-A)}{4\sigma}} = \left(\frac{1367\,\mathrm{W\,m^{-2}} \times (1-0.3)}{4 \times 5.67 \times 10^{-8}\,\mathrm{W\,m^{-2}\,K^{-4}}} \right)^{\frac{1}{4}} = 255\,\mathrm{K} \ \ \mathrm{or} \ -18^{\circ}\mathrm{C} \quad (3.9)$$

A global average temperature of -18°C is a frozen Earth's surface that could have an albedo considerably greater than 30%, leading to an even lower average temperature. Here, the role of the atmosphere as a modifier of the temperature and global climate comes into play. The phenomenon known as the greenhouse effect is discussed in the next section.

3 Greenhouse Effect

The analogy of the Earth's atmosphere to a greenhouse has an appeal of a familiar concept but it is a misnomer, even though it will continue to be used everywhere, including this book. A greenhouse or, in its older English name, a glasshouse lets sunlight in, where it is used by growing plants, and it must have a warm interior. Sunlight warms the interior of a greenhouse because air absorbs infrared radiation. However, a greenhouse stays warm because of good construction that prevents convective heat loss (no drafts allowed) and, in colder climates, by additional heating. Before proceeding further, the three main mechanisms of heat transport may be briefly summarized here. Heat transport by *radiation* is through emission or absorption of electromagnetic radiation that is converted to heat. *Convection* or *advection* is a process of heat transport by moving masses of the medium material, such as wind or water flow. *Conduction* of heat is a process driven by molecular motions analogous to the transport of matter by molecular diffusion. Heat or material transport in a turbulent medium may be considered as a special case of conduction, with the difference that it is not molecular motions but rather turbulent eddies of macroscopic dimensions that transfer heat or matter by exchanging it between them. Eddy diffusion or eddy conduction of heat is treated as a process analogous to molecular diffusion or conduction, but where the values of the transport coefficients are much greater, up to several orders of magnitude, than in the transport driven by forces on the molecular scale.

Glass provides some resistance to heat loss because of its relatively low thermal conductivity (lower than, for example, the thermal conductivities of metals but much greater than that of air that is, in general, a very good thermal insulator). In fact, heat loss from a greenhouse by thermal conduction of the glass may be greater than heat loss by radiation to the external environment. This aspect of a greenhouse is discussed below.

3.1   A Model Greenhouse

One may consider a greenhouse that is maintained at a comfortable plant growing temperature of 25°C inside while it is freezing cold outside, 0°C. The net radiative heat flow from the greenhouse to the surroundings, F_r, assuming both are ideal black body radiators, is:

$$F_r = \sigma(T^4 - T_0^4) \quad (\text{W m}^{-2}) \tag{3.10}$$

where $T = 298$ K is the greenhouse temperature and $T_0 = 273$ K is the outside air temperature. Then the rate of heat loss by the greenhouse is:

$$F_r = 5.67 \times 10^{-8}(298^4 - 273^4) \approx 130 \text{ W m}^{-2} \tag{3.11}$$

Heat will also flow from the greenhouse by thermal conduction of the glass in a direction perpendicular to the glass surface:

$$F_c = -K\frac{\Delta T}{\Delta z} \quad (\text{W m}^{-2}) \tag{3.12}$$

where K is the coefficient of thermal conductivity of the glass ($\text{W m}^{-1}\text{K}^{-1}$), ΔT is the temperature difference across the glass walls, and Δz is the glass thickness. If the inside temperature is maintained at a constant value and the outside air is also at a constant temperature and well mixed, then the temperature difference between the inside and outside glass surfaces is constant at 25°C, and equation (3.12) describes a steady-state flow of heat. Thermal conductivity coefficients for glasses of various composition, in the range from 0 to 100°C, are between 0.75 and 1.25 $\text{W m}^{-1}\text{K}^{-1}$ (Lide, 1994). The glass thickness, for this example, can be 1 cm. With these values, conductive heat flow from the greenhouse is:

$$F_c = (1 \text{ W m}^{-1}\text{K}^{-1}) \times \frac{298 \text{ K} - 273 \text{ K}}{0.01 \text{ m}} \approx 2500 \text{ W m}^{-2} \tag{3.13}$$

The much higher rate of heat loss from the greenhouse surface by conduction than by thermal radiation should not be confused with the transfer of heat within the greenhouse by air circulation or, by analogy, within the Earth's atmosphere where air convection is the main mechanism of heat distribution. In practice, one would try to reduce heat loss from a greenhouse by additional insulation at night and, perhaps, by using glass of lower thermal conductivity or double-glazed construction, with the space between the glass panes filled with air or even evacuated. Thermal conductivity of air is about 40 times smaller than that of glass at room temperature, so that an air-filled layer between double glass panes and its thickness greater than 1 cm might reduce the conductive heat flow to a few tens of watts per square meter, making it smaller than the radiative heat flux. This is of course the principle behind double-paned windows.

Heat transport or, in particular, the rate of cooling by convective flow of air is known from experimental evidence to be approximately proportional to the temperature difference between the cooling object and the surrounding air. However, many factors affect this rate, including the geometric shape of the object, velocity of air flow, and relative humidity. Numerous studies of convective cooling of different materials in different shapes and natural systems under different conditions have produced many empirical equations describing specific systems, such as the cooling rates by wind of exposed or vegetation-covered soils, or lake and ocean water surfaces. In the present case of a model greenhouse, with the two principal modes of heat loss, by radiative and conductive cooling, it is the radiative part of the heat budget that lends the name "greenhouse warming" to the Earth's surface and the atmosphere.

3.2   Greenhouse Warming of the Earth

The Earth's surface system that includes the solid Earth's surface, the oceans, and the atmosphere is too heterogeneous and complex for its temperature to be accounted for by simple equations of black body radiation balance, such as those given in the preceding section and those discussed in Box 3.2. However, another simple greenhouse model, even though it also oversimplifies the picture, is that some fraction B of the energy received by the Earth is used not in direct reradiation but in "something else" or it is "trapped" in the atmosphere, so that the energy balance between the incoming solar radiation and black body radiation outgoing from the Earth's surface and lower atmosphere can be written as:

$$\frac{S_0(1 - A)}{4} = \sigma T^4(1 - B) \quad (W\,m^{-2}) \tag{3.14}$$

Box 3.2   Some simple radiation balance models

Simple models trying to explain the mean temperature of the Earth's surface in terms of black body radiation balance between the incoming solar radiation and the outgoing terrestrial radiation (Fig. 3.1) start with the balance equation in the following form:

$$\frac{S_0(1 - A)}{4} = \sigma T^4 \quad (W\,m^{-2}) \tag{3.2.1}$$

where $S_0/4$ is the solar constant distributed over the Earth's surface area ($1367/4 \approx 342\,W/m^2$) and $A \approx 0.30$ is the albedo or the fraction of solar radiation reflected by the Earth as a whole. From the preceding equation, the equilibrium temperature of the Earth is 255 K or $-18°C$, as was shown in the text, equation (3.9). The examples in this box deal with a one-layer atmosphere and radiation in one direction. A discussion of radiation balance and temperature in a multilayer atmosphere can be found in Harte (1988).

(a)   If solar radiation heats a layer of the atmosphere above the Earth's surface, then the energy balance may be taken as in the figure below (after Kump et al., 2004):

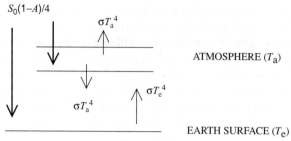

Figure 3.2.1

Because the atmosphere reradiates solar energy over two surfaces of the layer, up and down, the energy balance for a one-layer atmosphere is:

$$\frac{S_0(1 - A)}{4} = 2\sigma T_a^4 \tag{3.2.2}$$

The Earth's surface receiving solar radiation would have an equilibrium temperature T_e:

$$\frac{S_0(1 - A)}{4} = \sigma T_e^4 \tag{3.2.3}$$

From the preceding two equations, it follows that the Earth's surface temperature is greater by a factor of $2^{1/4}$ than the temperature of the atmospheric layer:

$$\sigma T_e^4 = 2\sigma T_a^4$$

and
$$T_e = 2^{1/4} T_a = 1.19\, T_a \tag{3.2.4}$$

For $T_a = 255$ K, this makes the Earth's surface temperature 303 K, about 15 K higher than at present.

(b)   A slightly more accurate balance for the atmosphere layer and the Earth's surface that are shown in Fig. 3.2.1 is the following. Atmosphere receives incoming solar radiation and the radiation emitted by the Earth, and it reradiates it across the upper and lower surfaces:

$$\frac{S_0(1 - A)}{4} + \sigma T_e^4 = 2\sigma T_a^4 \tag{3.2.5}$$

The Earth's surface receives solar radiation and the radiation from the atmosphere across its lower boundary, and it reradiates upward:

$$\frac{S_0(1 - A)}{4} + \sigma T_a^4 = \sigma T_e^4 \tag{3.2.6}$$

Subtraction of (3.2.6) from (3.2.5) and rearrangement of the terms gives a relationship between the Earth's surface and atmosphere temperature:

$$T_e = \left(\frac{3}{2}\right)^{1/4} T_a = 1.11 T_a \tag{3.2.7}$$

The latter result gives the Earth's surface temperature of 283 K, about 5 K lower than now.

(c) Based on the schematic models in **(a)** and **(b)**, the fractions of solar radiation absorbed by the atmosphere and by the Earth's surface may be used to estimate the difference between their temperatures. The Earth absorbs 44% of incoming solar radiation, $S_0/4$, or 63% of net radiation after the reflected fraction of 30% has been subtracted, $S_0(1 - A)/4$ (Figs. 3.1 and 3.2.1). The remaining 37% is the energy fraction absorbed by the atmosphere, as shown schematically in Fig. 3.2.2 below.

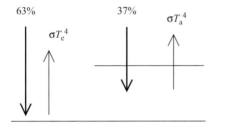

ATMOSPHERE (T_a)

EARTH SURFACE (T_e)

Figure 3.2.2

The balance equations in this case are:

For the atmosphere, $\qquad 0.37 \cdot \dfrac{S_0(1 - A)}{4} = \sigma T_a^4$ $\qquad\qquad$ (3.2.8)

For the Earth's surface, $\qquad 0.63 \cdot \dfrac{S_0(1 - A)}{4} = \sigma T_e^4$ $\qquad\qquad$ (3.2.9)

The ratio of the temperatures is, therefore, $\qquad \sigma T_e^4 = \dfrac{0.63}{0.73} \sigma T_a^4$

and $\qquad\qquad\qquad\qquad\qquad T_e = 1.14 T_a$ $\qquad\qquad$ (3.2.10)

The latter result falls between those of **(a)** and **(b)**, 291 K.

(d) The last simple model combines the reasoning of **(b)** and **(c)**. Energy balance of incoming and outgoing radiation is partitioned into 63% received by the Earth's surface and 37% received by the atmosphere. The radiation balance is therefore:

for the atmosphere, $\quad 0.37 \cdot \dfrac{S_0(1 - A)}{4} + \sigma T_e^4 = 2\sigma T_a^4$ $\qquad\qquad$ (3.2.11)

and for the Earth's surface, $\quad 0.63 \cdot \dfrac{S_0(1 - A)}{4} + \sigma T_a^4 = \sigma T_e^4$ $\qquad\qquad$ (3.2.12)

Subtracting (3.2.11) from (3.2.12): $0.26 \cdot \dfrac{S_0(1 - A)}{4} + 3\sigma T_a^4 = 2\sigma T_e^4$ $\quad$ (3.2.13)

From the latter equation, the temperatures of the Earth's surface and atmosphere are interrelated by:

$$T_e^4 = \frac{3}{2}T_a^4 + \frac{0.13}{\sigma} \cdot \frac{S_0(1-A)}{4}$$

$$= (1.5 + 0.13) \cdot \frac{S_0(1-A)}{4\sigma} \qquad (3.2.14)$$

The Earth surface is warmer at $T_e = 288$ K than the atmosphere.

The similarity of equations (3.14) and (3.7) may be noted, despite the difference in the physical significance of the terms $(1 - B)$ and the coefficient of emissivity ε.

Because the incoming flux of solar energy must be equal to the flux emitted by Earth, absorption of some fraction of the incoming radiation requires the temperature of the Earth-atmosphere system to rise in order to increase the outgoing flux. The equilibrium temperature then becomes:

$$T = \sqrt[4]{\frac{S_0(1-A)}{4\sigma(1-B)}} = \left(\frac{1367\,\mathrm{W\,m^{-2}} \times (1-0.3)}{4 \times 5.67 \times 10^{-8}\,\mathrm{W\,m^{-2}\,K^{-4}} \times (1-0.39)}\right)^{\frac{1}{4}} = 288\,\mathrm{K}\ \text{or}\ 15°C$$

$$(3.15)$$

The value of absorbed radiation fraction, $B = 0.39$, gives the temperature of 15°C for the present Earth's surface. This fraction is slightly higher than $0.26/0.70 = 0.37$ of the incoming solar radiation absorbed by the atmosphere, as discussed in Section 2. The same temperature of 288 K is obtained from equation (3.15) with the values of the radiation absorbed fraction $B = 0.37$, as given in Fig. 3.1, and an average emissivity coefficient $\varepsilon = 0.97$. The magnitude of the *greenhouse warming* is the difference between 288 K and 255 K or +33°C.

In Fig. 3.2A are shown the equilibrium temperatures that correspond to different combinations of the albedo and absorption. The relatively narrow band of temperatures of environmental concern (Fig. 3.2B) lies between about 282 K (9°C) at the Last Glacial Maximum 18,000 years ago and 300 K (27°C) that is near the upper limit of shallow ocean water during the Phanerozoic Eon of the last 540 million years (Wallmann, 2004). The trend of continuing emissions from fossil fuel burning and land-use change to the year 2100 may make the mean temperature 2 to 6°C higher than the approximate present-day 15°C (Section 5). Because the radiation absorption fraction, B, depends on the chemical composition of the atmosphere, it is one of the main forcings on the Earth's surface temperature and climate.

The radiation balance of the Earth's surface and atmosphere (Figs. 3.1, 3.3A) shows that 239 W m^{-2} are received from the incoming solar radiation and lost by emission from the upper atmosphere. The solar constant of 1385 W m^{-2} in Fig. 3.3A is slightly greater than the value of 1367 W m^{-2} used here. Absorption of radiation by different atmospheric gases is shown in Fig. 3.3B. In the shortwave, UV range, incoming solar radiation is absorbed by molecular oxygen (O_2) and ozone (O_3) in the stratosphere. In the longwave, infrared range, the main absorbing gases are water vapor (H_2O), carbon

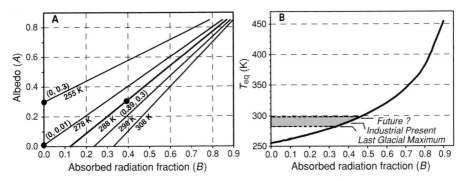

Figure 3.2   A. Temperature corresponding to different combinations of the Earth's albedo (A) and fraction of incoming solar radiation absorbed by the atmosphere (B), equation (3.13) (after Dutton, 1995). B. Earth's surface temperature as a function of the radiation fraction absorbed by the atmosphere (B).

dioxide (CO_2), methane (CH_4), nitrous oxide (N_2O), tropospheric ozone (O_3), and in the modern world, the synthetic chorofluorocarbon gases. In particular, the occurrence of CO_2 absorption peaks should be noted in the infrared region of wavelengths between about 2 and 15 μm. The effect of increasing concentrations of CO_2, water vapor, and other gases on the temperature of the Earth's surface depends on the characteristics of their absorption lines at the different wavelengths of the longwave radiation. For example, CO_2 has a strong absorption band near the wavelength of 15 μm. This band is essentially saturated and further increases in atmospheric CO_2 would not affect absorption at that wavelength (Seinfeld and Pandis, 1998; Shine et al., 1995). At other wavelengths, however, such as near 10 μm, CO_2 absorption bands are not saturated and an increase in CO_2 concentration will result in additional absorption and warming of the atmosphere. Computations of radiation balance in atmospheres of changing chemical composition require detailed knowledge of the absorption spectra of the individual gases and such computations form the basis of predictive climate change models (Hansen et al., 1988; and reports in Houghton et al., 2001; Wigley, 1994, with references to earlier editions of the Intergovernmental Panel on Climate Change volumes). The visible light portion of the solar radiation contained between 0.4 and 0.7 μm (Fig. 3.3A) accounts for about 37% of the total energy spectrum. Estimation of the energy fraction contained between two wavelengths is given in Box 3.3.

The radiative energy balance that is given in equations (3.14) and (3.15) means that of 239 W m^{-2} of the reradiated energy, a fraction

$$\frac{1}{1 - (0.39 \text{ to } 0.37)} - 1 = 0.64 \text{ to } 0.59$$

equivalent to 153 to 141 W m^{-2} does not escape directly into space but contributes to the warming of the atmosphere and Earth's surface. This absorbed energy is a forcing on the Earth-atmosphere system and without it the black body equilibrium temperature

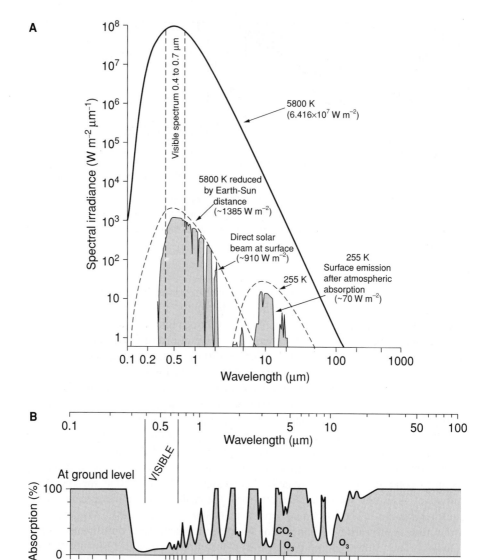

Figure 3.3   A. Radiation absorption spectrum of the Earth's atmosphere in wavelength range from 0.1 to 100 µm (modified from Henderson-Sellers and Robinson, 1986, Fig. 2.5(b), by permission of the authors). B. Absorption of spectra of atmospheric gases in the shortwave and longwave range. After Henderson-Sellers and Robinson (1986, Fig. 2.3), by permission of Pearson Education, and after Goody (1995, Fig. 1.3(b)), by permission of Oxford University Press.

Box 3.3   Energy fractions of emission spectrum

The Stefan-Boltzmann law of radiation emitted by an ideal black body is an integral of the Planck radiation law over all wavelengths, from $\lambda = 0$ to infinity. Total energy emitted is:

$$E = \frac{2\pi k^4}{h^3 c^2} T^4 \int_0^\infty \frac{u^3}{e^u - 1} du \qquad (3.3.1)$$

where the variable u is:

$$u = \frac{hc}{\lambda k T} \qquad (3.3.2)$$

and h is the Planck constant, c the velocity of light, k the Boltzmann constant, and T is temperature. Integration in equation (3.3.1) gives an explicit form of the Stefan-Boltzmann law (e.g., Weisstein, 2005):

$$E = \frac{2\pi^5 k^4}{15 h^3 c^2} T^4$$
$$= \sigma T^4 = 5.67 \times 10^{-8} \times T^4 \, \mathrm{W \, m^{-2}} \qquad (3.3.3)$$

Energy fraction of a black body radiation spectrum (Fig. 3.3A) that is contained below some wavelength λ is:

$$E_{<\lambda} = \frac{2\pi k^4}{h^3 c^2} T^4 \int_u^\infty \frac{u^3}{e^u - 1} du$$
$$= \sigma T^4 f_1(u) \qquad (3.3.4)$$

where $f_1(u) = 0$ when $\lambda = 0 \, (u = \infty)$, and $f_1(u) = 1$ when $\lambda = \infty \, (u = 0)$. A close approximation of $f(u)$ is given by the following polynomial (Weisstein, 2005, equation (16)):

$$f_1(u) = e^{-u} \left[1 + u + \frac{u^2}{2} + \left(\frac{1}{6} - \frac{5}{\pi^4} \right) u^3 + \cdots \right] \qquad (3.3.5)$$

The cumulative energy fraction curve as a function of the wavelength, $E_{<\lambda}$, is plotted in Fig. 2.7 and it is calculated from equation (3.3.5). The fraction of energy spectrum contained in the visible light, between the wavelengths $\lambda = 400 \, \mathrm{nm}$ and $\lambda = 700 \, \mathrm{nm}$, is

$$\frac{E_{<700} - E_{<400}}{\sigma T^4} = f_1(u = 2.055 \times 10^4 / T) - f_1(u = 3.597 \times 10^4 / T) \qquad (3.3.6)$$

where all the parameters in u are in c.g.s. units. At the solar radiation temperature of $T = 5780 \, \mathrm{K}$, the energy fraction of the visible light from equation (3.3.6) is 0.351 or 35% (Figs. 3.3A and 2.6). For the young Earth and faint young Sun that emitted only about 75% of its present-day energy, the emission temperature would be $T = 5379 \, \mathrm{K}$ and the visible light band would be 0.326 or about 33% of the total energy spectrum.

A complete solution of $E_{<\lambda}$ in equation (3.3.4) is given by the explicit series of the integral term (Abramowitz and Stegun, 1972, equation 27.1.2):

$$\int_u^\infty \frac{u^3}{e^u - 1} du = \sum_{k=1}^\infty e^{-ku} \left[\frac{u^3}{k} + \frac{3u^2}{k^2} + \frac{6u}{k^3} + \frac{6}{k^4} \right] \qquad (3.3.7)$$

The value of the preceding integral between the limits of 0 and infinity is:

$$\int_0^\infty \frac{u^3}{e^u - 1} du = \frac{\pi^4}{15}$$

Therefore a fractional value of the energy spectrum can be defined as $f_2(u)$ from (3.3.7):

$$f_2(u) = \frac{15}{\pi^4} \sum_{k=1}^\infty e^{-ku} \left[\frac{u^3}{k} + \frac{3u^2}{k^2} + \frac{6u}{k^3} + \frac{6}{k^4} \right] \qquad (3.3.8)$$

From (3.3.4) and (3.3.8), we have an explicit form of $E_{<\lambda}$:

$$E_{<\lambda} = \sigma T^4 f_2(u)$$

$$= \sigma T^4 \times \frac{15}{\pi^4} \sum_{k=1}^\infty e^{-ku} \left[\frac{u^3}{k} + \frac{3u^2}{k^2} + \frac{6u}{k^3} + \frac{6}{k^4} \right] \qquad (3.3.9)$$

The energy fraction of the visible radiation, computed from the complete relationship (3.3.9), is only slightly higher than the approximate results given above: 36.7% for the present and 34.5% for the faint young Sun.

The fraction of the solar constant of 1370 W m^{-2} contained between the wavelengths 400 and 700 nm is 38.15% (cited by Houghton, 1991).

would have been 33°C lower. This forcing magnitude should be kept in perspective with respect to the global warming during the past 200 to 300 years that is attributed to anthropogenic emissions of CO_2 and other greenhouse gases from fossil fuel burning and changing land-use practices. The anthropogenic forcing has increased from zero to about 2.4 W m^{-2} (not including the forcing from tropospheric O_3) since the late 1700s to the year 2000, as is shown in Fig. 3.5A.

3.3   The Milankovitch Atmosphere Model

Radiation absorption properties of the atmosphere depend on its chemical composition and, specifically, on the concentrations or partial pressures of radiation absorbing gases. Water vapor and carbon dioxide are the two most important greenhouse gases of the pre-industrial atmosphere, and even at present they account for the major part of absorption of longwave infrared radiation. The first attempt to explain quantitatively the role of carbon dioxide as a warming gas in the atmosphere containing also water vapor was that of Svante Arrhenius (1896). Earlier, by the early 1860s, John Tyndall recognized the role of water vapor in the atmosphere as a strong absorber of infrared radiation (Tyndall, 1863). A mathematical model of the equilibrium temperature for the Earth's surface and atmosphere was developed by Milankovitch (1920) as a relationship between incoming solar radiation, absorptive properties of the atmosphere for the shortwave and longwave radiation, and a small departure of the Earth's surface from the characteristics of an ideal black body. The Milankovitch atmosphere consists of one "average" gas that absorbs differently the incoming shortwave and outgoing longwave radiation and it is characterized by one scale height.

The relative simplicity of the Milankovitch atmosphere model as a forerunner of the later, much more detailed temperature and climate models, makes it a useful intro-duction to the radiation and temperature balance of the Earth's surface and atmosphere. For the Earth's surface, the radiation energy balance depends on the solar constant (S_0), the planetary albedo (A), the emissivity of the Earth's surface (ε), and the fractions of the transmitted shortwave (p_{sw}) and longwave (p_{ir}) radiation (Milankovitch, 1920, p. 140; 1930, p. 80):

$$\varepsilon \sigma T^4 = \frac{1}{2} \frac{S_0(1-A)}{4} \left[1 + p_{sw} + \frac{\ln p_{ir}}{\ln p_{sw}}(1 - p_{sw}) \right] \quad (\text{W m}^{-2}) \qquad (3.16)$$

It is more convenient to recast equation (3.16) in terms of absorbed radiation fractions, a_{sw} denoting the absorbed fraction of shortwave radiation and a_{ir} the absorbed fraction of longwave radiation:

$$a_{sw} = 1 - p_{sw} \quad (0 \le a_{sw} \le 1) \qquad (3.17)$$

$$a_{ir} = 1 - p_{ir} \quad (0 \le a_{ir} \le 1) \qquad (3.18)$$

Equation (3.16), written in terms of the absorbed rather than transmitted radiation fractions, and with the emissivity coefficient ε disregarded, becomes:

$$\sigma T^4 = \frac{1}{2} \frac{S_0(1-A)}{4} \left[2 - a_{sw} + \frac{\ln(1 - a_{ir})}{\ln(1 - a_{sw})} a_{sw} \right] \quad (\text{W m}^{-2}) \qquad (3.19)$$

If the atmosphere is completely transparent to the incoming shortwave radiation, then there is no absorption of the shortwave radiation ($a_{sw} = 0$) and the quotient term in brackets, on the right-hand side of (3.19), $a_{sw}/\ln(1 - a_{sw})$, would be 0/0. The limiting value of this quotient is -1 as a_{sw} tends to 0, which is obtained by differentiation of the numerator and denominator with respect to a_{sw}, and taking the limit of $a_{sw} \rightarrow 0$. In this limiting case of no absorption of shortwave radiation, the equilibrium temperature would be given by the relationship:

$$\sigma T^4 = \frac{1}{2} \frac{S_0(1-A)}{4} [2 - \ln(1 - a_{ir})] \quad (\text{W m}^{-2}) \qquad (3.20)$$

Furthermore, if there is no infrared absorption in the atmosphere, then $a_{ir} = 0$ and equation (3.20) becomes the simple radiation balance as given in equations (3.9) and (3.2.1).

Because the composition of the atmosphere during some periods of the geologic past was likely different from its present-day composition, the absorbed fractions of the shortwave and longwave radiation would have been also different. The Earth's surface temperature can be estimated from relationships (3.19) and (3.20), using the following parameter values. The fraction of the shortwave radiation absorbed by the atmosphere is $a_{sw} = 0.37$ (Fig. 3.1). The fraction of the longwave radiation absorbed by the lower atmosphere, a_{ir}, is near 0.90 and cited often in the literature as being about 90%: this is the difference between the absorption of infrared radiation at the Earth's surface and at the tropopause, at an altitude of 11 km (Schneider and Kellogg, 1973). Using $a_{ir} = 0.87$ in equation (3.19) gives 288 K for the equilibrium temperature:

$$
\begin{aligned}
T &= \sqrt[4]{\frac{S_0(1 - A)}{8\sigma} \left[2 - a_{sw} + \frac{\ln(1 - a_{ir})}{\ln(1 - a_{sw})} a_{sw} \right]} \\
&= \left(\frac{1367 \, \text{W m}^{-2} \times (1 - 0.3)}{8 \times 5.67 \times 10^{-8} \, \text{W m}^{-2} \, \text{K}^{-4}} \left[2 - 0.37 + \frac{\ln(1 - 0.87)}{\ln(1 - 0.37)} \times 0.37 \right] \right)^{\frac{1}{4}} = 288 \, \text{K}
\end{aligned}
$$

$$(3.21)$$

The radiative forcing in this Milankovitch atmosphere model of equation (3.19) is essentially the same as the forcing given by the term $(1 - B)$ in equation (3.14):

$$
\frac{1}{1 - B} = 1 - \frac{a_{sw}}{2} + \frac{a_{sw}}{2} \cdot \frac{\ln(1 - a_{ir})}{\ln(1 - a_{sw})}
$$

$$(3.22)$$

and, with the same parameter values as used in equation (3.21),

$$
B = 0.39
$$

The preceding discussion shows how several simple models of radiation balance give a similar result for the present-day average temperature of the Earth's surface and lower troposphere. The main natural absorbers of infrared radiation in the atmosphere are water vapor, carbon dioxide, clouds, methane, nitrous oxide, and tropospheric ozone (Fig. 3.3B). The latter is a minor component in the present-day atmosphere, but among the former three, clouds may account for 50% of the trapped longwave radiation and H_2O is by far more important as a greenhouse gas than CO_2: among the two, H_2O has been estimated to account for about 88% and CO_2 for 12% of the joint longwave absorption by these two gases (Goody, 1995; Ramanathan and Coakley, 1978; Wayne, 2000). It should be noted that one of the consequences of an atmospheric temperature rise due to increasing atmospheric CO_2 concentrations may be an increase in the mass of water vapor in the atmosphere that adds to its warming. Such an increase has recently been noted by modeling and in observations made by instrumentation aboard orbiting satellites (Minschwaner and Dessler, 2004).

4   Temperature of a Prebiotic Atmosphere

The most pronounced change in the solar constant has probably been the increase in the luminosity and energy emission since the formation of the Sun (Gough, 1981). Because of the burning of hydrogen to helium in the Sun's interior, its density and luminosity have increased. The luminosity and energy emission about 4.5 Ga ago were about 71 to 75% of the present, depending on the physical model of the Sun's interior. The Sun at that stage is also known as the faint young Sun. A decrease of 30% in the present-day solar constant would have the same effect on the equilibrium temperature of the

Earth as a 30% albedo, as shown in equation (3.9), making the Earth a frozen planet if its atmosphere were not different from the present. However, the Earth's prebiotic atmosphere 4.5 to 3.9 billion years ago was probably very different from the present, and of the two most important greenhouse gases, water vapor and carbon dioxide, at least CO_2 was likely to be present in the atmosphere at a much higher concentration than now (Chapter 2). Changes in the solar constant, Earth's albedo, and radiation absorptive power of the atmosphere would affect the temperature as described in a simplified form in equation (3.15): a lower value of the solar constant might be compensated by a higher value of the radiation absorption in the atmosphere.

After the Earth cooled sufficiently for the water to condense and make the hydrosphere, the prebiotic atmosphere likely contained carbon dioxide and nitrogen as its main constituents. Although we do not know what was the cooling path of such a CO_2-N_2 atmosphere, a range from about 280°C down to 25°C is plausible (Fig. 2.2). In terms of the Milankovitch atmosphere model, Fig. 3.4A shows the temperatures for the Earth's surface for different combinations of the absorbed fractions of the shortwave and longwave radiation. For any given value of shortwave absorption, greater absorption of the longwave radiation results in a higher temperature of the Earth's surface. The present-day absorption conditions, $a_{sw} = 0.36$ and $a_{ir} = 0.87$ that were used in equation (3.21), are indicated by a black dot on the iso-temperature curve of 288 K. If the longwave radiation absorption is not too great then an increase in shortwave absorption results in less radiation reaching the Earth's surface and a lower temperature. In Fig. 3.4A, for a given value of a_{ir}, increasing values of a_{sw} correspond to lower equilibrium temperatures. This relationship very much diminishes at the higher values of longwave absorption, $a_{ir} \gtrsim 0.90$, where the slopes of the iso-temperature curves are less sensitive to an increase in shortwave absorption, a_{sw}, and the longwave radiation absorption exerts a dominating role in the control of the equilibrium temperature.

The increase in temperature as a function of the longwave radiation absorption is shown in Fig. 3.4B for a limiting case of an atmosphere that is transparent to the solar shortwave radiation: $a_{sw} = 0$, the relationship given in equation (3.20). The highest equilibrium temperatures shown in Fig. 3.4B are for 99% longwave absorption or $a_{ir} = 0.99$: 344 to 376 K or about 70 to 100°C. However, despite the limitations of this simple model, we may still conclude that if a prebiotic atmosphere consisting predominantly of CO_2 was characterized by strong absorption of longwave radiation, then the Earth's surface temperature would have been closer to 100°C rather than 25°C. The agreement should be noted between the Milankovitch model of radiation absorption that results in temperatures higher than on the present-day Earth's surface and the notions that such higher temperatures existed on the prebiotic Earth.

The views and hypotheses that were cited in Chapter 2 address the temperature of the primordial and later atmosphere variably from about 4000 Ma ago through the different stages of the Proterozoic, up to 750 Ma. The focus there is to identify the concentration ranges of such greenhouse gases as methane and ammonia that could supplement carbon dioxide at the time of the faint young Sun and later, and also provide a shield from the ultraviolet radiation that is now absorbed by molecular oxygen and ozone in the stratosphere (Fig. 3.3B). As the occurrence of carbon dioxide and other greenhouse

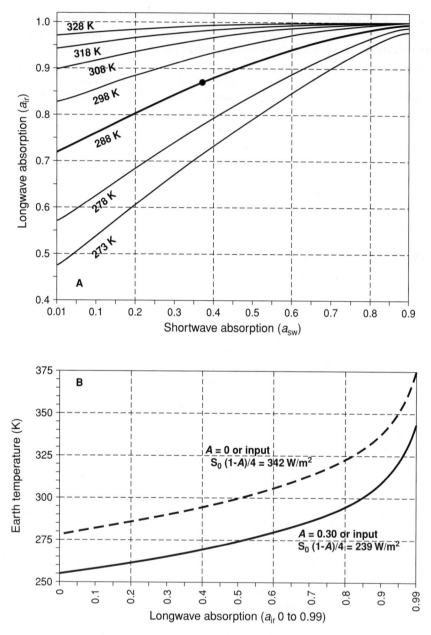

Figure 3.4   The Milankovitch model of Earth's surface temperature as a function of longwave and shortwave radiation absorption in the atmosphere. A. Earth's surface temperature at different values of shortwave and longwave radiation absorption in the atmosphere, equation (3.20). Black dot: model present-day conditions, $a_{sw} = 0.36$, $a_{ir} = 0.87$. B. Dependence of Earth's surface temperature on absorption of longwave radiation in the atmosphere (no absorption of shortwave solar radiation). Curves for albedo $A = 0$ and 30%.

gases is only one of the several factors that affect the Earth's surface temperature, it should be noted that theoretical response times of the atmosphere to changes in the radiation balance are very short and the computed times of temperature change are measured in days (Houghton, 1991). The Earth's surface environment, however, comprises in addition to the atmosphere also the upper part of the continental crust, weathered regolith, and the global ocean, all exchanging heat with the atmosphere. Seasonal temperature variations penetrate into the soil and crustal rocks to depths of 6.5 m to 10 m, where the amplitude of annual temperature variation between the coldest and warmest periods at the surface is attenuated to 0.05 of its value (Carslaw and Jaeger, 1959; Hillel, 1998; Marshall et al., 1996). The heat content of a soil or granite layer of such a thickness is comparable within a factor of about 2 to that of the atmosphere, if both are at 15°C. The world ocean, at a mean temperature taken as 4°C, is a much bigger heat reservoir and its heat content is of an order of 10^3 times greater than that of the atmosphere. Because the internal circulation time of the ocean is 10^3 years, a response time of the entire Earth-surface system to changes in the radiation balance is likely to be longer than for the atmosphere alone. Thus longer-term physical and climatic changes in the environment, such as those driven by major tectonic forces, probably had a more determinative effect on the geologic history of the Earth's surface and its life.

5 CO$_2$ and Climate Change

When CO_2 concentration increases in the atmosphere, some of the outgoing radiation is absorbed by the added amount, and the incoming and outgoing radiation are no longer at equilibrium, as given in equation (3.6) or (3.14). To reestablish the balance, the temperature of the atmosphere has to rise to make the outgoing radiation equal to the incoming. The imbalance, called the radiative forcing (ΔE, in units of W m^{-2}), is usually calculated at the level of the top of the atmosphere. In the Industrial Age, most of the radiative forcing has been due to the carbon dioxide accumulation, and a global mean temperature rise from 1861 to 2000 is about 0.7°C. This temperature change is reported relative to the land-surface air and ocean surface mean temperature of the period 1961 to 1990, when it increased from -0.4°C below the mean in 1861 to $+0.3$°C above the mean in 2000 (Houghton et al., 2001), the total change reaching 0.8°C in 2005. A temperature change ΔT is related to a change in the input of radiative energy ΔE by a coefficient χ that is also called the *climate sensitivity parameter*, in units of K W^{-1} m^2:

$$\Delta T = \chi \Delta E \qquad (3.23)$$

The climate sensitivity parameter is obtained from equation (3.1) or (3.14) as:

$$\frac{dT}{dE} = \frac{1}{4\sigma T^3} \equiv \chi \qquad (3.24)$$

An increment of radiative energy ΔE may be caused by a change in any of the

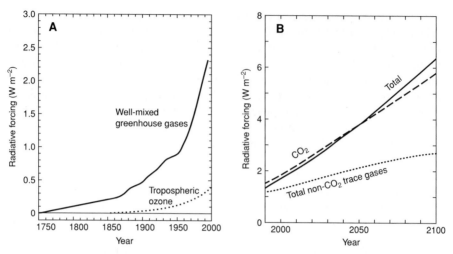

Figure 3.5   Radiative forcing on the atmosphere in the past (A, after Ramaswamy *et al.*, 2001) and in the future (B, after Kattenberg *et al.*, 1996). Note that the forcing values for the 1990s differ in the two models: no water vapor contribution is given in B. The dominant role of CO_2 in radiative forcing is shown in B. Total forcing until about 2050 is slightly smaller than that of CO_2 alone because of the negative contributions of the sulfate and biomass-burning aerosols.

energy-balance parameters that are given in equation (3.14), such as the Earth's albedo *A* or the forcing due to a change in the chemical composition of the atmosphere as represented by parameter $1-B$ that is also temperature dependent (Raval and Ramanathan, 1989). The simple form of the climate sensitivity parameter in (3.24) gives $\chi = 0.27$ to 0.19 K W^{-1} m^2 for a temperature range from 255 to 288 K. Estimates of χ, without any feedbacks from an increase in the water vapor content of a warmer atmosphere, give a value of 0.3 K W^{-1} m^2 (Seinfeld and Pandis, 1998; Shine *et al.*, 1990). For a radiative forcing increase to $\Delta E = 2.3$ W m^{-2} by the year 2000 (Fig. 3.5A), the temperature increase would be $\Delta T = 0.3 \times 2.3 = 0.7°$C.

A change of 1% in the mean solar constant would produce a change of 0.6 to 0.7°C in the equilibrium temperature of the Earth, using the same values of the albedo and atmospheric absorption as in equations (3.9), (3.15), or (3.21). Statements can be encountered in the scientific literature that a change of 1% in the mean solar constant would produce about a 1°C change in the equilibrium temperature, as derived from Global Circulation Models (GCM). However, it should be kept in mind that a 1% change is quite large, amounting to about 14 W m^{-2}. The closeness of the reported GCM result to the very simple energy balance computation in this section is probably coincidental, but it nevertheless confirms the value of simple, even if inexact, models in obtaining answers of a reasonably correct magnitude.

Feedbacks due to an increased concentration of water vapor in the atmosphere may increase the climate sensitivity parameter χ to 0.43. Projections to the year 2100

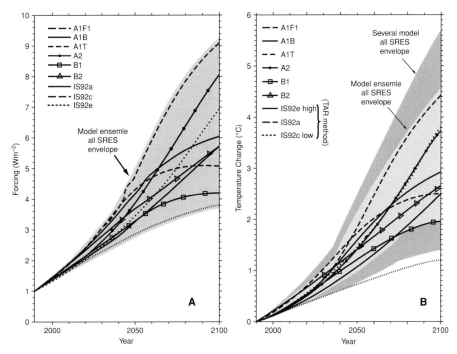

Figure 3.6   Radiative forcing (A) and temperature increase (B) according to different model scenarios projected to the year 2100. SRES are Special Reports on Emission Scenarios (from Cubasch *et al.*, 2001).

(Figs. 3.5B and 3.6A) show a mean of the radiative forcing estimates of about $\Delta E = 6.5$ W m^{-2} and a corresponding rise in the mean temperature of 2.8°C (Fig. 3.6B). This rise is from equation (3.23): $\Delta T = 0.43 \times 6.5 = 2.8$°C.

At the present, the Earth's atmosphere is at one of the low-CO$_2$ stands of its geologic history, whereas higher and even much higher atmospheric CO$_2$ levels might have been the rule during most of Phanerozoic time, as shown in Fig. 3.7. The Earth's surface temperature, as estimated for the shallow oceans, was also higher than now during most of the Phanerozoic. The dips in the temperature curve in the Paleozoic approximately coincide with the glaciations in the Ordovician and Permian, and the cooling trend of the Late Tertiary is the latest geologic stage of declining temperature before industrial time. The mean global temperature during the Phanerozoic varied within a band of about 13°C, with the higher values of 20 to 25°C in the Early Phanerozoic. As these temperatures are similar to those in the present-day tropical ocean, it is not possible to infer from the calculated temperature curve that calcifying organisms in Early Paleozoic seas lived at temperatures very different from those of the warmer parts of the modern ocean. However, the main calcifying benthonic groups in the Paleozoic, that include mollusks, brachiopods, corals, echinoderms, and later foraminifera, emerged

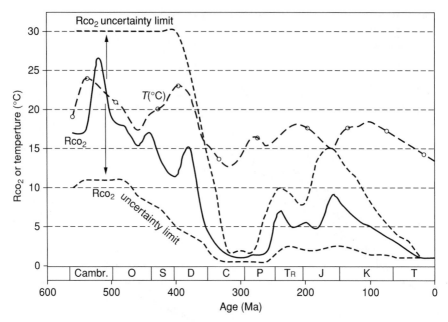

Figure 3.7 R_{CO_2}, ratio of the atmospheric CO_2 concentration in the past to the Holocene value of 300 ppmv (Berner and Kothavala, 2001). Temperature based on the oxygen-isotope record of shallow-water biogenic marine carbonates (Wallmann, 2004).

at mean temperatures higher than those of the Holocene. The calculated atmospheric CO_2 concentrations might have been 20 to 25 times higher near the beginning of the Phanerozoic about 500 Ma ago than the Holocene value of 300 ppmv. The decline in atmospheric CO_2 from the Devonian to the Carboniferous is attributed to the emergence of higher plants on land, with a consequent increase in the weathering rate, and the pronounced minimum in the Carboniferous and Permian, around 330 to 270 Ma ago, is likely to be the result of the enhanced storage of carbon in trees that became massive coal deposits. Relationships between the atmospheric CO_2 trends during the last 550 Ma and the major changes in the sedimentary and tectonic processes relevant to them are discussed in Chapters 7 and 10.

At the time scale of the last 18,000 years, since the end of the last glaciation, and about 100 years into the future, the atmospheric CO_2 content may increase by a factor of four. An increase of 100 ppmv, from 180 to 280 ppmv, occurred since the Last Glacial Maximum to the end of pre-industrial time about 200 to 300 years ago, while the Earth's surface warmed from about 9° or 10° to 15°C. In industrial time, atmospheric CO_2 increased by an additional 100 ppmv, from 280 to 380 ppmv by the year 2005, and the doubling of 380 ppmv, or even a greater increase, is one of the projected scenarios for the coming century. A functional relationship between the atmospheric CO_2 concentration and radiative forcing indicates that the doubling of the atmospheric CO_2 concentration would increase the radiative forcing to about 4.4 W m^{-2} (Shine *et al.*, 1990):

$$\Delta E = 6.3 \ln(C/C_0) = 4.4 \text{ W m}^{-2}, \quad \text{for } C < 1000 \text{ ppmv} \qquad (3.25)$$

where C is CO_2 concentration in ppmv and C_0 is initial concentration. As the coefficients of radiative forcings of CO_2 and other greenhouse gases are being continuously revised, the estimated contributions to radiative forcing of each gas per unit of its concentration in the atmosphere are also changing (Ramaswamy *et al.*, 2001; Shine *et al.*, 1990). Tabulated data of radiative forcings of CO_2, methane (CH_4), numerous carbon-halogen trace gases, and other atmospheric gases can be found in these references.

The upper limit of 1000 ppmv for atmospheric CO_2 concentration in equation (3.25) should be taken in perspective of the rising CO_2 due to emissions from fuel burning and land-use practices. Relative to the 380 ppmv CO_2 in the year 2005, 1000 ppmv is an increase by a factor of about 2.6 or by about 1316 Gt C, from 807 to 2123 Gt C. The proven mass of fossil fuel reserves (Chapter 1) at the end of the year 2003 is about 800 Gt C, but an estimate of total potentially recoverable fossil fuels is much larger, 5000 Gt C. Of the total of anthropogenic CO_2 emissions since the beginning of the Industrial Age, about 43% remain in the atmosphere (Fig. 11.5). Thus if the same partitioning fraction holds in the future, an increase in atmospheric CO_2 by 620 ppmv or 1316 Gt C would require consumption of about 3100 Gt C, which is much more than the known reserves in the year 2003 or 60% of all the estimated fossil fuel reserves. This would indicate a temperature increase due to CO_2 of 2.7°C that is within the range of the estimates for the 21st century (Fig. 3.6B).

In the last approximately 500,000 years, the Earth went through four glacial and four interglacial periods, the last one starting about 18,000 years ago. As discussed in more detail in Chapter 10, the Milankovitch theory of periodic changes in the Earth's insolation (Imbrie *et al.*, 1984; Milankovitch, 1941) due to periodic variations in the Earth's orbital parameters, finds parallels in the variations of the atmospheric carbon dioxide and methane concentrations during the glacial and interglacial periods. Changes in the orbital parameters affect the distribution of heat on the Earth's surface and seasonal insolation of the northern and southern hemispheres. For example, if the Earth's axis of rotation were perpendicular to the plane of the orbit (obliquity 0°), the poles would be always cold, the lower latitudes uniformly warmer closer to the equator, and both hemispheres would be in the same seasonal phase. At present, with the axial tilt at 23.5°, the poles are still cold. A tilt of about 50° would result in a more uniform temperature distribution than at present, making the equator colder and the poles warmer, as was deduced by Ward (1974; also cited in Kump *et al.*, 1999) in a general analysis of solar energy distribution on a planet. Although it is not well known how far into the geologic past did the Milankovitch orbital periodicities extend, there are numerous studies that attribute cyclical changes in the older, pre-Pleistocene sedimentary record to environmental changes occurring with the periodicities of the Milankovitch cycles, those at the time scales of about 23 (time of perihelion or precession of the Earth's axis of rotation), 41 (axial tilt or the obliquity), and 100 ka (eccentricity of the Earth's orbit around the Sun).

The temperature of the Earth's atmosphere, water, and land surface is a major factor that controls many of the inorganic and biologically-mediated processes occurring in the surface environment of the Earth. The essential relationships that are discussed in this chapter form the background for the variety of temperature-dependent processes in the carbon cycle addressed in subsequent chapters, such as, for example, weathering rates of rock minerals, air-sea CO_2 exchange, photosynthesis and respiration, and isotopic exchange.

Chapter 4

Mineralogy, Chemistry, and Reaction Kinetics of the Major Carbonate Phases

Among the carbon-containing minerals, the carbonates are by far the most abundant in the outer shell of the Earth. Elemental carbon, in the mineral forms of diamond and graphite, forms at higher temperatures and pressures. Occurrences of mineral carbides, such as silicon carbide moissanite (SiC), in terrestrial rocks are rare, but several metal carbides have been found in meteorites (Bernatowicz *et al.*, 1996; Di Pierro *et al.*, 2003). There are salts of organic acids classified as minerals, either containing metal cations or without them, that occur as alteration products in organic-matter-rich sediments or on other minerals that were probably subject to bacterial alteration. Two oxalate minerals containing calcium, whewellite ($CaC_2O_4 \cdot H_2O$) and weddellite ($CaC_2O_4 \cdot 2H_2O$), are more common, occurring in soils where they probably form in the presence of oxalic acid of plant origin. In this chapter, we consider the chemical composition, mineralogy, and stability of the important carbonate phases calcite (including the variety magnesian calcite), dolomite, and aragonite. In addition, the kinetics of dissolution and precipitation of these minerals is discussed, preparing the reader for discussion of their solubility behavior in natural waters (Chapter 5). We show in this chapter that experimental work has produced sensible kinetic expressions for carbonate dissolution and precipitation that can be used to constrain models of the carbon cycle.

Furthermore, these considerations of the geochemistry of the carbonate minerals are important in any discussion of the use of proxies (chemical, mineralogical, biological, and isotopic records from various sources that are used to interpret climate and environmental changes of the past) in carbonate phases to determine the environmental state of the ocean-atmosphere system through geologic time, as discussed in Chapter 7. One must be satisfied that the chemical or mineralogical proxies are primary and have not been altered by secondary processes of diagenesis and metamorphism. Applications of the kinetic mineral data to the formation and diagenesis of carbonate sediments are given in Chapter 9. An understanding of the basic thermodynamics and kinetics of carbonate phases and their aqueous solutions provides constraints on the interpretation of the geologic record and the future, as affected by the likely continuing rise of atmospheric CO_2.

1   Carbonate Minerals

Sediments and sedimentary rocks contain a variety of carbonate minerals. Calcite (trigonal rhombohedral $CaCO_3$, Fig. 4.1A) and dolomite (hexagonal $CaMg(CO_3)_2$, Fig. 4.1C) are by far the most abundant carbonate minerals in ancient sedimentary rocks. Aragonite (orthorhombic $CaCO_3$, Fig. 4.1B) is rare and other carbonate minerals are largely confined to special deposits such as evaporites ("salt deposits" of NaCl, $CaSO_4$, $MgSO_4$, KCl, and more complex minerals), iron formation deposits containing iron compounds with the elements O, S, Si, or C, or are present as minor minerals, such as ankerite, $CaFe(CO_3)_2$, in other sedimentary deposits. Calcitic and dolomitic

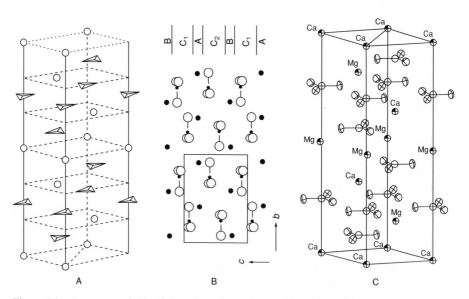

A   B   C

Figure 4.1   Structures of (A) calcite, trigonal rhombohedral $CaCO_3$; (B) aragonite, orthorhombic $CaCO_3$; and (C) dolomite, hexagonal $CaMg(CO_3)_2$, showing the arrangement of carbonate anion groups (triangular shapes) and the atoms of Ca and Mg (small open or filled circles). For orientation in (A) and (C), the c-axis of the minerals is perpendicular to the planes of the carbonate anions and metal element cations.

sedimentary rocks comprise 25% of the mass of sedimentary rocks. Carbonate rocks are important reservoirs of oil and gas and contain commercial ore bodies, and important information, in the form of biological, chemical, mineralogical, and isotopic signatures, concerning the evolution of the Earth. Understandably, the mineralogy and geochemistry of the carbonate minerals constituting carbonate sediments and rocks have been investigated in great detail.

There are three polymorphs (minerals of the same chemical composition but different crystal structure) of $CaCO_3$ (Fig. 4.1) that occur in sediments and in the structures of organisms. The rhombohedral mineral calcite is the most abundant and is thermodynamically stable near the Earth's surface (Box 4.1). Aragonite, the orthorhombic form, is also abundant but found primarily in young sediments and the skeletal structures of marine organisms. Aragonite is the denser form and hence is the $CaCO_3$ phase stable

Box 4.1   Calcite and aragonite equilibrium

For calcite and aragonite, some of the main thermodynamic and physical parameters at 25°C and 1 atm total pressure are listed below:

Mineral phase	ΔG_f° (kJ mol^{-1})	$V^{\circ(1)}$ (cm^3 mol^{-1})	$\rho^{(1)}$ (g cm^{-3})	$\beta^{(3)}$ (bar^{-1})	$\alpha^{(4)}$ (K^{-1})
$CaCO_3$ calcite	$-1128.8 \pm 1.4^{(1)}$ $-1129.07^{(2)}$	36.934	2.71	1.367×10^{-6}	1.88×10^{-5}
$CaCO_3$ arag.	$-1127.8 \pm 1.5^{(1)}$ $-1128.3^{(2)}$	34.15	2.93	1.55×10^{-6}	5.53×10^{-5}

[1] Robie *et al.* (1978), density from V° and gram-formula mass of mineral 100.089 g/mol.
[2] Drever (1997).
[3] Birch (1966), for pressure from 1 to 10,000 bar.
[4] Skinner (1966, p. 82), mean of three expansion values from 20° to 100°C, 200°C, and 400°C for each mineral.

In the preceding table and elsewhere (*see* Chapter 5), ΔG_f° is the Gibbs free energy of formation from elements in the standard state (25°C, 1 atm total pressure), V° is the mineral molar volume, ρ is mineral density, β is the coefficient of volume compressibility at constant temperature, and α is the coefficient of volume expansion at constant pressure.

At the standard conditions of 25°C and 1 atm total pressure, pure calcite and aragonite can not be at equilibrium because of the difference in their ΔG_f° values. However, at a higher pressure $P > 1$ bar, an equilibrium reaction is thermodynamically possible:

$$CaCO_{3\,cal}\,(25°C, P) = CaCO_{3\,arag}\,(25°C, P) \qquad (4.1.1)$$

An equilibrium can exist if the Gibbs free energies of formation of the two minerals (pure phases) at a higher pressure P become equal:

$$\Delta G_{f\,cal}\,(25°C, P) = \Delta G_{f\,arag}\,(25°C, P) \qquad (4.1.2)$$

From the general thermodynamic relationship between the Gibbs free energy and pressure at constant temperature, $(\partial \Delta G/\partial P)_T = V$, the free energy of formation of each mineral is:

$$\Delta G_f(25°C, P) - \Delta G_f^\circ(25°C, P = 1) = \int_{P=1}^{P} V dP \qquad (4.1.3)$$

One may think of the physical significance of equation (4.1.3) as an increase in the Gibbs free energy of formation of a mineral by work done on it by pressure, $V\Delta P$. Molar volume or density of a mineral depends on pressure through the coefficient of compressibility, $\beta = -(1/V)(\partial V/\partial P)_T$ (bar^{-1}):

$$V = V^\circ - V^\circ \beta (P - 1) \qquad (4.1.4)$$

From (4.1.4) and (4.1.3), the change in the Gibbs free energy at 25°C for an increase in pressure from $P = 1$ bar to $P > 1$ is:

$$\Delta G_{f,P} = \Delta G_{f,P=1}^\circ + V^\circ(P-1)\left[1 - \frac{1}{2}\beta(P-1)\right] \qquad (4.1.5)$$

To obtain the equilibrium pressure at 25°C between calcite and aragonite, we substitute (4.1.5) in (4.1.2), disregard the term in $(P-1)^2$, and use the data from the table in this box:

$$P_{eq} \approx \frac{\Delta G_{f\,arag}^\circ - \Delta G_{f\,cal}^\circ}{V_{cal}^\circ - V_{arag}^\circ} + 1$$

$$\approx \frac{(0.8 \text{ to } 1.0) \times 10^3 \text{ J mol}^{-1}}{2.78 \text{ cm}^3 \text{ mol}^{-1}} \times 10 \text{ bar cm}^3 \text{J}^{-1} = 2900 \text{ to } 3600 \text{ bar} \qquad (4.1.6)$$

In the continental crust, a pressure of about 3000 bar is attained at the depth of about 10 km, where the temperature is also higher, about 300°C. In old marine limestones preserved on land, the constituent minerals are calcite and dolomite, but not aragonite. The reasons behind this are that the thickness of sedimentary overburden was usually less than 10 km, the formation of dolomite is favored as a diagenetic replacement of calcite, and the generally less abundant aragonite in young oceanic sediments is converted to calcite by dissolution and reprecipitation. Also, at higher temperatures, the difference between the ΔG_f° values of aragonite and calcite increases (at 600 K it is about 2.5 kJ/mol, Robie *et al.*, 1978). This makes the numerator in equation (4.1.6) larger and the denominator slightly smaller because of the slightly greater coefficient of thermal expansion of aragonite. The result is that the equilibrium pressure at about 300°C would be greater and this translates into a deeper burial of the sediment.

at higher pressure and temperature, but metastable relative to calcite at low pressure and temperature. It is about 1.5 times more soluble than calcite. Vaterite is the third anhydrous $CaCO_3$ phase, has a hexagonal structure, and is metastable relative to aragonite and calcite under the environmental conditions that characterize sediments and sedimentary rocks. It is approximately 3.7 times more soluble than calcite and 2.5 times more soluble than aragonite. It rarely is observed in natural systems (e.g., Albright, 1971, gives a summary of its natural occurrences and stability), but can be produced in the laboratory under experimental conditions that give rise to high precipitation rates for carbonate phases. In addition to the anhydrous $CaCO_3$ minerals found in sediments, there are scarce occurrences of hydrated $CaCO_3$ minerals; an example is ikaite ($CaCO_3 \cdot 6H_2O$) observed in Antarctic shelf sediments (Suess et al., 1982).

Dolomite is one of the main carbonate phases, in addition to calcite and aragonite that make up carbonate rocks. However, it does not occur as a skeletal component of organisms as do calcite and aragonite. Even after many years of study, the mode of formation of dolomite remains controversial. Its properties under the environmental conditions at and near the surface of the Earth are less well known than those of calcite and aragonite. This is partly a reflection of the fact that the synthesis of dolomite in low temperature laboratory experiments has proven to be difficult. Table 4.1 lists the variety of carbonate minerals found in nature and some of their properties. Because of the predominant abundance of calcite and dolomite, and the lesser but important abundance of aragonite in the sedimentary rock record, the mineralogy and geochemistry of these phases are discussed in some detail.

2   Calcites

Calcites are major and very important biogenic and inorganic constituents of modern marine sediments, and Pleistocene and older rocks. They are complex minerals that can contain up to 30 mol % $MgCO_3$. Calcites containing more than a few percent $MgCO_3$ are usually referred to as magnesian calcites. Generally, for the same $MgCO_3$ content, the biogenic calcites have greater concentrations of sodium, sulfate, water, hydroxide, and bicarbonate and tend to have larger cell volumes and greater carbonate positional disorder in their structure than natural or synthetic inorganic phases. Chemical and microstructural heterogeneities characterize biogenic magnesian calcites, and dislocations and plane defects are common in their crystals. All of these characteristics may affect the thermodynamic and kinetic properties and the reactivity of the phases in aqueous solution.

In contrast to younger sediments, calcites found in Paleogene (Early Tertiary) and older rock units commonly contain only a few percent $MgCO_3$. This contrast in composition and its interpretation have been the principal driving force behind studies of the basic crystal chemistry, thermodynamic, and kinetic properties of calcites. The literature concerning these properties is substantial and is summarized in Arvidson and Mackenzie (1999), Mackenzie et al. (1983), Morse and Mackenzie (1990), and Tribble et al. (1995).

Table 4.1   Carbonate minerals and some of their physical and chemical properties[1]

Mineral	Formula	Formula wt ($g\ mol^{-1}$)	Density ($g\ cm^{-3}$)	Crystal System	$\Delta G_{f,298}$ ($J\ mol^{-1}$)	Various estimates of $-\log K_{sp}$
Calcite	$CaCO_3$	100.09	2.71	Trig.	−1128842	8.30, 8.35, 8.48, 8.46
Aragonite	$CaCO_3$	100.09	2.93	Ortho.	−1127793	8.12, 8.22, 8.34, 8.30
Vaterite	$CaCO_3$	100.09	2.54	Hex.	−1125540	7.73, 7.91
Monohydrocalcite	$CaCO_3 \cdot H_2O$	118.10	2.43	Hex.	−1361600	7.54, 7.60
Ikaite	$CaCO_3 \cdot 6H_2O$	208.18	1.77	Mono.	—	7.12
Magnesite	$MgCO_3$	84.32	2.96	Trig.	−1723746	8.20, 7.46, 5.10, 8.10
Nesquehonite	$MgCO_3 \cdot 3H_2O$	138.36	1.83	Trig.	−1723746	5.19, 4.67
Artinite	$Mg_2CO_3(OH)_2 \cdot 3H_2O$	196.68	2.04	Mono.	−2568346	18.36
Hydromagnesite	$Mg_4(CO_3)_3(OH)_2 \cdot 3H_2O$	359.27	—	Mono.	−4637127	36.47, 30.6
Dolomite	$CaMg(CO_3)_2$	184.40	2.87	Trig.	−2161672	17.09
Huntite	$CaMg_3(CO_3)_4$	353.03	2.88	Trig.	−4203425	30.46
Strontianite	$SrCO_3$	147.63	3.70	Ortho.	−1137645	8.81, 9.03, 9.13, 9.27
Witherite	$BaCO_3$	197.35	4.43	Ortho.	−1132210	7.63, 8.30, 8.56
Barytocalcite	$CaBa(CO_3)_2$	297.44	—	Trig.	−2271494	17.68
Rhodochrosite	$MnCO_3$	114.95	3.13	Trig.	−816047	10.54, 9.30, 10.59
Kutnohorite	$CaMn(CO_3)_3$	215.04	—	Trig.	−195058	55.79
Siderite	$FeCO_3$	115.85	3.80	Trig.	−666698	10.50, 10.68, 10.91
Cobaltocalcite	$CoCO_3$	118.94	4.13	Trig.	−650026	11.87, 9.68
	$CuCO_3$	123.56	—	Trig.	—	9.63, 11.51
Gaspeite	$NiCO_3$	118.72	—	Trig.	−613793	7.06, 6.87
Smithsonite	$ZnCO_3$	125.39	4.40	Trig.	−731480	9.87, 10.00
Otavite	$CdCO_3$	172.41	4.26	Trig.	−669440	11.21, 13.74
Cerussite	$PbCO_3$	267.20	6.60	Ortho.	−625337	12.80, 13.13, 12.15
Malachite	$Cu_2CO_3(OH)_2$	221.11	4.00	Mono.	—	33.78, 33.46
Azurite	$Cu_3(CO_3)_2(OH)_2$	344.65	3.88	Mono.	—	45.96
Ankerite	$CaFe(CO_3)_2$	215.95	—	Trig.	−1815200	19.92
Natronite	$Na_2CO_3 \cdot 10H_2O$	285.99	—	Ortho.	−3428997	1.03
Thermonatrite	$Na_2CO_3 \cdot H_2O$	124.00	—	Ortho.	−1286538	0.403, 0.54
Trona	$NaHCO_3 \cdot Na_2CO_3 \cdot 2H_2O$	229.00	2.25	Mono.	−2386554	2.07, 1.00
Nahcolite	$NaHCO_3$	84.01	2.16	Mono.	−851862	0.545, 0.39
Natron	$NaHCO_3 \cdot H_2O$	102.03	—	Mono.	—	0.80

[1] Table modified from Morse and Mackenzie (1990).

2.1   Crystal Structure and Chemistry

During recent years, a number of new observations concerning the crystal structure and chemistry of calcites has been made. These observations have proved useful in interpreting the experimental data on the solubility of calcites in aqueous solution and the behavior of calcites during diagenesis, and their environmental interpretation. The following is a summary of the more important recent findings.

(1) Unit cell parameters of synthetic calcites vary smoothly as a function of $MgCO_3$ content, up to at least 20 mol % $MgCO_3$. These phases have larger c/a axial ratios than those obtained from a straight line connecting the c/a axial ratios of calcite and magnesite or disordered dolomite (Fig. 4.2A). Furthermore, the synthetic phases exhibit slightly negative excess cell volumes in the composition range 0–20 mol % $MgCO_3$ and perhaps slightly positive excess volumes for $MgCO_3$ contents greater than 20 mol % (Fig. 4.2B). Excess volume refers to the departure in cell volume from that anticipated for a simple mixture of calcite and magnesite (or disordered dolomite). In contrast, the unit cell parameters of biogenic calcites do not vary smoothly as a function of $MgCO_3$

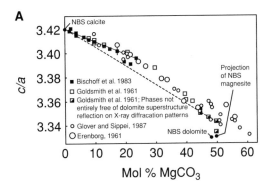

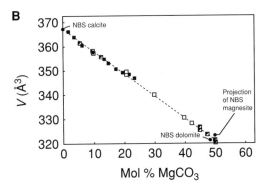

Figure 4.2   Unit cell axial ratios (A) and volumes (B) versus composition for synthetic magnesian calcites. Solid line is quadratic, least-squares fit for synthetic phases. The dashed line connects pure calcite and least-ordered dolomite (after Bischoff *et al.*, 1983).

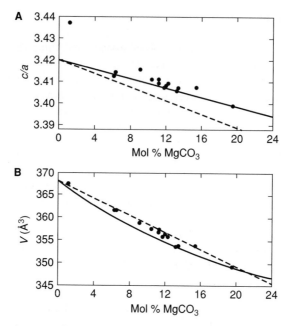

Figure 4.3   Biogenic magnesian calcites. A. The c/a axial ratios of the unit cell. B. Unit cell
volume in cubic ångströms ($Å^3$). The dashed lines are straight-line connections for cell-edge c/a
ratio and unit cell volume between synthetic calcite and synthetic disordered dolomite. Solid lines
fit the synthetic data from Fig. 4.2. The calcites were formed in the biomineralization process of
various skeletal carbonate organisms of different magnesium contents (after Bischoff et al., 1983).

content, and their axial ratios and cell volumes are larger and smaller, respectively, than
those of synthetic solids of the same $MgCO_3$ content (Fig. 4.3).

(2) The excess c/a axial ratios in synthetic calcite phases are the result of the positional disorder of the CO_3^{2-}-anion group in the vicinity of Mg^{2+} ions. In essence, the
anion group becomes progressively more inclined to the c crystal axis with increasing
magnesium concentration in the calcite phase. This phenomenon has been documented
in Raman spectroscopic studies of calcites that show an increase in the half-widths
of Raman spectral bands with increasing magnesium concentration (Fig. 4.4). Single
crystal X-ray refinements of two biogenic magnesian calcites have confirmed the hypothesis of positional disorder in these phases with increasing Mg content and have
shown that this disorder is not simply limited to the CO_3 group but also affects the
cation positions (Paquette and Reeder, 1990).

(3) In general, biogenic calcites exhibit a greater degree of heterogeneity with respect to the distribution of magnesium in their structure than synthetic phases. Furthermore, some biogenic calcites and synthetic calcites precipitated at high supersaturations
from multicomponent aqueous solutions contain trace amounts of Na^+ and SO_4^{2-} owing
to point defects and dislocations. It is likely that substitutions of Na^+ for Ca^{2+} in the
calcite lattice are balanced at least in part by incorporation of HCO_3^- into the structure.

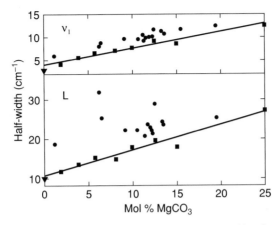

Figure 4.4   Halfwidths of the ν_1 and L Raman modes versus composition for synthetic (squares) and biogenic (dots) magnesian calcites. The triangle represents pure calcite. The straight line is the least-squares fit to the synthetic phase data (after Bischoff *et al.*, 1985).

(4) Busenberg and Plummer (1989) modeled the solid solution behavior of calcites based on experimentally obtained dissolution data. Compositionally pure binary solid solutions of $CaCO_3$ and $MgCO_3$ that include metamorphic and hydrothermal phases, synthetic phases prepared at high temperatures and pressures, and synthetic phases prepared from aqueous solution at low temperatures and very low calcite supersaturations were modeled as sub-regular solid solutions between calcite and dolomite or disordered dolomite. For the biogenic calcites and synthetic calcites synthesized at high calcite supersaturations from aqueous solution, a sub-regular solid solution between defective calcite and disordered dolomite appeared to fit best the experimental data (solid solutions are discussed in Box 4.2).

The solubility of a calcite in aqueous solution depends significantly on its physical and chemical characteristics. These characteristics must be defined and quantified before one can interpret the laboratory solubility or kinetic data necessary to studies of the reactivity of calcites in natural environments.

Box 4.2   Solid Solutions

Solid solutions are by definition nonelectrolyte solutions consisting of uncharged particles. The theory of nonelectrolyte solutions was developed initially for treatment of molecular solutions but has been extended to include solid solutions in minerals, cation exchangers, molten magma, and organic materials, both solid and liquid. The only major class of solutions of geological and biogeochemical concern for which this theory is not used is aqueous solutions of electrolytes. In order to understand solid solutions, we need to take a step back and look first at the behavior of liquid nonelectrolyte solutions.

For an ideal solution, it has been shown with various degrees of accuracy that a simple relationship exists between the vapor pressure due to each component of the solution and the mol fraction of each of the components:

$$P_j = P_j^* N_j \tag{4.2.1}$$

where P_j is the equilibrium vapor pressure of component j of the solution, N_j is the mol fraction in solution of that component, and P_j^* is the equilibrium vapor pressure of the pure substance j. This relationship is known as Raoult's law. A solution obeying Raoult's law and for which the vapors behave as perfect gases is defined as an ideal solution. The pressure of an ideal gas has been given a special name, the fugacity, and in the notation of equation (4.2.1):

$$P_j = f_j \tag{4.2.2}$$

where f_j is the fugacity of component j in the solution. Substituting (4.2.2) in (4.2.1), we obtain

$$f_j = f_j^* N_j \tag{4.2.3}$$

for the ideal solution, where f_j^* is the fugacity of the pure substance j. Although the concept of fugacity has been developed above through considerations of the ideal solution and the perfect gas, the fugacity of a dissolved component in any condensed phase of liquid or solid is equal to the fugacity of that component in the vapor phase in equilibrium with the condensed phase.

The fundamental definition of the activity (a_j) of a dissolved component of any solution is:

$$a_j = f_j / f_j^0 \tag{4.2.4}$$

where f_j is the fugacity of the dissolved component j, and f_j^0 is its fugacity in the standard state, usually taken as 1 atmosphere total pressure and at a specified temperature for a pure liquid or solid. For these standard state conditions, the standard state fugacity is the fugacity of the pure substance

$$f_j^0 = f_j^* \tag{4.2.5}$$

and equation (4.2.4) becomes

$$a_j = f_j / f_j^* \tag{4.2.6}$$

For a pure substance $f_j = f_j^*$ and thus a_j^*, the activity of the pure substance, is 1. Combining equations (4.2.3) and (4.2.6), we obtain

$$a_j^* = N_j \tag{4.2.7}$$

where N_j is the mol fraction of the dissolved component j. Figure 4.2.1 shows how the activities of two components vary as a function of composition expressed as mol fraction for an ideal binary solution. For volatile substances vapor pressure determinations can be a practical way of determining activities of substances having appreciable vapor pressures since the fugacity can be calculated from the vapor pressure. For less volatile components, like $MgCO_3$ dissolved in $CaCO_3$

(the magnesian calcities), for which the vapor pressures of the two components are very low, other indirect methods for determining activities must be employed.

Most actual solutions are not ideal and the activity of the j-th component does not equal its mol fraction in solution. The activities of the two components in the binary solution of Fig. 4.2.1 would not plot as linear lines as a function of composition expressed as mol fraction but as curves, where as the mol fraction of a component approaches unity, its activity approaches unity.

There are various types of nonideal solutions. For such solutions over all compositional ranges, the activity of the j-th component is related to its mol

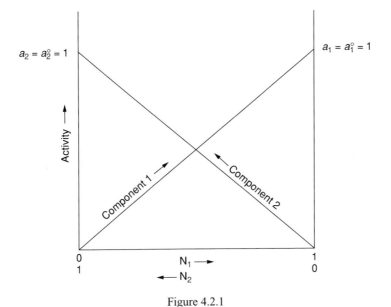

$a_2 = a_2^o = 1$ $a_1 = a_1^o = 1$

Activity

Component 1

Component 2

$0 \atop 1$ $N_1 \longrightarrow$ $1 \atop 0$

$\longleftarrow N_2$

Figure 4.2.1

fraction by $a_j = \lambda_j N_j$, where λ_j is termed the activity coefficient. One example of nonideal solution behavior is the regular solution. It has been found that for many binary regular solutions, including minerals, the activity of each component can be expressed by an equation of the type $\log \lambda_j = b'(1 - N_j)^2$, where b' is an empirically determined constant. The magnesian calcites have been modeled as types of regular solutions (Busenberg and Plummer, 1989).

2.2   Solubility and Solid Solution Behavior

The solubilities of the calcites depend on their $MgCO_3$ content, in addition to other chemical and environmental parameters, and the solubility trend with composition of these phases at 25°C and 1 atmosphere total pressure appears reasonably well defined (Fig. 4.5) owing to the work of Bischoff *et al.* (1987, 1993) and Busenberg and Plummer

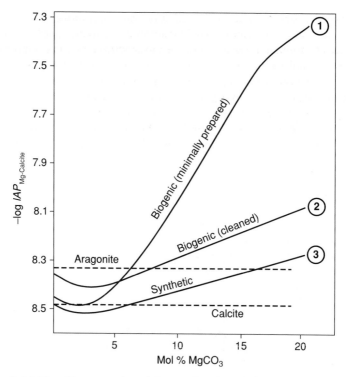

Figure 4.5   Solubility of the magnesian calcites as a function of the $MgCO_3$ in the phase, at 25°C. Curve 2, biogenic (cleaned), includes some data for impure, structurally disordered, synthetic precipitated phases. Further explanation is given in the text (after Bischoff *et al.*, 1993).

(1989). The solubilities of magnesian calcites can be expressed as a function of their $MgCO_3$ content and the ion activity product (*IAP*) of a solution in equilibrium with the solid (e.g., Thorstenson and Plummer, 1977):

$$Ca_{1-x}Mg_xCO_3 = (1-x)Ca^{2+} + xMg^{2+} + CO_3^{2-}$$

$$IAP_{\text{calcite}} = K_{\text{sp}} = (Ca^{2+})^{1-x}(Mg^{2+})^x(CO_3^{2-}), \qquad (4.1)$$

where parentheses in the latter expression denote the activities of the ions in aqueous solution, x is the mol fraction of $MgCO_3$ in the solid, and K_{sp} is the solubility product.

The solubilities of calcites expressed in terms of their ion activity products exhibit a minimum around 2 mol % $MgCO_3$, beyond which they increase nearly linearly. The solubility product of pure calcite at 25°C and 1 atmosphere is about $10^{-8.46}$. There appear to be two trends in the solubility-composition data: one for inorganic solids that are well crystallized, compositionally homogeneous, and chemically pure (Fig. 4.5, curve 3), and the other for biogenic phases and inorganic solids that exhibit substantial lattice defects, carbonate positional disorder, and substitution of such foreign ions as sulfate and sodium (Fig. 4.5, curve 2). A third solubility-composition curve is shown in Fig. 4.5, that of Plummer and Mackenzie (1974) for biogenic materials subjected to minimal cleaning procedures in the laboratory before use in dissolution experiments designed to determine the solubilities of these phases (Fig. 4.5, curve 1). This curve probably reflects primarily kinetic rather than thermodynamic factors, a result of the lack of annealing of the biogenic materials and possibly the non-removal of fine-grained particles from the reactant solids used in these experiments.

As can be seen in Fig. 4.5, biogenic and some inorganic calcites (Fig. 4.5, curve 2) are unstable in aqueous solutions relative to synthetic phases of similar $MgCO_3$ content (Fig. 4.5, curve 3). Their solubilities differ by approximately 0.15 pIAP units (p$IAP = -\log IAP$), or about 4.6 kJ/mol (1.1 kcal/mol). This difference arises because of physical and chemical differences between the two types of solids. The complex microarchitecture, greater positional disorder of the CO_3^{2-} anion group, and chemical impurity of the biogenic phases and some inorganic calcites result in higher solubilities of these solids. It is likely that the solubility-composition curve 3 (Fig. 4.5) for pure, compositionally homogeneous and structurally well-ordered calcite phases represents the true metastable equilibrium solubility of calcites in aqueous solution at 25°C and 1 atmosphere. This difference between the equilibrium and biogenic solubilities of calcite phases of the same $MgCO_3$ content has important implications for diagenesis of the metastable assemblage of carbonate minerals deposited on the seafloor today and in the geologic past (Bischoff *et al.*, 1993, and Chapter 9).

3   Dolomite

The mineral dolomite, $CaMg(CO_3)_2$, is a major constituent of carbonate rocks. Dolomite was named in honor of Dieudonné Dolomieu (also known as Déodat Dolomieu), a geologist who worked in the Pyrenees and Alps in the 18th century and first described its characteristics and occurrence. For geologists dolomite has been the center of an enduring debate regarding its mode of formation in surface environments, its past abundance relative to other sedimentary carbonates, and its overall significance in the geologic record (Hardie, 1987; Machel and Mountjoy, 1986; Mackenzie and Morse, 1992; McKenzie, 1991). This debate has motivated the search for a modern analogue to the surface or near-surface environment that is capable of producing large quantities of dolomite, whether as a primary precipitate, or (as is more commonly interpreted in ancient carbonate rocks) an early diagenetic phase replacing primary $CaCO_3$. Although this work has yielded possible scenarios in which certain sedimentary dolomites no doubt form, such as in the Persian Gulf sabkha environments (Chapter 7), it has failed to offer a reasonable model for the formation of *massive* dolomites in the geologic record. This lack of a modern analogue for such an important mineral has challenged experimentalists to observe and measure directly the physicochemical conditions and rates under which dolomite precipitation and replacement reactions occur.

3.1   Mineralogy and Phase Relations

Dolomite is distinguished from calcite and the other rhombohedral carbonates by its stoichiometry; ideal dolomite has equal numbers of Ca and Mg atoms, and the Ca and

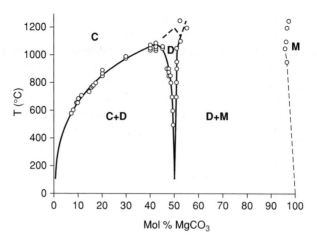

Figure 4.6   Subsolvus relationships for the $CaCO_3$-$MgCO_3$ join. The top of the calcite-dolomite immissibility gap occurs at 1075°C, at this temperature, the phase composition is $Ca_{0.575}Mg_{0.425}(CO_3)_2$. Dotted line represents limit of detectable ordering in the dolomite structure. C = calcite; M = magnesite; and D = dolomite (after Tribble *et al.*, 1993; data from Goldsmith and Graf, 1958, and Goldsmith and Heard, 1961).

Mg are segregated in distinct lattice planes. These planes are oriented normal to the *c*-axis, each alternating with *c*-normal trigonal CO_3 groups that are also in essentially planar orientation (Fig. 4.1C). This cation ordering results in a coupled rotation of the carbonate groups and a reduction of symmetry relative to calcite from R3c to R3. Each cation is octahedrally coordinated by six oxygen atoms, which are themselves threefold coordinated with a calcium, magnesium, and carbon atom.

The univariant temperature-CO_2 curve for the calcite-dolomite-magnesite system was determined by Harker and Tuttle (1955a) at high to moderate temperatures and pressures. The order of decomposition with increasing temperature was shown to be magnesite, dolomite, and calcite. The location of calcite-dolomite and dolomite-magnesite solvi along the $CaCO_3$-$MgCO_3$ binary join (Fig. 4.6) was established through the collective work of Harker and Tuttle (1955b), Graf and Goldsmith (1955, 1956, 1958), and Goldsmith and Heard (1961). Lattice constants for Ca-Mg carbonates have also been determined by Goldsmith and Graf (1958), Goldsmith *et al.* (1961), and Graf (1961). Those of Goldsmith and Graf (1958) have been abandoned.

The top of the calcite-dolomite solvus was located by Goldsmith and Heard (1961) at 1075°C, giving a composition of $Ca_{0.575}Mg_{0.425}CO_3$ (Fig. 4.6). Above this temperature, complete solid solubility is seen between calcite and a carbonate of dolomite composition. However, a phase change must still separate the disordered Mg calcite (R3c space group) from ordered dolomite (R3). To distinguish this phase change as discontinuous (first order) or continuous (second order) is problematic: a first-order transition would demand coexistence of ordered and disordered phases at the temperature of transition. Goldsmith and Heard (1961) observed a continuous increase in disorder from $\sim 1000°$ through $1150°C$, based on the decreasing intensity of cation ordering reflections. The intensity of ordering reflections depends on the difference between *average* scattering factors for the atoms occupying distinct sites. Thus Goldsmith and Heard concluded that the phase change is of a higher order, with the caveat that a two-phase field, if present, may be so narrow as to have eluded detection.

With decreasing subsolvus temperatures, the amount of excess $CaCO_3$ that can be accommodated in stable dolomite is greatly reduced (Fig. 4.6). Dolomites formed at temperatures of $\sim 500°C$ or lower with more than one mol percent excess $CaCO_3$ must thus be regarded as metastable. Goldsmith and Graf (1958) recognized the difficulty in accommodating this excess Ca in Mg sites, and showed that natural dolomites whose compositions deviate from stoichiometric proportions also showed evidence of some cation disorder.

3.2 Early Work on Synthesis and Mechanism

The work of Graf and Goldschmidt (1956) is one of the first experimental attempts to obtain information on the rate and mechanism of dolomite formation as a function of temperature and reactant composition. Important conclusions of this work are:

(1) Ordered, stoichiometric dolomite similar to natural samples can be readily synthesized under dry conditions during brief reaction times at temperatures of 500° to 800°C using a variety of reactants. However, the reaction below 400°C is so slow that no dolomite product is identifiable after reaction times of thousands of hours.

(2) Dolomite can be synthesized in hydrothermal (wet) conditions over a range of temperatures. The phases formed at a given temperature depend on the partial pressure of CO_2 and H_2O. At temperatures down to 200°C, ordered, stoichiometric dolomite can be synthesized. At lower temperatures or for shorter reaction times, the product dolomites may exhibit 1:1 Ca:Mg ratios, but cation ordering is usually incomplete.

(3) At temperatures less than 100°C, regardless of reaction time, dolomite-like materials are formed having compositions that deviate significantly from ideal and are typically calcium rich. These Ca-rich dolomites completely lack evidence of ordering. The term *protodolomite* was introduced by Graf and Goldschmidt (1956) to describe Ca-Mg carbonates that deviate from stable, ordered, stoichiometric dolomite. Protodolomite can transform to dolomite with sufficient time or temperature.

Since Graf and Goldschmidt's pioneering work, subsequent experimental work focused on the stability relations in the $CaCO_3$-$MgCO_3$ system under hydrothermal conditions in chloride solutions (Gaines, 1974, 1980; Katz and Matthews, 1977; Rosenberg and Holland, 1964; Rosenberg *et al.*, 1967). Gaines (1974) was able to produce ordered, stoichiometric dolomite from calcite and aragonite at 100°C reacted in Ca-Mg-chloride solutions for a relatively short time of about 200 hours. He demonstrated qualitatively that the dolomite reaction rate depended on temperature, ionic strength, Mg:Ca ratio in solution, and the mineralogy of the reactant carbonate solid (calcite, magnesian calcite, and aragonite increasing in reactivity, respectively). The activation energy of the formation reaction of an ordered dolomite recalculated from Gaines' experiments by Arvidson and Mackenzie (1997) was found to be 42.1 kcal/mol or 176.1 kJ/mol.

3.3   More Recent and Current Work in Mineralogy and Synthesis

During the past quarter century, there has been a considerable amount of effort devoted to studies of dolomite mineralogy and synthesis under experimental conditions to shed light on the dolomite formation mechanism(s). Some highlights of this work follow.

First of all, Schultz-Guttler (1986) explored the relationship between disordering in dolomite and the mol fraction of $MgCO_3$ in coexisting calcite at relatively high temperatures and pressures, using pure dolomite and calcite reacted in water with Mg-oxalate over long reaction times ($\sim$100 hours). Schultz-Guttler documented a positive relationship between isothermal $MgCO_3$ solubility in calcite and the disorder of coexisting dolomite, and suggested that discrepancies among previous data in the location of the calcite limb of the solvus may reflect the time required to achieve a completely ordered structure, even at high temperatures and pressures. One would expect this effect to be exacerbated at lower temperatures.

Later, Lumsden et al. (1989) synthesized ordered, stoichiometric dolomite from calcite in Mn-doped $MgCl_2$ solutions at 192° and 224°C, with magnesite ($MgCO_3$) and brucite [$Mg(OH)_2$] appearing as reactive intermediates. At a given temperature, the addition of Mn apparently decreased the reaction rate. Cation ordering in product dolomites appeared to decrease with increased residence time of the solutions (co-incident with an increase in the mol fraction of $CaCO_{3, dol}$); however, these data are not well constrained, and there are problems inherent in intensity ratio measurements. Thus it is difficult to ascertain whether this trend in order/disorder actually reflects Mn substitution in dolomite (Peacor et al., 1987).

In a following important paper, Sibley (1990) produced stoichiometric, well-ordered dolomite from calcite at 218°C. Intermediate products included various fractions of magnesian calcite and non-stoichiometric dolomite whose compositions and rate of formation were functions of the Mg:Ca ratio in solution. Sibley concluded that the reaction sequence of magnesian calcite → Ca-rich dolomite → dolomite is controlled by the relative (decreasing) nucleation and growth rates of these phases.

Although cation disorder in dolomite has typically been estimated from intensity ratios of ordered versus non-ordered reflections in X-ray diffraction, structural disorder resulting from other defects may play a critical role in reactivity. Relevant to the above dolomite ordering studies is the work of Wenk et al. (1983). They summarized the occurrence of a diverse set of features known collectively as *modulated structures*. These are seen as diffuse intensity contrasts in transmission electron microscopic (TEM) images of a variety of carbonates, including dolomite (particularly Ca-rich varieties, although the structures are often absent in stoichiometric dolomite). The most common microstructure is a modulation parallel to (1014) having a wavelength of 200 Å. The role of these microstructures and their relationship to cation ordering is not yet fully understood, but this area holds some potential in terms of documenting the relationship between dolomite of replacement origin and the carbonate precursor, and thus yielding insight into the reaction mechanism.

Of importance to the question whether dolomite is a primary precipitate or a replacement product of preexisting carbonate phases is the further work of Wenk *et al.* (1993). They reported the possible evidence for the direct precipitation of ordered dolomite in a modern sabkha facies in Abu Dhabi, on the Persian Gulf. TEM images and electron diffraction data were used to differentiate coexisting aragonite, disordered magnesian calcite (grading to partially ordered calcian dolomite), and ordered dolomite. Despite the coexistence of these phases, Wenk and co-workers argued that there is no evidence to suggest that the ordered dolomite formed as a result of ordering of a magnesian calcite precursor, and thus must have formed by direct precipitation, possibly with concurrent aragonite dissolution.

Finally, Arvidson and Mackenzie (1997, 1999, 2000) showed from theoretical considerations and continuous flow, dolomite-seeded, reactor experiments that the dolomite precipitation reaction rate is strongly dependent on temperature and moderately dependent on the saturation state of the solution from which the dolomite is forming. The dolomite produced in their experiments was variable in composition but typically was a calcium-rich protodolomite that formed syntaxial overgrowths on the seed material. The activation energy for this protodolomite precipitation reaction was found to be 31.9 kcal/mol (133.5 kJ/mol), lower than that for reactions involving ordered dolomite. The energy required to convert a calcium-rich protodolomite to an ordered dolomite is about 5.5 kcal/mol or 23 kJ/mol; the solubility product for ordered dolomite at $25°C$ and 1 atmosphere is about 10^{-17}.

The above experimental, observational, and analytical work demonstrates that both protodolomite and dolomite growth can occur at low to moderate temperatures on relatively short time scales, given an appropriate solution chemistry. Further aspects of dolomite formation and its relevance to ocean-atmosphere conditions are discussed later in this chapter and in Chapter 7.

4 Aragonite

The most abundant orthorhombic carbonate considered in this book is aragonite, the dimorph of calcite. Other rare isomorphs (similar structure, different cation composition) include strontianite ($SrCO_3$), witherite ($BaCO_3$), and cerrusite ($PbCO_3$) that are given in Table 4.1. At near Earth's surface pressure and temperature, aragonite occurs as the inorganic component of many common invertebrate skeletons (Table 4.2), as sediments derived from the physical and biological disintegration and erosion of these skeletons, as cements in modern marine sediments and Neogene limestones, in some carbonate cave deposits and travertine precipitated by hot springs, and often as a replacement mineral in igneous and metamorphic rocks and ore deposits (Speer, 1983).

Although aragonite has larger cation sites, the phase is denser than calcite and hence is stable relative to calcite at elevated pressure and temperature (Fig. 4.7; Box 4.1). The solubility product of aragonite at $25°C$ and 1 atmosphere is approximately $10^{-8.30}$.

Table 4.2   Occurrences of $CaCO_3$ minerals, including magnesian calcites, in autotrophic and heterotrophic organisms that secrete calcareous skeletons.

Mineral[1]	Kingdom and Phylum[1],[2]	Trophic state or production mode[2]	Common name, ecological habitat[2], comments
	Bacteria (Monera)		
Calcite	Cyanobacteria	Photoautotrophs	Blue-green algae or bacteria
Aragonite	Cyanobacteria	Photoautotrophs	Blue-green algae or bacteria
	Pseudomands	Heterotrophs	Also classified as Protobacteria or purple bacteria
	Protoctista		
Calcite	Haptophyta	Photoautotrophs	Coccoliths, predominantly marine plankton
	Chlorophyta	Photoautotrophs	Green algae, marine and freshwater
	Rhodophyta	Photoautotrophs	Red or coralline algae, vast majority marine
	Rhizopoda	Heterotrophs	Amoebas, ocean and fresh water, soils
	Dinoflagellata	Some photoautotrophs, some heterotrophs	Mostly marine plankton
	Zoomastigina	Heterotrophs	Zooflagellates, largely fresh water
	Foraminifera	Heterotrophs	Marine plankton and benthos
	Ciliophora	Heterotrophs	Ciliates, marine and freshwater
	Myxomycota	Heterotrophs	Plasmodial slime molds
Aragonite	Haptophyta	Photoautotrophs	Coccoliths
	Chlorophyta	Photoautotrophs	Green algae, marine and freshwater
	Rhodophyta	Photoautotrophs	Red or coralline algae, vast majority marine
	Phaeophyta	Photoautotrophs	Brown algae, littoral and pelagic
	Foraminifera	Heterotrophs	Superfamily *Robertinacea*, benthonic
Vaterite	Rhodophyta	Photoautotrophs	In genus *Galaxaura*
	Fungi	Heterotrophs	
Calcite	Ascomycota		Lichens, formerly Mycophycophyta
	Plantae	Photoautotrophs	
Calcite			Mosses
			Ferns
	Angiospermophyta		Flowering plants, also classified as Anthophyta
Aragonite	Angiospermophyta		same
Vaterite	Angiospermophyta		same

(*continued*)

Table 4.2   (*Continued*)

Mineral[1]	Kingdom and Phylum[1],[2]	Trophic state or production mode[2]	Common name, ecological habitat[2], comments
	Animalia	Heterotrophs	
Calcite	Cnidaria		Tabulate and rugose corals (extinct); Octocorallia (except Helioporida)
	Porifera		Sponges, Calcarea (calcareous sponges)
	Platyhelminthes		Flat worms, freshwater, marine, soils
	Sipuncula		Peanut worms, mostly benthonic
	Annelida		Marine worms, Polychaeta, serpulid worms
	Ectoprocta		Bryozoans, mostly marine
	Arthropoda		Carapace and cuticle: Trilobitomorpha (trilobites, extinct), ostracods, barnacles (Cirripedia), Malacostraca
	Mollusca		Shells only: Bivalvia and Gastropoda (snails)
	Brachiopoda		Fossil and extant lamp shells
	Echinodermata		Plates, spicules, spines of Mg-calcite in all groups: sea urchins, sea stars, brittle starts (ophiuroids), sea lilies (crinoids), sea cucumbers (holothurians).
Aragonite	Cnidaria		Madreporian corals and groups Milliporina, Helioporida, Scleractinia
	Porifera		Sponges, Calcarea (calcareous sponges)
	Platyhelminthes		Flat worms, freshwater, marine, soils
	Annelida		Marine worms, Polychaeta, serpulid worms
	Ectoprocta		Bryozoans, mostly marine
	Mollusca		Shells only: Aplacophora, Monoplacophora, Polyplacophora, Bivalvia, Gastropoda (including planktonic Pteropoda), Cephalopoda
	Arthropoda		Cuticle in some Malacostraca, base plate in some barnacles
Vaterite	Platyhelminthes		Flat worms, freshwater, marine, soils

[1] Lowenstam (1986), Lowenstam and Weiner (1989), Mackenzie *et al.* (1983), Runnegar and Bengston (1990).
[2] Margulis and Schwartz (1998).

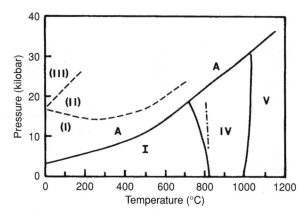

Figure 4.7   Stability fields of aragonite and different polymorphs of calcite as a function of temperature and pressure. A: aragonite; I through V: calcite polymorphs with metastable fields shown by parentheses; dash-dot line at 800°C represents aragonite-calcite transition encountered in experimental cooling runs; solid line at lower temperature represents transition encountered in experimental heating runs (after Carlson, 1980).

The crystal structure of aragonite was first determined by Bragg in 1924 (Bragg, 1924). Refinements of the structure were done nearly 40 years later (in Speer, 1983). The structure of orthorhombic aragonite compared to those of calcite and dolomite is shown in Fig. 4.1.

Aragonites exhibit only limited solid solution, primarily with Pb and Sr. Plumbian aragonites with up to 2.5 mol % PbCO, are common, and strontian aragonites occurring in hot springs may contain up to 14 mol % $SrCO_3$ (Speer, 1983). The latter limited solid solution is of importance here because of the fact that the degree of solid solution varies among different groups of organisms having aragonitic shells (Fig. 4.8A). In certain organisms, the Sr content of the skeleton has been used to obtain the Sr:Ca ratio or the temperature of the water from which the organism precipitated its skeleton (Figs. 4.8B and 4.9). However, the Sr content is a function of the mechanism by which the organism precipitates its skeleton and it varies from organism to organism producing a "species or vital effect." For example, biomineralization in mollusks is different from that in more primitive phyla in that the chemical constituents for skeletal formation in mollusks come from the extrapallial fluid through the blood that in turn is acquired from the water. In these organisms the concentration of Sr in the invertebrate skeleton is not that predicted to be in equilibrium with the water, and hence the skeleton of the organism in terms of its Sr concentration is not a good indicator of the environmental conditions of the water from which the skeleton was precipitated. On the contrary, for more primitive organisms like certain corals and algae, biomineralization is extracellular and the shell composition is more like inorganic aragonite precipitated under similar environmental conditions as the skeletal aragonite (Lowenstam, 1986). Smith *et al.* (1979) found that in large-scale incubation experiments, the Sr:Ca ratio of three genera of scleractinian coral skeletons varied with temperature and the Sr:Ca ratio of the incubation water. At a nearly constant Sr:Ca ratio in surface ocean water, the Sr:Ca ratios of the coralline

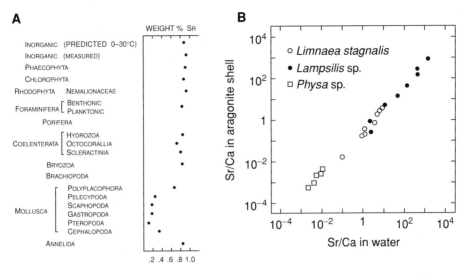

Figure 4.8   A. The extent of solid solution of Sr in wt % in skeletal carbonates of different organisms. B. An example of a good correlation between Sr/Ca in water and the Sr/Ca ratio in aragonitic shells. One could use this information to predict from shell chemistry, the chemistry of the water (after Speer, 1987).

aragonite exhibited an approximate negative correlation with temperature (Fig. 4.9). The slope of the curve was offset from that of inorganic aragonite showing that the distribution coefficient for Sr between the inorganic aragonite and coralline aragonite and water were different but nevertheless demonstrating that corals could be used for obtaining the temperature of the ambient seawater, if the Sr distribution coefficient were

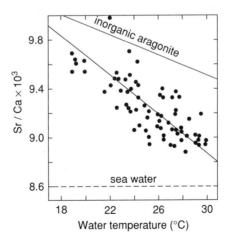

Figure 4.9   The relationship between the Sr/Ca ratio (by weight) in corals and the water temperature in °C. For comparison, the ratio in inorganically precipitated aragonite as a function of water temperature is also shown (after Speer, 1987).

known for the particular coral species. This finding has had considerable application in determination of paleotemperatures of ocean water. This type of thermometry is known as sclerothermometry.

Because aragonite is unstable relative to calcite at near Earth's surface pressures and temperatures (Box 4.1), with time and changes in environmental conditions, inorganic and biogenic aragonite can be converted to calcite or dolomite. Thus the original texture of the aragonite can be altered and the chemical and isotopic information contained in the phase lost. Rocks older than Neogene contain very little aragonite because of diagenetic reactions that lead to leaching of the components of the phase from the rock or its recrystallization to another mineral, as discussed in the next section. Scant aragonite has been found in rocks as old as Pennsylvanian in age (Smith *et al.*, 1994).

5   Carbonate Dissolution and Precipitation Kinetics

Carbonate minerals are common, chemically moderately reactive minerals under Earth's surface conditions. They are not as "easy" to dissolve or precipitate as simple salts like NaCl (halite) or as "difficult" to dissolve or precipitate as the aluminosilicate minerals. An understanding of their reactivity (dissolution and precipitation) is important in consideration of a range of topics in geochemistry, including early and late-stage diagenesis, fate of fossil fuel CO_2, formation of modern carbonate sediments and their transformation over time, and the global biogeochemical cycles of carbon and associated elements. The vast majority of kinetic studies has focused on calcite and aragonite with less attention being given to the third most important carbonate mineral, dolomite, probably because of its slow reactivity under low temperature and pressure conditions. The topic of carbonate mineral dissolution is covered in detail in Morse and Arvidson (2002) and that of precipitation in Morse and Mackenzie (1990). Here we discuss the results of laboratory studies and their applicability to natural systems.

The most commonly used equation to describe the rate of carbonate mineral dissolution is (e.g., Morse, 2004; Morse and Arvidson, 2002; Sjöberg, 1976):

$$R = -\frac{dm_{\text{carbonate}}}{dt} = k^*(1 - \Omega)^n \tag{4.2}$$

where R is the rate in such units as, for example, $\mu\text{mol m}^{-2}\text{ h}^{-1}$, m is moles of carbonate mineral per unit of its surface area, and t is time. k^* is an empirical rate constant in units that must agree with those of R. Ω is the ratio (dimensionless; defined in Chapter 5, Section 4) of the ion activity product (IAP) of the carbonate mineral in solution to its solubility product (K_{sp}), and n is a positive constant known as the order of reaction. The units of R and k^* may be either mass area^{-1} time^{-1} or they may be %/time. The ratio of $IAP/K_{sp} = \Omega$ is a measure of disequilibrium between the carbonate solid and solution; that is, a measure of the Gibbs free energy drive for dissolution (or precipitation). For undersaturation it must range from 0 to 1 and for supersaturation from 1 to higher values (Chapter 5).

It follows from equation (4.2) that by taking the logarithm of both sides of the equation we have

$$\log R = n \log(1 - \Omega) + \log k^* \qquad (4.3)$$

Thus a plot of $\log R$ vs. $\log(1 - \Omega)$ yields a straight line with the intercept $\log k^*$ and the slope n. This equation and variations of it have been used to obtain the reaction order n for calcite, aragonite, and dolomite. For calcite, where $\log R$ was determined in simple solutions as a function of pH at 25°C and atmospheric pressure in the pH region where diffusion is the major control on the kinetics of the reaction (low pH range), Plummer *et al.* (1978) obtained $n = 1$ and a k of about 50×10^{-3} cm/s, whereas Sjöberg and Rickard (1984) found $n = 0.9$ and a k of about 3×10^{-3} cm/s. It should be noted that the rate parameter k in cm/s is usually used in dissolution rate equations of the form $k(C_{eq} - C)$, where C is a concentration in solution or at the mineral surface and C_{eq} is a concentration at equilibrium with the solid, and the product kC_{eq} has the same dimensions as k^* in equation (4.2). At a pH of 4, the different results are equivalent to predicted rates of dissolution that differ by a factor of 7. Morse and Arvidson (2002) attributed this difference to be due to the choice by the authors of the thickness of the boundary layer on the surface of the mineral through which diffusion occurs. If the layer is assumed to have a thickness of a lower limiting value of 4 μm, the results are nearly identical.

Figure 4.10 shows a compilation of calcite dissolution rates as a function of pH from several investigators. Notice that the rates decrease by several orders of magnitude from

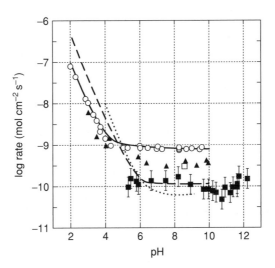

Figure 4.10   Logarithm of the rate of dissolution of calcite in mol cm^{-2} s^{-1} from various experimental investigations as a function of pH (after Morse and Arvidson, 2002).

low pH values to high, and that the rates appear to reach a nearly constant value at a pH $\approx$ 6.

For aragonite the laboratory-determined reaction order has been found to be close to 3 in both seawater and more dilute solutions (Morse et al., 1979). The reaction order of 2.93 for aragonite in seawater is close to that observed for calcite determined in seawater under similar conditions (Morse and Arvidson, 2002). Laboratory determination of dolomite dissolution kinetics has been quite limited reflecting the general difficulty in low temperature experiments involving dolomite of both slow reaction rates, and the large uncertainties involving the realationship between the mineral's reactivity, structure, and composition. In the seminal study of Busenberg and Plummer (1982), these authors proposed the following overall rate equation for the dissolution of dolomite in dilute solutions at varying P_{CO_2} levels (hence pH) from 0 to 1 bar and temperatures from 5° to 65°C:

$$R = k_1 a_{H^+}^{1/2} + k_2 a_{H_2CO_3}^{1/2} + k_3 a_{H_2O}^{1/2} - k_4 a_{HCO_3^-} \tag{4.4}$$

where the ks represent rate constants of the same dimensions as the dissolution rate R, a is the activity of the dissolved subscripted chemical species, and the first three terms in the equation represent three parallel forward reactions with a common order of reaction of $1/2$ at 25°C. The fourth term represents a "backward" precipitation term. The values of the rate constants at 25°C for k_1 through k_4 are 1.7×10^{-8}, 4.3×10^{-10}, 2.5×10^{-13}, and 2.8×10^{-8}, respectively. Since Busenberg and Pummer's work (1982), there have been several dolomite dissolution studies using different experimental approaches and measurement techniques (e.g., Chou et al., 1989; Gautelier et al., 1999; Lüttge et al., 2003; Pokrovsky et al., 1999). Although there is controversy concerning the exact surface reaction mechanisms involved and the extent to which the experimentally determined total surface of the dolomite used participates in the various dissolution experiments, a compilation of most of the rate vs. pH experiments shows reasonable agreement (Fig. 4.11). As with most carbonate minerals, the dolomite rate of dissolution decreases with increasing pH and may approach a near constant rate at higher pH values (Fig. 4.11).

In summary, as can be seen in Fig. 4.12 over a range of pH, aragonite, calcite, and witherite ($BaCO_3$) have about the same rate of dissolution whereas dolomite and magnesite ($MgCO_3$) dissolve at rates one to three orders of magnitude slower, respectively. The experimental data in Fig. 4.12 were fit by equations of the type shown below for calcite:

$$R = k_1 a_{H^+} + k_3 + k_6 m_{Ca^{2+}} + m_{CO_3^{2-}} \quad (\text{mol cm}^{-2}\,\text{s}^{-1}) \tag{4.5}$$

where k_1, k_3, and k_6 are first-order rate constants that are temperature dependent. The experimental results confirm the geological evidence discussed above that the reactivity of dolomite is considerably less than that of calcite or aragonite.

The kinetics of carbonate mineral precipitation in chemically simple solutions has received considerably less attention than that of dissolution reactions. However, the work of Busenberg and Plummer (1982, 1986) dealing with a comparative study of the dissolution and crystal growth kinetics of calcite and aragonite, and dissolution of

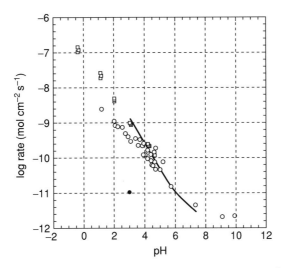

Figure 4.11   Logarithm of the rate of dissolution of dolomite in mol cm^{-2} s^{-1} from various experimental investigations as a function of pH (after Morse and Arvidson, 2002).

dolomite is of particular interest. They investigated the dissolution and crystal growth kinetics of calcite and aragonite in pure water, and in the presence of HCl, $CaCl_2$, $KHCO_3$, and KOH at various CO_2 partial pressures at 25°C. The forward dissolution rate of reactions for the carbonate minerals is described by equations similar to that of (4.4). The rate of crystal growth of calcite and aragonite that was studied in $Ca(HCO_3)_2$

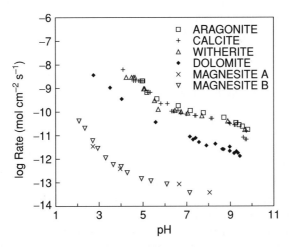

Figure 4.12   Comparison of the logarithm of the rates of dissolution of several carbonate minerals in mol cm^{-2} s^{-1}. Note that magnesite dissolves at a rate about three orders of magnitude slower and dolomite at a rate about two orders of magnitude slower than that of calcite (after Chou *et al.*, 1989).

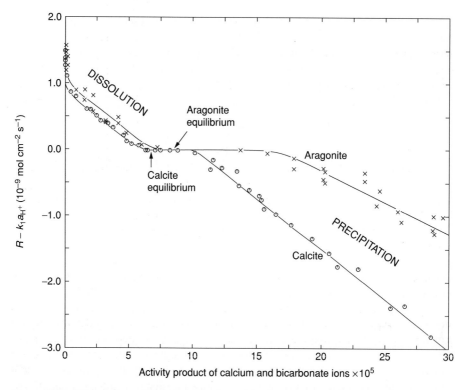

Figure 4.13   Comparison of dissolution and precipitation rates of calcite and aragonite single crystals as a function of the activity product of Ca^{2+} and HCO_3^- at 0.96 atm P_{CO_2}. Note that the rates of precipitation of calcite are significantly faster than those for aragonite; however, aragonite dissolves at a rate only slightly faster than calcite (after Busenberg and Plummer, 1986).

solutions at CO_2 partial pressures of $10^{-3.5}$ (atmospheric) to 1 atmosphere, can be described by the following rate equation:

$$R_{\text{precipitation}} = k_4(cE - a_{Ca^{2+}}a_{HCO_3^-}) \qquad (4.6)$$

where k_4 and c are constants and E is the equilibrium activity product of Ca^{2+} and HCO_3^- at equilibrium with the specific $CaCO_3$ dimorph at the CO_2 partial pressure of the experiment. Figure 4.13 is a summary of their results. It can be seen from the figure that the rates of dissolution of aragonite are slightly greater than those for calcite in solutions of similar composition.

However, for supersaturated $Ca(HCO_3)_2$ solutions, the rates of precipitation of calcite are significantly faster than those of aragonite. Since soil and groundwater solutions are mainly Ca^{2+}-HCO_3^- solutions, these results to some extent explain why aragonite when exposed to freshwater environments is dissolved and its chemical components leached from the environment or re-precipitated as calcite.

A similar forward rate equation as that of (4.6) can be used to describe dolomite dissolution rates (Busenberg and Plummer, 1982), except that the rates are half-order with respect to H^+, H_2CO_3, and H_2O. One overall conclusion is that the rates of dissolution of dolomite in dilute solutions are much slower than those of calcite and aragonite, as might be anticipated from geologic field observations. In extremely dilute solutions, the rate of dissolution of dolomite is 100 times slower that of calcite and aragonite.

6   Carbonate Precipitation and Dissolution in Marine Ecosystems

Throughout this book we discuss processes and mechanisms, reservoirs and fluxes involving the biogeochemical cycles of the elements. In this section we provide kinetic equations that form the foundation of a model of the global carbonate cycle in the coastal ocean discussed in Chapter 9. The kinetic discussion is given here as it provides some insight at this time of how one can introduce kinetic parameterizations from field or experimental data into global biogeochemical cycle models. The discussion focuses on carbonate calcification and dissolution rates.

In one case, the biogenic calcification rate, as a relative rate value, was shown to depend on carbonate saturation state and can be expressed as a linear relationship derived from multiple experiments on corals and coralline algae (Gattuso et al., 1999):

$$R_\Omega = 21.3\Omega + 12, \tag{4.7}$$

In a second case, the calcification rate is given by a curvilinear relationship based on the experimental results for the coral, *Stylophora pistillata* (Gattuso et al., 1998; Leclerq et al., 2002):

$$R_\Omega = 228(1 - e^{-\Omega/0.69}) - 128, \tag{4.8}$$

where R_Ω is the relative rate of calcification (approximately equal to 100 at the initial conditions) and Ω is the surface water aragonite saturation state. The dependence of biogenic calcification on temperature can also be expressed by two different relationships. In one, a negative parabolic relationship can be used, as obtained from experimental results for a red coralline alga, *Porolithon gardineri* (Agegian, 1985; Mackenzie and Agegian, 1989) and normalized to the maximum rate of calcification:

$$R_T = 100 - 1.32\Delta T^2 \tag{4.9}$$

where R_T is the relative rate of calcification expressed as a percentage of the rate at the initial temperature in the year 1700 and ΔT is the temperature change in degrees Celsius. In a second case, a positive linear relationship can be obtained from the observed rates of calcification of multiple coral colonies from the Great Barrier Reef, Hawaii, and Thailand (Grigg, 1982, 1997; Lough and Barnes, 2000; Scoffin et al., 1992):

$$R_T = 100 + 28\Delta T \tag{4.10}$$

Table 4.3   Adopted constants for inorganic $CaCO_3$ precipitation and dissolution rates (R)

Precipitation $R_p = k_p(\Omega - 1)^{n_p}$ (Zhong and Mucci, 1989)	Rate constant (k) ($\mu mol \ m^{-2} \ h^{-1}$)	Reaction order (n)
Calcite	$10^{-0.20}$	2.80
15 mol % Mg-calcite*	$10^{-0.20}$	2.80
Aragonite	$10^{1.00}$	2.36
Dissolution $R_d = k_d(1 - \Omega)^{n_d}$ (Walter and Morse, 1985)	($\mu mol \ g^{-1} \ h^{-1}$)	
Calcite	$10^{2.82}$	2.86
15 mol % Mg-calcite	$10^{2.62}$	3.34
Aragonite	$10^{2.89}$	2.48

*Assuming same rate as calcite

Inorganic dissolution (R_d) and precipitation (R_p) of carbonate minerals within the pore water-sediment model system can be described by kinetic rate equations of the general form:

$$R_d = k_d(1 - \Omega)^{n_d},$$ (4.11)

and,

$$R_p = k_p(\Omega - 1)^{n_p},$$ (4.12)

where k_d is the rate constant for dissolution and k_p is the rate constant for precipitation, Ω is the pore water carbonate saturation state, and n_d and n_p are the reaction orders for dissolution and precipitation, respectively (Table 4.3). The constant parameters were obtained from the experimental results of Zhong and Mucci (1989) for precipitation and Walter and Morse (1985) for dissolution. Reaction rates can be calculated based on the total mass of calcium carbonate in the sediments by converting the calculated rates per unit area to mass per unit time using average specific surface areas of the minerals (0.1–0.5 m^2/g) and the ratio between reactive surface area and total area (0.003–0.66) for typical shallow-water biogenic sediment components (Walter and Morse, 1984, 1985). Inhibition by dissolved phosphate (10 μmol/L) and dissolved organic matter (10 mg/kg) can also be taken into consideration by modifying the uninhibited rates by factors derived from the relationship between the concentrations of these constituents and the rates (Morse et al., 1985). The application of the above relationships can be found in Chapter 9 that discusses air-sea exchange of CO_2 and effects of changing carbonate saturation state and temperature on future biogenic calcification and carbonate dissolution rates in the coastal ocean.

7   Some Geological Considerations

Questions concerning the diagenesis of calcites in carbonate sediments and rocks, and the occurrence and mode of formation of dolomite have plagued geologists for more than a century. Our knowledge of the mineralogy, chemistry, phase relations, and reaction kinetics of these minerals acquired from experimental work has formed the necessary foundation to obtain answers to some of the questions. For example, the finding that calcites rich in magnesium are not stable relative to pure calcite and dolomite is the principal reason that the former phases convert at near Earth's surface temperatures and pressures to the latter with increasing age of the sedimentary rock in which the minerals occur. The fact that aragonite is unstable relative to calcite is also the primary reason this phase converts to calcite (or dolomite) with increasing age. Modern calcareous sediments of shallow-water origin are characterized by a metastable assemblage of calcites and aragonite, primarily of biogenic origin, and sparse inorganic dolomite and inorganic carbonate cements. Table 4.2 shows the classes of organisms that produce carbonate minerals as skeletons, tests, or shells and their mineralogy. With time and burial, this metastable assemblage generally will be converted to the stable assemblage of calcite and dolomite; aragonite will alter to calcite, the stable dimorph, and magnesium-rich calcite will lose magnesium and be converted to nearly pure calcite and contemporaneously, or later in the history of the sedimentary deposit, to dolomite. Evidence for the diagenetic stabilization by the replacement of aragonite and magnesian calcites is ample, particularly from limestones subjected to freshwater alteration reactions above the groundwater table (vadose zone diagenesis) and below it (phreatic zone diagenesis).

In general, the alteration of magnesian calcite to calcite is described using petro-graphic thin section techniques as a pseudomorphic replacement reaction, in which the original skeletal or non-skeletal materials retain their original microarchitecture and structure (e.g., Land, 1967). The original fabrics are retained because the chemical reactions that transform magnesian calcites to calcite probably occur along a microscopic diagenetic front. Incongruent reaction is the reaction mechanism most commonly invoked to explain the retention of microarchitecture in magnesian calcite skeletal materials or other substrates. The reaction involves dissolution of the magnesian calcite accompanied by precipitation of calcite with a lower concentration of magnesium than that of the initial solid. Based mainly on experimental and theoretical arguments, Bischoff et al. (1983) showed that there are several potential stabilization pathways by which a magnesian calcite is converted to calcite. However, these pathways are poorly documented in the rock record because of a lack of adequate field studies.

Figure 4.14 illustrates an example of a very well documented pathway for the diagenesis of a metastable mixture of the carbonate minerals aragonite, calcite, and magnesian calcite to a stable monomineralic rock, calcitic limestone. The figure shows what happens to the mineralogy, chemistry, and isotopic composition of sediments originally deposited as biogenic debris in Holocene and Pleistocene environments on the Bermuda platform. With time and contact with mainly meteoric vadose (subsurface environment above the groundwater table) and phreatic (subsurface environment below the groundwater table) waters, the metastable mineral assemblage is altered to limestone. Note that the aragonite and magnesian calcite components in the original sediments are leached from the rock or altered to calcite. In general, the elements

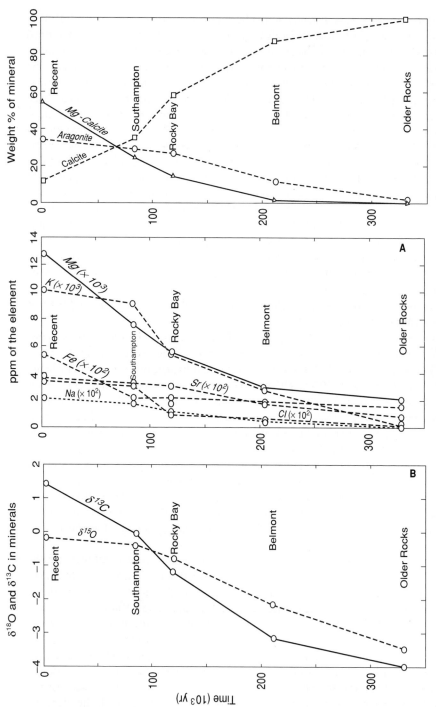

Figure 4.14   Changes in the mineralogy, chemistry, and oxygen and carbon isotopic compositions of Holocene and Pleistocene carbonate rocks as a function of time of exposure to vadose and phreatic waters on the Bermuda islands (details in the text; after Morse and Mackenzie, 1990).

Mg, K, Fe, Sr, Na, and Cl are lost from the rock during this conversion process. In addition, both the δ^{13}C and δ^{18}O isotopic signatures of the rock become lighter (depleted in the heavy isotopes ^{13}C and ^{18}O) with increasing rock age as the original carbon and oxygen of the solid phases are exchanged with the subsurface meteoric waters that are depleted in the heavy isotope of carbon and oxygen (isotopic exchange and reactions are discussed in Chapter 6, and long-term trends in Chapter 7). This overall conversion pathway of diagenesis is not unusual and has been observed in many Holocene and Pleistocene sedimentary sequences, such as those along the Mediterranean coast of Israel (Gavish and Friedman, 1969), in Barbados (Pingitore, 1976), in Florida and the Bahamas (Stehli and Hower, 1961), and on Majora atoll, Marshal Islands (Anthony *et al.*, 1989). The important point here is that as we go back in time in the geologic record, much of the mineralogical, chemical, and biological information contained in the original carbonate sediment can be lost or transformed (Chave, 1962). Thus when proxy information is used to gain insight into past environmental conditions, one must be very careful to select materials for analysis that have not undergone significant alteration since deposition. Any proxies used to decipher changes in Earth's surface environmental conditions must be evaluated in light of these changes that can bias the proxy record (e.g., Brand and Veizer, 1980, 1981).

In the case of dolomite, the initial dolomitic phase that forms is usually a protodolomite with excess calcium and lack of structural ordering. With increasing time since formation, the original metastable dolomite phase can go through successive recrystallization reactions (transformation of one phase to another generally by processes of microsolution and precipitation) leading to a more ordered and stoichiometric phase (Fig. 4.15). Massive accumulations of dolomite as found in the rock record cannot form simply by conversion of the metastable phases to dolomite in a closed system. There

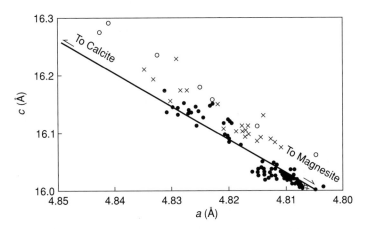

Figure 4.15   Unit cell measurements in ångströms of synthetic (open circle), Holocene ($\times$), pre-Holocene (filled black circles), and compositionally and structurally "ideal" dolomite ($+$). Notice the decreasing cell dimensions as the dolomite phase becomes more ordered and compositionally homogeneous (after Land, 1985).

is not sufficient magnesium present in the phases to make large quantities of dolomite; magnesium must be supplied from another source. This source generally is thought to be seawater or modified seawater circulating in the pore space of recently deposited sediments or subsurface fluids, including modified seawater solutions, relatively rich in magnesium found in sedimentary basins.

The core of the dolomite problem is the apparent paradox posed by the paucity of dolomite in modern marine depositional environments and its relative abundance in the sedimentary rock record. There are two major contrasting paradigms that have been promoted to explain this observation. The first is that most dolomite is not primary in nature but forms over time from the alteration of the other principal carbonate phases of calcite and aragonite. The second is that the atmospheric and oceanic environmental conditions of today are not favorable to dolomite precipitation but were favorable at times during the geologic past. This latter paradigm requires changes in seawater chemistry and perhaps atmospheric CO_2 concentrations during geologic time.

To obtain further insight into this latter paradigm, the experimental study of Arvidson and Mackenzie (1999; and Section 3.3) determined that the rate of dolomite precipitation is a strong function of temperature and a function of the degree of saturation of the solution with respect to dolomite, as has also been shown with calcite and aragonite. The dependence on temperature and saturation state observed in the experiments led the authors to conclude that it is the overall rate of dolomite precipitation relative to competing carbonate phases of calcite and aragonite at surface temperatures that may determine to some extent the abundance of dolomite in the sedimentary regime. They also concluded that small changes in temperature substantially affect the dolomite precipitation rate, increasing it with increasing temperature. Thus it is possible that the abundance of dolomite in the sedimentary rock record reflects its absolute precipitation rate in the environment, the rate being controlled in part by the environmental variables of temperature and solution composition, such as the saturation state and Mg/Ca ratio of seawater. There is little doubt that the global mean temperature of the planet has varied during geologic time as has the chemistry of seawater (Chapters 3, 7, and 10). Thus the abundance of dolomite might reflect to some degree these changes. In addition, abundant dolomite in the sedimentary rock record is commonly associated with sedimentary deposits accumulated in relatively shallow-water shelf and associated environments. During extended periods of relatively high sea level in the geologic past, such environments were probably more abundant than during periods of low sea level (Hallam, 1984; Vail et al., 1977). Flooding of ancient continental shelf areas may have helped to provide the environmental conditions necessary for substantial dolomite formation. These assertions are currently being debated and further investigated, as are the chemical mechanisms and the role of bacteria in the dolomite precipitation reaction (Holland, 2004; Veizer and Mackenzie, 2004; Warthmann et al., 2000).

The material of this chapter provides the necessary thermodynamic and kinetic background for discussion of calcite and aragonite in natural waters as functions of temperature and atmospheric CO_2 concentration that are given in Chapters 5 and 9.

In addition, it is relevant to an understanding of the material on isotopes and carbon-isotope fractionation that are the subject of Chapter 6. The dolomite problem and its relevance to ocean-atmosphere evolution is discussed in more detail in Chapter 7 that deals with changes in Earth's surface environment during geologic time as recorded in the sedimentary rock sequences.

Chapter 5

Carbon Dioxide in Natural Waters

This chapter presents some of the basic physical and chemical relationships that are relevant to an understanding of the behavior of CO_2 in the Earth's surface environment. It would be an understatement to say that the literature on this subject is extensive—it is prodigious! In Chapter 2 we dealt with the Earth's atmosphere in the past and present that is based mostly on the model of an ideal gas. However, CO_2 on the Earth's surface interacts with other gases, it dissolves in water, and reacts with water, other dissolved species, and minerals. Many of the chemical reactions involving CO_2 are either directly or indirectly mediated by biological processes, as is briefly discussed in Chapter 1 and further addressed in Chapters 6 and 9. This chapter deals with some of those fundamental aspects of CO_2 and oceanic and continental waters that are reasonably closely related to the global-scale natural systems and processes, and it includes the following subjects: dissolution of CO_2 in water at the Earth's surface conditions; a summary of chemical thermodynamic equilibrium relationships; transfer of CO_2 between atmosphere and water, and the rising atmospheric CO_2 concentration; solubilities of calcite and aragonite and the saturation of natural waters with respect to these minerals; CO_2

in the liquid and hydrate phases as related to the potential disposal of industrial CO_2 in the deep ocean; and sea-air exchange of CO_2 due to the formation and storage of calcium carbonate and organic carbon in sediments.

1 Dissolution and Dissociation of CO_2 in Water

1.1 Simple CO_2-Water System

Dissolution of carbon dioxide in water is the first step that enables photosynthetic production of organic matter in aquatic systems, precipitation of carbonate minerals, and chemical weathering of the Earth's crust and sediments. Carbon dioxide dissolves in water and reacts with it producing negatively charged bicarbonate and carbonate ions, the electric charges of which are balanced in pure water by the hydrogen ion or in a natural solution by other metal cations and the hydrogen ion. Dissolution and dissociation of the gaseous species $CO_2(g)$ in water are represented by the following reactions:

$$CO_2(g) = CO_2(aq) \tag{5.1}$$

$$CO_2(aq) + H_2O = H^+ + HCO_3^- \tag{5.2}$$

$$HCO_3^- = H^+ + CO_3^{2-} \tag{5.3}$$

The aqueous uncharged CO_2 includes the species H_2CO_3 that constitutes only a small fraction, about 1/400, of CO_2 in solution. Alternative use of either CO_2 or H_2CO_3 for all of the aqueous species $CO_2(aq)$ is acceptable without a loss of correct meaning (e.g., Millero, 1996). Because of the formation of the aqueous ions HCO_3^- and CO_3^{2-} from CO_2 in solution, the total concentration of dissolved inorganic carbon is the sum of all the carbon species, $CO_2(aq)$, HCO_3^-, and CO_3^{2-}. This parameter is also called dissolved inorganic carbon, to distinguish it from dissolved organic compounds, and it is usually denoted ΣCO_2 or DIC:

$$DIC = CO_2(aq) + HCO_3^- + CO_3^{2-} \quad (mol\,kg^{-1}) \tag{5.4}$$

Concentration units of DIC in mols of carbon per 1 kg solution—that is, 1 kg of solution consisting of water and dissolved species—are conventionally and widely used for natural waters[1].

In this chapter relationships in the CO_2-carbonic acid-carbonate system are mainly described in terms of apparent or stoichiometric constants that are dependent not only on temperature and pressure, as are true thermodynamic equilibrium constants, but also on composition (e.g., salinity). In addition, detailed carbon system calculations

---

[1] The units of DIC and of the CO_2 solubility, and dissociation constants K that are given in (5.4)–(5.6), can be converted to *molal* concentrations, in units of mols per 1 kg H_2O. For a solution of known composition, a value in mol/kg solution divided by a fraction $(1 - $ kg of solutes in 1 kg solution) converts the concentration to mols per 1 kg H_2O. It may also be noted that the units of ppm, parts per million, are mass/mass, such as milligrams per 1 kg solution. *Molar* concentrations are in mol/liter, making them temperature dependent, and for their conversion to mass/mass units the density of the solution must be known.

are given to provide the reader with a feeling for how one can treat aquatic systems quantitatively to resolve scientific problems.

1.2   Dissolved Inorganic Carbon

The concentration of dissolved inorganic carbon (DIC) in solution is from (5.4) and the carbonate dissociation constants given in Box 5.1:

$$[DIC] = P_{CO_2} K_0 \left(1 + \frac{K_1}{[H^+]} + \frac{K_1 K_2}{[H^+]^2}\right) \quad (mol\,kg^{-1}) \qquad (5.5)$$

Box 5.1   Carbonate equilibria

---

1. Solubility of CO_2 in seawater, K_0' in Table 5.1.1, temperature range $-1°C$ to $40°C$, salinity range from 0 to 40 g/kg (Weiss, 1974).

$$\ln K_0' = -60.2409 + 9345.17/T + 23.3585 \ln(T/100)$$
$$+ S[0.023517 - 0.00023656T + 0.0047036(T/100)^2] \quad (5.1.1)$$

Temperature T is in kelvin. To convert K_0' from units of mol (kg solution)$^{-1}$ atm^{-1} to units of mol kg^{-1} bar^{-1}, the value should be divided by 1.01325 bar atm^{-1} or 0.0132 subtracted from $\ln K_0$.

Table 5.1.1   Upper part: apparent dissociation constants K_i' of the carbonate system at ocean water salinity of 35, 1 bar total pressure. Lower part: dissociation constants in pure water (K_i, at zero ionic strength) based on ion activities and fugacity of CO_2 gas, and solubility constants are for pure calcite and aragonite.

	Parameter	At 25°C	At 15°C	At 5°C
1	$K_0' = [CO_2]/P_{CO_2}$	2.839×10^{-2}	3.746×10^{-2}	5.213×10^{-2}
2	$K_1' = [H^+][HCO_3^-]/[CO_2]$	1.392×10^{-6}	1.119×10^{-6}	8.838×10^{-7}
3	$K_2' = [H^+][CO_3^{2-}]/[HCO_3^-]$	1.189×10^{-9}	7.970×10^{-10}	5.191×10^{-10}
4	$K_w' = [H^+][OH^-]$	6.063×10^{-14}	2.380×10^{-14}	8.549×10^{-15}
5	$K_B' = [H^+][B(OH)_4^-]/[B(OH)_3]$	2.526×10^{-9}	1.921×10^{-9}	1.431×10^{-9}
6	B_T (mol/kg)	4.160×10^{-4}	4.160×10^{-4}	4.160×10^{-4}
7	$K_{cal}' = [Ca^{2+}][CO_3^{2-}]$	4.273×10^{-7}	4.315×10^{-7}	4.309×10^{-7}
8	$K_{arag}' = [Ca^{2+}][CO_3^{2-}]$	6.482×10^{-7}	6.719×10^{-7}	6.824×10^{-7}
1	$K_0 = a_{CO_2}/f_{CO_2} = a_{CO_2}/P_{CO_2}$	3.388×10^{-2}	4.571×10^{-2}	6.457×10^{-2}
2	$K_1 = a_{H^+} a_{HCO_3^-}/a_{CO_2}$	4.467×10^{-7}	3.802×10^{-7}	3.020×10^{-7}
3	$K_2 = a_{H^+} a_{CO_3^{2-}}/a_{HCO_3^-}$	4.677×10^{-11}	3.715×10^{-11}	2.818×10^{-11}
4	$K_w = a_{H^+} a_{OH^-}/a_{H_2O}$	1.000×10^{-14}	4.467×10^{-15}	1.862×10^{-15}
7	$K_{cal} = a_{Ca^{2+}} a_{CO_3^{2-}}$	3.313×10^{-9}	3.714×10^{-9}	4.034×10^{-9}
8	$K_{arag} = a_{Ca^{2+}} a_{CO_3^{2-}}$	4.969×10^{-9}	5.667×10^{-9}	6.270×10^{-9}

2. The dissociation constants (Table 5.1.1) of the carbonate, borate, and water species, and calcite and aragonite in normal seawater are from the equations of Mucci (1983), DOE (1994), and other sources, as summarized in Zeebe and Wolf-Gladrow (2001, pp. 255–258 and 266–267). For pure water, at the limit of ionic strength of zero, the data are from Plummer and Busenberg (1982) and Drever (1997). Brackets [ ] denote concentrations of aqueous species in mol/kg, P_{CO_2} is partial pressure of CO_2 in the atmosphere (bar), and B_T is total dissolved boron concentration (mol B/kg). The pH is used on the total scale (Zeebe and Wolf-Gladrow, 2001).

3. The pressure dependence of the mineral and aqueous species is calculated from the parameters given in Millero (1995) and Zeebe and Wolf-Gladrow (2001).

4. Solubility of calcite and aragonite
A reaction between a pure mineral and its aqueous ions at equilibrium,

$$CaCO_3 = Ca^{2+} + CO_3^{2-},$$

is characterized by an equilibrium constant, also called a solubility or dissociation constant:

$$K_{sp} = \frac{a_{Ca^{2+}} a_{CO_3^{2-}}}{a_{CaCO_3}}$$

$$= [Ca^{2+}]\gamma_{Ca^{2+}}[CO_3^{2-}]\gamma_{CO_3^{2-}} = IAP \qquad (5.1.2)$$

where a_i are thermodynamic activities ($a_{solid} = 1$ for a pure mineral), brackets denote concentrations (e.g., mol/kg-solution or mol/kg H_2O), γ_i are the individual species' ionic activity coefficients, and IAP stands for ion activity product. Because estimation of the activity coefficients in solutions of higher concentration of dissolved solids, such as seawater, is subject to some uncertainties, a practical definition of K'_{sp} is an apparent solubility product based on the ionic concentrations:

$$K'_{sp} = \frac{K_{sp}}{\gamma_{Ca^{2+}}\gamma_{CO_3^{2-}}}$$

$$= [Ca^{2+}][CO_3^{2-}] = ICP \qquad (5.1.3)$$

where ICP denotes ion concentration product.

The thermodynamic equilibrium constant K_{sp} depends on temperature and pressure, but not on the solution composition that affects the values of the activity coefficients γ_i. However, the concentrations of the aqueous ions depend on a solution composition and the apparent equilibrium constant K'_{sp} also varies with the solution composition and ionic strength, as well as with temperature and pressure.

4.1 Temperature Dependence of K

The dependence of an equilibrium constant K on temperature T, at constant pressure, is:

$$\left(\frac{\partial \ln K}{\partial T}\right)_P = \frac{\Delta H^\circ}{RT^2} \quad (\mathrm{K}^{-1}) \tag{5.1.4}$$

where R is the gas constant and ΔH° is the enthalpy change in the reaction in the standard state. In an exothermal reaction, heat is released and the change in the reaction enthalpy is positive ($\Delta H^\circ > 0$). This makes the equilibrium constant increase with increasing temperature ($T > T_0$) and the mineral solubility increases. In the opposite case of an endothermal reaction ($\Delta H^\circ < 0$), such as at the equilibrium between calcite and its ions in an aqueous solution at $25°C$, a decrease in temperature makes the equilibrium constant larger and increases the mineral solubility. The heat capacity or enthalpy of a reaction is also temperature dependent and, by definition, its change with T at constant pressure is the heat capacity change in the reaction:

$$\left(\frac{\partial \Delta H^\circ}{\partial T}\right)_P = \Delta C_P^\circ \quad (\mathrm{J\,mol^{-1}\,K^{-1}}) \tag{5.1.5}$$

Heat capacities at constant pressure of minerals and aqueous species, C_P, are usually given in textbooks as polynomials in T with constant coefficients that can be integrated in a straightforward manner. The value of K_T as a function of its value at some reference temperature, K_{T_0}, and other parameters is obtained by integration of (5.1.5) and (5.1.4):

$$\ln\frac{K_T}{K_{T_0}} = \frac{\Delta H_0^\circ}{R}\left(\frac{1}{T_0} - \frac{1}{T}\right) + \frac{1}{R}\int_{T_0}^{T}\left(\int_{T_0}^{T}\Delta C_P^\circ dT\right)\frac{dT}{T^2} \tag{5.1.6}$$

If the reaction enthalpy can be considered as constant over some temperature range, then only the first term on the right-hand side of (5.1.6) may be used.

4.2 Pressure Dependence of K

An equilibrium constant K depends on pressure, at constant temperature, by the following relationship:

$$\left(\frac{\partial \ln K}{\partial P}\right)_T = -\frac{\Delta V^\circ}{RT} \quad (\mathrm{bar}^{-1}) \tag{5.1.7}$$

where ΔV° is the volume change of the reaction in the standard state ($\mathrm{cm^3\,mol^{-1}}$). Volumes of solids and aqueous species may change only slightly over a pressure range from atmospheric to a few hundred bars. When the reaction products are denser than the reactants, ΔV° of the reaction is negative, and the equilibrium constant increases with increasing pressure. The physical significance of this is that at a higher pressure the product has a smaller molar volume and higher density that counteract the effect of the increased pressure.

If the molar volume change in the reaction is not constant, then the dependence of K on pressure may be improved by inclusion of the first derivative of the molar volume that is the coefficient of compressibility at constant temperature of each reactant and product:

$$\beta_i = -\frac{1}{V_i}\left(\frac{\partial V_i}{\partial P}\right)_T \qquad (\text{bar}^{-1}) \qquad (5.1.8)$$

The compressibility coefficient is a positive quantity ($\beta > 0$) if the volume decreases with an increasing pressure, as is usually the case. The molar compressibility of the reaction, denoted Δk°, is the algebraic sum of the individual terms:

$$\Delta k^\circ = \sum_i \beta_i V_i^\circ = -\left(\frac{\partial \Delta V^\circ}{\partial P}\right)_T \qquad (\text{cm}^3\,\text{bar}^{-1}) \qquad (5.1.9)$$

The volume change due to an increase in pressure from P_0 to P is:

$$\Delta V^\circ = \Delta V_0^\circ - \Delta k^\circ (P - P_0) \quad (\text{cm}^3) \qquad (5.1.10)$$

Substitution from (5.1.10) in (5.1.7) and integration of the latter give:

$$\ln \frac{K_{T,P}}{K_{T,P_0}} = -\frac{\Delta V_0^\circ}{RT}(P - P_0) + \frac{1}{2}\Delta k^\circ (P - P_0)^2 \qquad (5.1.11)$$

where subscript T indicates a constant temperature and the molar volume change for the reaction, ΔV_0°, is at the initial pressure P_0. The compressibility of the common mineral-solution reactions is small and the contribution of the second-power term $(P - P_0)^2$ to the change in the equilibrium constant is also usually very small.

It should be noted that equations (5.1.6) and (5.1.11) describe the temperauture and pressure dependence of the thermodynamic equilibrium constant or the IAP, not of the apparent solubility product or the ICP.

At any temperature and a fixed solution composition, which in the present case means a fixed hydrogen-ion concentration or a constant pH (definition: pH $= -\log_{10}[\text{H}^+]$, and the pH is on the so-called total scale; Box 5.1), the DIC concentration is directly related to the partial pressure of atmospheric CO_2. However, as DIC also increases with a decrease in the hydrogen-ion concentration, $[\text{H}^+]$, this indicates that the higher the pH of the solution and the more alkaline it is, the greater is the solubility of CO_2 in it. Conversely, acidic solutions having high values of the hydrogen-ion concentration, $[\text{H}^+]$, or low pH, have a limiting equilibrium DIC value of:

$$[\text{DIC}] \approx K_0' P_{\text{CO}_2} \quad (\text{mol kg}^{-1}) \qquad (5.6)$$

In acidic solutions of pH $\lesssim$ 4, the sum of the three terms in parentheses in (5.5) is close to unity, making relationship (5.6) a reasonably good approximation to CO_2 solubility in acidic waters.

Each of the three species comprising dissolved inorganic carbon represents some fraction of DIC that depends on the pH value of the solution. As the pH of natural waters is often determined by combinations of many different biogeochemical factors, it is useful to consider it as an environmental variable that determines the distribution of the dissolved inorganic carbon species. DIC can be written as a sum of the fractions of the undissociated CO_2 (α_0), bicarbonate ion HCO_3^- (α_1), and carbonate ion CO_3^{2-} (α_2):

$$\alpha_0 + \alpha_1 + \alpha_2 = \frac{[CO_2]}{[DIC]} + \frac{[HCO_3^-]}{[DIC]} + \frac{[CO_3^{2-}]}{[DIC]} = 1 \qquad (5.7)$$

The individual fractions can be expressed in terms of the dissociation constants and the hydrogen-ion concentration as follows (Stumm and Morgan, 1981):

$$\text{Fraction of } CO_2 \text{ (aq)} \qquad \alpha_0 = \frac{1}{1 + K_1/[H^+] + K_1 K_2/[H^+]^2} \qquad (5.8)$$

$$\text{Fraction of } HCO_3^- \qquad \alpha_1 = \frac{1}{[H^+]/K_1 + 1 + K_2/[H^+]} \qquad (5.9)$$

$$\text{Fraction of } CO_3^{2-} \qquad \alpha_2 = \frac{1}{[H^+]^2/K_1 K_2 + [H^+]/K_2 + 1} \qquad (5.10)$$

The individual values of the three fractions of DIC are shown in Fig. 5.1 for different sets of conditions. In Fig. 5.1A are shown the fractions of CO_2, HCO_3^- and CO_3^{2-} in pure water as functions of the solution pH, between pH $= 2$ and 12, at two temperatures of $5°$ and $25°C$, at 1 bar total pressure. In a solution of zero ionic strength, meaning a very dilute solution, lower temperature favors relatively higher concentrations of undissociated CO_2 and lower concentrations of the carbonate ion, CO_3^{2-}. The same temperature effect is also shown for seawater, in Fig. 5.1B. In seawater, the relative abundance of the bicarbonate ion, HCO_3^-, decreases at the expense of the carbonate ion. However, the effect of pressure is small at a mean ocean water salinity of 35: a total pressure increase from atmospheric (pressure-gage $P = 0$) to 300 bar has only a small effect on the shift in the relative abundances of the carbonate species.

1.3 CO_2 and the pH of Fresh and Sea Waters

A change in the CO_2 partial pressure affects the hydrogen-ion concentration, $[H^+]$ or the pH, of a solution. Thus unless the pH of a natural water is maintained at a constant value, changes in atmospheric CO_2 would affect the DIC and hydrogen-ion concentrations. Pre-industrial pure rain water would have a pH of about 5.7 if it were equilibrated with atmospheric CO_2 gas and contained no other acids and dissolved mineral solids. Although the latter conditions are questionable because the sulfuric and nitric acids from natural sources occur in the atmosphere (Chapter 8) and there are dissolved salts in rain from ocean-water spray and continental dust particles, they

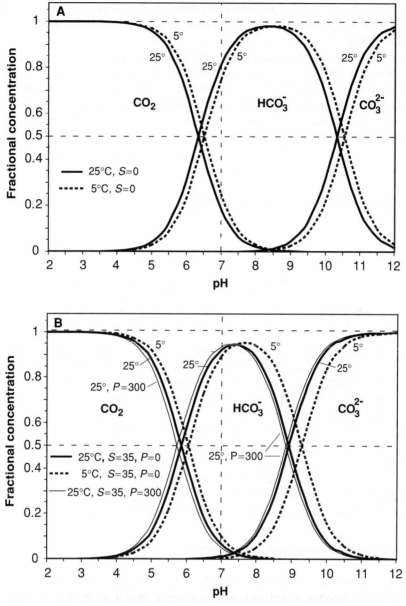

Figure 5.1   A. Fractional concentrations of dissolved inorganic carbon species in pure water (salinity $S = 0$) as a function of the pH at 5° and 25°C. B. Fractional concentrations of dissolved inorganic carbon species in seawater ($S = 35$) at 5° and 25°C at atmospheric pressure, and at the pressure of 300 bar. Carbonate equilibrium constants for pure and seawater are given in Box 5.1.

may be disregarded for the purpose of the present discussion. From Fig. 5.1A, at a pH $\approx$ 5.5, DIC contains about 10% bicarbonate ion and 90% undissociated CO_2. In the more acidic rains of the modern Industrial Age where the pH can be 4.0 to 4.5 or lower, the bicarbonate fraction in it would be down to about 1%, while undissociated CO_2 would account for most of the dissolved inorganic carbon, 99%. The relationships plotted in Fig. 5.1B indicate that in alkaline seawater, the dissociation of the bicarbonate ion is greater and, consequently, the concentration of CO_3^{2-} is higher and may promote precipitation of $CaCO_3$ in solution.

The effect of a changing atmospheric CO_2 content on the dissolved inorganic carbon concentration in continental and ocean waters is at the heart of the problem of the CO_2 transfer between the atmosphere and natural waters. A rise in the atmospheric CO_2 concentration from the pre-industrial value of 280 to 375 ppmv produces changes in the dissolved inorganic carbon concentration and pH of a water solution that is at equilibrium with the atmosphere. These changes depend on temperature and composition of the solution, as explained below, and they are shown in Table 5.1 for pure water and seawater at the mean global surface temperature of 15°C. In fresh water, the concentration of dissolved inorganic carbon (DIC) is much more sensitive to changes in atmospheric CO_2: an increase of about 30% in CO_2 is matched by a similar increase in DIC. In seawater, however, more of the CO_2 dissolves in 1 kg of solution than in fresh water, but the initially higher DIC concentration in seawater makes the fractional increase smaller, by only 2%.

Table 5.1 Computed pH and dissolved inorganic carbon (DIC) concentration at equilibrium with two partial pressures of atmospheric CO_2, for seawater and fresh water at 15°C. Parameters from Table 5.2 and equations (5.5) and (5.15).

[Alk] (mol-equivalent/kg)	P_{CO_2} (ppmv)	pH	DIC (mol/kg)
2.3×10^{-3}	280	8.195	2.078×10^{-3}
	375	8.087	2.124×10^{-3}
0	280	5.656	1.500×10^{-5}
	375	5.593	1.969×10^{-5}

Changes in a solution pH are related to changes in atmospheric CO_2 at equilibrium with the solution through a quantity known as the *alkalinity*. In a natural water containing positively charged cations and negatively charged anions, an electrical charge balance must be maintained and the sum of charges of all the positive species must be equal to that of all the negatively charged. Among the cations and anions, there are those that are conservative and their concentrations, at least to a first approximation, do not depend on the hydrogen-ion concentration, $[H^+]$. The conservative species include such common cations as Ca^{2+}, Mg^{2+}, Na^+, K^+, and the anions Cl^- and SO_4^{2-}. On the other hand, there are also cations and anions, the concentrations of which depend on H^+. In general, these include H^+ and OH^-, the ions HCO_3^- and CO_3^{2-}, the borate ion $B(OH)_4^-$, the phosphate ions $H_2PO_4^-$ and HPO_4^{2-}, ammonium NH_4^+, in some types of acidic waters the aluminum hydroxide ionic species $Al(OH)_2^+$ and $AlOH^{2+}$, and in anoxic waters HS^-. For seawater, the difference between the equivalent concentrations of the conservative cations and anions is equal to the algebraic sum of the concentrations of the H^+-dependent ions, and this difference is called the *total alkalinity* of the solution, denoted $[A_T]$:

$$2[Ca^{2+}] + 2[Mg^{2+}] + [Na^+] + [K^+] - [Cl^-] - 2[SO_4^{2-}]$$
$$= [HCO_3^-] + 2[CO_3^{2-}] + [B(OH)_4^-] + [OH^-] - [H^+] \quad \text{(mol-equivalent/kg)}$$
$$(5.11)$$

$$[A_T] = [HCO_3^-] + 2[CO_3^{2-}] + [B(OH)_4^-] + [OH^-] - [H^+]$$
$$= \frac{[CO_2]K_1'}{[H^+]}\left(1 + \frac{2K_2'}{[H^+]}\right) + \frac{B_T}{1 + [H^+]/K_B'} + \frac{K_w'}{[H^+]} - [H^+]$$
$$= [A_C] + [A_B] + [A_w] \quad \text{(mol-equivalent/kg)} \quad (5.12)$$

where the three subscripted A terms denote carbonate, borate, and water alkalinity, respectively.

As long as the concentrations of the species on the left-hand side of (5.11) are not affected by such processes as changes in input or precipitation or dissolution of minerals, their algebraic sum is constant and total alkalinity is constant. In fresh waters, where the borate concentration is low, about 1.1×10^{-6} mol B/kg (Harris, 1969) as compared to 416×10^{-6} mol B/kg in seawater, the bicarbonate ion is usually the main H^+-dependent anion contributing to alkalinity, and the relationship for A_T can be approximated by:

$$[Alk] \approx [HCO_3^-] + 2[CO_3^{2-}] + [OH^-] - [H^+]$$
$$\approx [A_C] + [A_w] \quad \text{(mol-equivalent/kg)} \quad (5.13)$$

For mean ocean water, carbonate alkalinity accounts for 94 to 97% of the total alkalinity. Alkalinity can be written as a function of the partial pressure of CO_2 at equilibrium with the solution and the $[H^+]$-ion concentration:

$$[Alk] \approx P_{CO_2}K_0'\left(\frac{K_1'}{[H^+]} + \frac{2K_1'K_2'}{[H^+]^2}\right) + \frac{K_w'}{[H^+]} - [H^+] \quad (5.14)$$

Total alkalinity of a solution, $[A_T]$, can be determined by an acidimetric titration (Drever, 1997; Edmond, 1970; Stumm and Morgan, 1981) and the pH of the solution measured at the time of sampling. From these two measurements, the P_{CO_2} value at equilibrium with the solution is calculated. Often the result differs from the atmospheric CO_2 partial pressure because of such processes as biological production or respiration of organic matter that cause the solution to be out of an instantaneous equilibrium with the atmosphere, or because not all the pH-dependent species were taken into account. As long as alkalinity is constant, a change in atmospheric CO_2 that is represented by its partial pressure P in (5.14) would require a change in $[H^+]$ that is determined from

Table 5.2 Solubility of CO_2 and dissociation constants at 15 °C for waters of two different salinities and alkalinities. Constants K for pure and seawater from Box 5.1.

[Alk] $= 2.3 \times 10^{-3}$ mol/kg	0 mol/kg
$S = 35$	0
$K_0' = 3.746 \times 10^{-2}$	4.571×10^{-2}
$K_1' = 1.119 \times 10^{-6}$	3.802×10^{-7}
$K_2' = 7.970 \times 10^{-10}$	3.715×10^{-11}
$K_w' = 2.380 \times 10^{-14}$	4.467×10^{-15}

equation (5.14), written as a cubic equation in $[H^+]$:

$$[H^+]^3 + [Alk][H^+]^2 - (K_0'K_1'P_{CO_2} + K_w')[H^+] - 2K_0'K_1'K_2'P_{CO_2} = 0 \quad (5.15)$$

where all the coefficients of $[H^+]$ and the last term on the left-hand side are constants at a given temperature and solution composition (that is, its salinity or ionic strength).

For the alkalinity that is representative of the present-day oceans, $[A_C] \approx 2.3 \times 10^{-3}$ mol-equivalent/kg. In dilute solutions of fresh waters, alkalinity may be up to 100 times lower, and in acidic solutions it becomes negative. For a limiting case of pure water at equilibrium with CO_2 gas, $[Alk] = 0$. The parameters shown in Table 5.2 are used in equation (5.15) to calculate the pH and then in (5.5) to calculate the DIC concentration.

2 CO_2 Transfer from Atmosphere to Water

2.1 Industrial-Age Increase in Atmospheric CO_2

Transfer of CO_2 by equilibration of the changing atmospheric concentration with surface water is of fundamental importance to many environmental issues. The physical dimensions of the atmosphere and water reservoirs determine to a large extent the transfer of CO_2 between them, as shown in this section that estimates the thickness of the ocean surface layer that absorbed the rising CO_2 from the atmosphere during the last 200 years of the Industrial Age.

The mass of CO_2 (n_{CO_2}) at a concentration of 280 ppmv in the pre-industrial atmosphere, that is a layer of an ideal gas 8.4 km thick at a temperature of 15°C (see Chapter 2), is:

$$n_{CO_2} = \frac{P_{CO_2}V_a}{RT} = \frac{4\pi r_0^2 P_{CO_2}}{gM_a}$$

$$= \frac{2.8 \times 10^{-4}\,\text{bar} \times 5.1 \times 10^{14}\,\text{m}^2 \times 8.4 \times 10^3\,\text{m}}{8.315 \times 10^{-5}\,\text{bar}\,\text{m}^3\,\text{mol}^{-1}\,\text{K}^{-1} \times 288.15\,\text{K}} = 5.0 \times 10^{16}\,\text{mol or } 600\,\text{Gt C}$$

$$(5.16)$$

The net increase in atmospheric CO_2 due to fossil fuel burning and land-use changes was from 280 ppmv CO_2 to 370 ppmv by the end of the 20th century. At the CO_2 concentration of 375 ppmv and the same temperature, the present-day mass is $375/280 = 1.34$ times greater, which makes it 6.7×10^{16} mol or 804 Gt C. The increase of 1.7×10^{16} mol CO_2 or 204 Gt C since pre-industrial time is the net increase after all the inputs and removal into the ocean and land sinks have been accounted for. A balance of CO_2 additions to the atmosphere, land, and ocean up to the year 2000 is shown in Fig. 11.5: of a total 480 Gt C (4×10^{16} mol C) of input, about 200 Gt C (1.7×10^{16} mol) accumulated in the atmosphere, and about 130 Gt (1.1×10^{16} mol) were taken up by the ocean. The thickness of the surface ocean layer that takes up CO_2 from the atmosphere depends on the mass of carbon transferred from the atmosphere $\Delta n_a \approx 1 \times 10^{16}$ mol C, and the increase in DIC concentration within a water layer of mass M_w:

$$\Delta n_a = \Delta\,[DIC] \times M_w \quad (mol\ C) \qquad (5.17)$$

Equation (5.17) is a general relationship that gives the mass change of DIC in a body of water, without consideration of its causes. However, the computed thickness of the surface water layer depends on an assumption that CO_2 was gained by the ocean in a gas-solution equilibrium while the total alkalinity of ocean water remained constant, and on the water temperature.

2.2   Partitioning of CO_2 between Atmosphere and Water

The simplest approximation to the distribution of CO_2 between the atmosphere and surface ocean water is a solubility equilibrium at a certain temperature:

$$K_0' = \frac{[CO_2]}{P_{CO_2}} \qquad (5.18)$$

where $[CO_2]$ is the concentration of dissolved CO_2 species (mol/kg), P_{CO_2} is atmospheric partial pressure of CO_2 (bar) at equilibrium with the solution, and K_0' is the temperature-dependent solubility coefficient (mol kg^{-1} bar^{-1}, Box 5.1). The concept of a chemical equilibrium is only an approximation to the atmosphere in contact with surface ocean water because of several factors: ocean surface temperature varies with latitude and CO_2 solubility increases at lower temperatures and lower salinities; oceanic regions of deeper water upwelling are often sources of CO_2 emissions to the atmosphere, whereas the downwelling areas, such as the high-latitude North Atlantic, are sinks of atmospheric CO_2 transporting it into the deeper waters; and the global balance of directions of CO_2 flows across the sea-air interface shows large parts of the Northern and Southern hemispheres as CO_2 sinks and regions in the lower, warmer latitudes as sources (Chester, 2000; Takahashi, 1989). In addition, mineralization or oxidation of organic matter brought from land to the surface ocean is a source of additional CO_2 in ocean water; deposition of $CaCO_3$ by inorganic or biological processes releases CO_2 into ocean water; and primary production in surface waters, consuming dissolved CO_2, may create at least temporarily a disequilibrium between the atmosphere and ocean water.

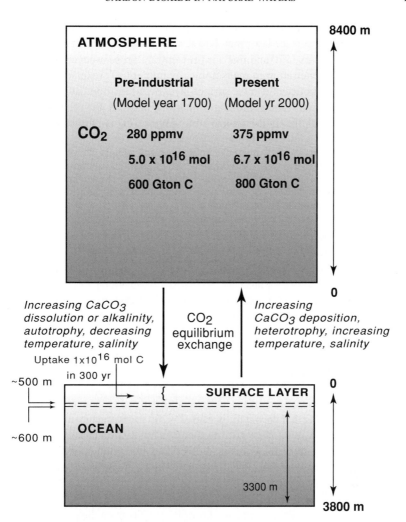

Figure 5.2 Carbon changes in the atmosphere and surface ocean in the last 300 years of industrial time. Thicknesses of the two reservoirs (8400 and 3800 m) shown approximately to scale. Processes that affect the direction of CO_2 exchange between the atmosphere and surface ocean are indicated next to the flux arrows. Thickness of surface ocean layer, 500 to 600 m, that is equilibrated with atmospheric CO_2 is explained in the text.

For a simplified model of CO_2 uptake by the surface ocean, a water layer of some thickness equilibrates with the atmosphere. This picture does not take into account the water exchange between the deep and surface ocean: the mixing time of the global ocean is usually taken as an order of magnitude of 10^3 years, and 1600 years is an often cited value derived from oceanographic models and, later, the measurements of ^{14}C activity (half-life 5715 years) in CO_2 of atmospheric origin in the surface and deep ocean water layers (Albarède, 1995; Broecker, 1974; Munk, 1966; Stumm and Morgan,

1981). Depending on the chosen thickness of the deep and surface layers, the water residence time in the surface layer may be a few centuries or shorter. At a 20-year time scale, between the years 1970 and 1991, in the Central Pacific water column over a wide range of latitudes, penetration of the isotopically lighter CO_2 (explained in Chapter 6, Section 5) was most noticeable in the surface layer between about 300 and 800 m depth (Quay et al., 1992). With the preceding limitations in mind, the thickness of a water layer equilibrated with the industrial-age atmospheric CO_2 is estimated below.

Table 5.3   Changes in the pH and DIC in seawater of salinity 35 at the pre-industrial and present-day atmospheric CO_2 concentrations, at 25 and 5 °C, at constant total alkalinity $A_T = 2.34 \times 10^{-3}$ mol-equivalent/kg. Note the strong effect of temperature on the carbonate ion concentration and the degree of ocean water saturation with respect to calcite (Ω), at the Ca^{2+}-ion concentration of 1.028×10^{-2} mol/kg.

Temperature	P_{CO_2} (ppmv)	$[CO_2]$ (mol/kg)	pH	$[DIC]$ (mol/kg)	$[CO_3^{2-}]$ (mol/kg)	Ω
25°C	280	7.950E-06	8.171	1.937E-03	2.890E-04	7.0
	375	1.065E-05	8.070	1.994E-03	2.430E-04	5.8
5°C	280	1.460E-05	8.181	2.126E-03	1.541E-04	3.7
	375	1.955E-05	8.070	2.175E-03	1.240E-04	3.0

For a rise in atmospheric P_{CO_2} from 280 to 375 ppmv, the dissolved $[CO_2]$ concentration also increases. The higher concentration of dissolved CO_2 results in an increase in total dissolved inorganic carbon (DIC), changes in the $[H^+]$-ion concentration, and changes in the concentrations of the individual dissolved carbonate species $[HCO_3^-]$ and $[CO_3^{2-}]$. Assuming that during the period of ≤ 200 years, the mean total alkalinity of surface ocean water $A_T = 2.34 \times 10^{-3}$ mol-equivalent/kg (Takahashi, 1989) remained constant (that is, no significant amounts of carbonate precipitation or dissolution or addition of alkalinity from land occurred), the new values of surface ocean water pH and DIC at the higher atmospheric P_{CO_2} of 375 ppmv are given in Table 5.3. An increase in dissolved CO_2 at 5°C is greater than at 25°C because of the higher solubility of CO_2 at lower temperatures. A 34% increase in atmospheric CO_2 from the pre-industrial value of 280 to 375 ppmv in the present corresponds to an increase of DIC by 2 to 3% and lowering of the pH by about 0.1 unit. This increase represents the total mass of carbon transferred from the atmosphere as CO_2 to ocean water, where the added CO_2 causes changes in the concentrations of dissolved carbonate species. Thus only a fraction of atmospheric CO_2 gas becomes dissolved $[CO_2]$ at equilibrium with the atmosphere, whereas most of the increase goes into the bicarbonate ion, increasing DIC by the reaction $CO_2 + CO_3^{2-} + H_2O = 2HCO_3^-$.

The increase in DIC at 25° and 5°C is, from Table 5.3:

$$\Delta[DIC] = [DIC]_{P_{CO_2}=370} - [DIC]_{P_{CO_2}=280}$$

$$= 1.994 \times 10^{-3} - 1.937 \times 10^{-3} = 5.69 \times 10^{-5} \text{ mol C/kg at } 25°C$$

$$= 2.175 \times 10^{-3} - 2.126 \times 10^{-3} = 4.97 \times 10^{-5} \text{ mol C/kg at } 5°C$$

Thus the thickness of the surface ocean layer, h, where dissolved CO_2 is equilibrated with the atmosphere, is from (5.17):

$$h_{25°C} = (1 \times 10^{16} \, \text{mol C})/(5.69 \times 10^{-5} \times 1027 \times 3.61 \times 10^{14}) \approx 470 \, \text{m}$$
$$h_{5°C} = (1 \times 10^{16} \, \text{mol C})/(4.97 \times 10^{-5} \times 1027 \times 3.61 \times 10^{14}) \approx 540 \, \text{m}$$

where the mean density of ocean water is taken as $1027 \, \text{kg/m}^3$ and ocean surface area is $3.61 \times 10^{14} \, \text{m}^2$. The above results for a 500 m to 600 m-thick surface ocean layer, shown in Fig. 5.2, are constrained by an external estimate of the carbon mass transferred to the ocean in industrial time, about $1 \times 10^{16} \, \text{mol C}$, and by our conceptual model that assumes an equilibrium between atmospheric and dissolved CO_2 in a water layer where total alkalinity remained constant. Observations of the CO_2-carbonic acid system in the world's oceans show that anthropogenic CO_2 has penetrated to these depths, and deeper due to oceanic circulation, actually leading to enhanced dissolution of sinking $CaCO_3$ particles (Feely et al., 2004; Orr et al., 2005).

3 Calcite and Aragonite in Natural Waters

The mineral properties of the two common polymorphs of $CaCO_3$, calcite and aragonite, are described in Chapter 4. Their fields of stability in the outer shell of the Earth, and dissolution and precipitation rates in solution are also discussed in that chapter. This and the following sections deal with the solubilities of calcite and aragonite in fresh and sea waters and with the degree of saturation of rivers, lakes, and ocean water with respect to these minerals.

The solubility of calcite and aragonite in pure water increases with decreasing temperature, as shown in Fig. 5.3A. In the range from 25° to 5°C, the increase in K_{sp} is 20 to 25%. However, in ocean water of normal salinity, the apparent solubility constants of the two minerals, K'_{sp}, change much less with temperature, as shown in Fig. 5.3B. The transition of the curve shape for the calcite solubility K'_{cal} at $S = 0$ to that at $S = 35$ is gradual, and from 25° to 5°C the solubility constant increases by a factor of only 1.004 or 0.4%. Furthermore, K'_{cal} shows a solubility maximum near 12.5°C that is by a factor of 1.01 or 1% higher than the solubility at 25°C, then declining from 12.5 to 5°C. The small solubility maximum also exists at salinities higher than normal ocean water, to the upper limit of $S = 44$ for the K'_{cal} equation (calcite solubility equations in Mucci, 1983, and Zeebe and Wolf-Gladrow, 2001; Box 5.1).

At a higher pressure of 500 bar, corresponding to a depth of about 5000 m in the ocean, the solubilities of the two carbonate minerals increase significantly (Fig. 5.3B). For calcite, K'_{cal} increases by a factor of 2 at 25°C and by a higher factor of 2.6 at 5°C. Thus the pressure increase with depth rather than the temperature decrease is one of the main factors controlling the increased calcite solubility with depth, as is discussed in more detail in the next section.

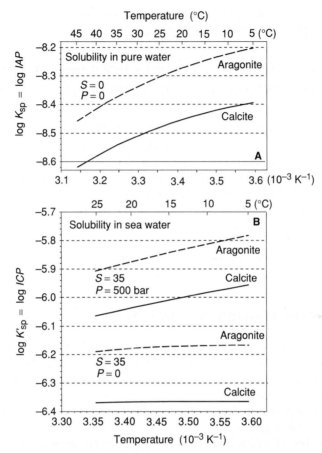

Figure 5.3   A. Solubility of calcite and aragonite in pure water (of salinity $S = 0$) between 5°
and 45°C (Plummer and Busenberg, 1982). Box 5.1. B. Solubility of calcite and aragonite in
normal sea water ($S = 35$) at atmospheric pressure ($P = 0$) and at depth of approximately 5000 m
in the ocean ($P = 500$ bar or 50 MPa) (Millero, 1995; Mucci, 1983; Zeebe and Wolf-Gladrow,
2001).

4   Degree of Saturation with Respect to Carbonate Minerals

A measure of the degree of saturation of a natural water with respect to calcite or
aragonite is a quotient

$$\Omega = \frac{[Ca^{2+}][CO_3^{2-}]}{K'_{min}} \tag{5.19}$$

where K'_{min} is the apparent solubility product of the mineral, K'_{cal} or K'_{arag}, when
concentrations of the ions are used in the numerator. A thermodynamic solubility
product (at zero ionic strength) replaces K'_{min} if thermodynamic activities rather than

concentrations are used:

$$\Omega = a_{Ca^{2+}} a_{CO_3^{2-}} / K_{min}$$

Values of $\Omega > 1$ indicate supersaturation and $\Omega < 1$ is undersaturation. For calcite, that is slightly less soluble than aragonite at Earth's surface conditions (Chapter 4), a measure of the degree of saturation known as the Calcite Saturation Index (CSI) is also used for fresh waters where the values of Ω are small:

$$CSI = \log \Omega \qquad (5.20)$$

At saturation, $\Omega = 1$ and CSI $= 0$. Supersaturation of oceanic and some continental waters with respect to calcite and aragonite can exist as a stable feature of the water body, as mentioned previously and shown in the next sections.

4.1 Rivers

Rivers vary in their saturation state with respect to calcite, from undersaturation to supersaturation, depending on their water composition that is in part a function of the drainage basin lithology. Fresh-water lakes, in particular those containing relatively high concentrations of calcium and carbonate ions and characterized by alkaline pH, can be supersaturated with respect to calcite (Fig. 5.4). The trend of the CSI values for a large number of cold- and warm-climate rivers holds for a variety of terranes, from sediments of the Paleozoic and Mesozoic platform to deformed rocks of a tectonic collision zone and crystalline shield. The mean undersaturation with respect to calcite of the rivers plotted in Fig. 5.4 is CSI $= -1.1$ and the mean pH is 7.6. At these values, the internal P_{CO_2} of a mean river is considerably higher than the atmospheric P_{CO_2} of about 380 ppmv, which is due to the respiration of organic matter in soils and the production of CO_2 (Chapter 6).

4.2 Ocean Water

Surface ocean water is generally supersaturated with respect to the minerals aragonite and calcite ($CaCO_3$), as defined in (5.19). As shown in Table 5.3, increasing temperature has a strong effect on the degree of calcite saturation, primarily by the lowering of the CO_3^{2-}-ion concentration. Deeper-ocean waters are undersaturated with respect to aragonite and calcite, and the saturation depth is shallower for aragonite than for calcite because aragonite is more soluble and $K'_{arag} > K'_{cal}$ (Box 5.1). The decrease in a degree of saturation with increasing depth is primarily due to a decrease in the CO_3^{2-}-ion concentration at the lower temperature of deep water, and an increase in the value of K'_{cal} with increasing pressure. The depth where supersaturation changes to undersaturation varies among the major oceans (Atlantic, Indian, and Pacific oceans), and with latitude and location within an ocean (Broecker and Peng, 1982; Morse and Mackenzie, 1990). A generalized example of a change with depth of the degree of saturation of ocean water with respect to aragonite and calcite is shown in Fig. 5.5. The saturation depth

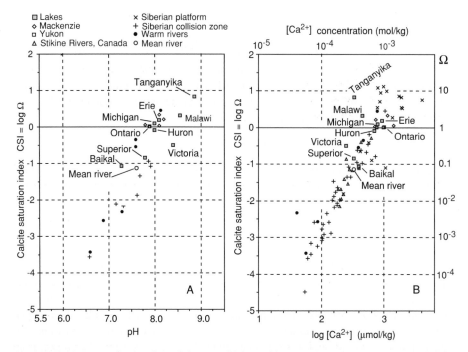

Figure 5.4   Degree of saturation of rivers and lakes with respect to calcite as a function of the pH (A) and calcium concentration (B). Northern rivers from Millot *et al.* (2003), Gaillardet *et al.* (2003), Huh and Edmond (1999), and Huh *et al.* (1998a, b). Warm rivers: Amazon (Gaillardet *et al.*, 1997), Orinoco (Edmond *et al.*, 1996), Ganges and Brahmaputra (Galy and France-Lanord, 1999), Indus (Karim and Veizer, 2000), Niger (Tardy *et al.*, 2004). Data for North American Great Lakes from Weiler and Chawla (1969), East African lakes from Oliver (2002), and Baikal from Votintsev (1993). Mean river values are for the data plotted in the figure.

of the more soluble aragonite is at about 500 m and that of calcite is at 3000 m. As mentioned previously, temperature decrease has little effect on the solubility of calcite in ocean water of normal salinity, but it has a strong effect on the lowering of the CO_3^{2-} concentration, and pressure also increases the calcite solubility. The undersaturation of ocean water with respect to aragonite begins at a shallower depth than undersaturation with respect to calcite (Fig. 5.5D), which accounts for the dissolution of $CaCO_3$ particles settling from the surface ocean layer and the return of calcium and carbonate to solution in this pump-like process. The dissolution rates of calcite and aragonite increase significantly at depth horizons in the ocean called the calcite and aragonite lysoclines, respectively.

The surface water in the Pacific is more strongly supersaturated with respect to calcite and aragonite in the latitudinal belt of N30° to S30° (Fig. 5.6) and the saturation declines in the colder waters of the higher latitudes in both hemispheres.

The carbonate saturation trends that are shown in Figs. 5.5 and 5.6 do not necessarily represent the shallow water regions of the coastal ocean and banks and reefs where

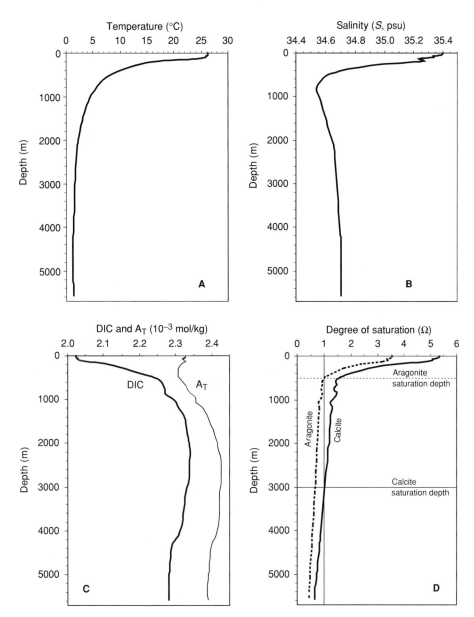

Figure 5.5   Central Equatorial Pacific, W168.75°, S0° (WOCE Section P15S/14S, NOAA CGC-96, 6 March 1996, http://cdiac.ornl.gov/oceans/RepeatSections/clivar_ p15s.html). Water depths calculated from pressure (Chapman, 2003). A. Temperature. B. Salinity. C. Dissolved inorganic carbon (DIC) and total alkalinity (A_T). D. Degree of calcite and aragonite saturation (Ω) from *T-S-P* and carbonate-system parameters (Zeebe and Wolf-Gladrow, 2001).

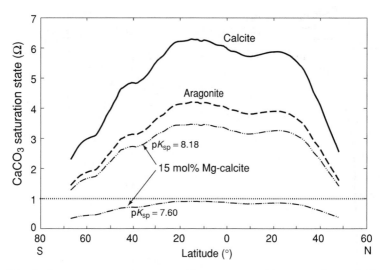

Figure 5.6   Surface water saturation state with respect to calcite, aragonite, and 15 mol % magnesian calcite as a function of latitude in the Pacific Ocean. Saturation states were calculated from *in situ* temperature and salinity at WOCE section P15 (CLIVAR and Carbon Hydrographic Data Office; http://whpo.ucsd.edu/index.htm). The different values of the magnesian calcite solubility result in significantly different estimates of the water saturation state with respect to $Ca_{0.85}Mg_{0.15}CO_3$ ($pK_{sp} = 8.18$ from Bischoff *et al.*, 1987, 1993; $pK_{sp} = 7.60$ from Plummer and Mackenzie, 1974; Mg-calcite solubility is at $25°C$ because of a lack of solubility data at other temperatures). After Andersson *et al.* (2005).

significant changes in the degree of carbonate saturation state can occur at a diurnal or diel time scale due to changes in the rates of photosynthesis and total respiration. During the day when photosynthesis is most active, the carbonate saturation state of the water will increase as CO_2 is removed from the water, whereas at night when respiration occurs and CO_2 is added to the water, the carbonate saturation state will fall. Such changes in water chemistry due to biological activity are important to the determination of the net calcification and remineralization rates of aquatic ecosystems (Chapter 9).

5   CO$_2$ Phases: Gas, Liquid, Hydrate, Ice

5.1   Phase Diagram

At the atmospheric pressure, CO_2 is gas over a wide range of temperatures, but at higher pressures it becomes solid at lower temperatures and liquid at higher temperatures (Fig. 5.7). Above the critical temperature and pressure of $31°C$ and 74 bar (Table 2.4), CO_2 exists as a superheated fluid. Occurrences of solid CO_2, also called dry ice when used in refrigeration, are mostly outside the conditions of the Earth's surface environment. Several solid CO_2 phases of higher density have been discovered at very high pressures (Yoo *et al.*, 1999, 2002).

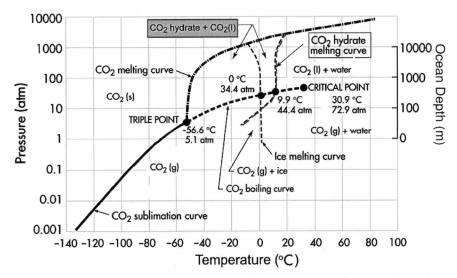

Figure 5.7 CO_2 and H_2O phase diagram in the field of temperature and pressure that includes the Earth's surface, atmosphere, oceans, and the upper lithosphere to depth of about 30 km. The phase boundaries of CO_2 hydrates are accentuated by dashed lines (from Murray et al., 1996, original part of the figure by permission of C. Nicholas Murray; for CO_2 gas-liquid-solid phase boundaries, see also phase diagrams by ChemicaLogic Corporation, Burlington, Massachusetts, 2004, http://www.chemicalogic.com/download/phase_diagram.html).

Liquid CO_2 is a stable phase between $-56.6°C$ and $+30.9°C$ and pressures between about 5 and 74 bar, and it can exist under the temperature and pressure conditions that occur in parts of the sedimentary lithosphere and ocean water. The phase boundary between CO_2 hydrate and gas (Fig. 5.7) shows that the hydrate decomposes to liquid water and gas in the temperature range from about 0 to 10°C and pressures between 10 and 44 atm. At greater pressures, the stability field of CO_2 hydrate overlaps that of liquid CO_2 below 10°C and the hydrate melts at approximately 10°C. The stability fields of the CO_2 and H_2O phases that are shown in Fig. 5.7 are those of the pure substances and do not take into account the chemical reactions between the gas, liquid, and hydrate with either pure or seawater.

CO_2 hydrates are compounds of composition $CO_2 \cdot nH_2O$, where the gas molecule is contained within the cavity of a cage-like crystal structure of the host H_2O molecules that form a regular and a slightly distorted dodecahedral framework (Fig. 5.8). In appearance, hydrates are crystals or fluffy masses as found in their natural occurrence in marine sediments. The first gas hydrate, chlorine hydrate, was discovered by Sir Humphrey Davy in 1810 (Sloan, 1998). In CO_2 hydrate, the hydration number n varies from $n = 5.75$ for the closest packed lattice to a theoretical limit of $n = 6.67$ when all the 20 vertices of the pentagonal dodecahedron are occupied by H_2O molecules. Six large cages and two small cages with 46 water molecules form the unit cell of the CO_2 hydrate, where some of the cages are unoccupied by CO_2, resulting in a mean composition of $CO_2 \cdot 6.5H_2O$ (Udachin, 2001).

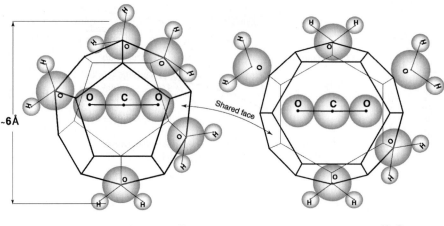

A Pentagonal dodecahedron (5^{12})   **B Tetrakaidecahedron (5^{12} 6^2)**

Figure 5.8   Structure of CO_2 hydrate made of small and large cages (after Sloan, 1998; Udachin *et al.*, 2001, 2002; and J. A. Ripmeester, *personal communication*, 2005). A. A small cage, a pentagonal dodecahedron (12 regular pentagons, 5^{12}), containing one CO_2 molecule. B. A large cage with one CO_2 molecule, a tetrakaidecahedron (12 pentagons and 2 hexagons, 5^{12}6^2). Water molecules are shown as occupying only some of the vertices of the two cages. Molecular dimensions of CO_2 gas and H_2O liquid are approximately to scale. The name tetrakaidecahedron is also used for other fourteen-faced polyhedra, such as the truncated cube (6 octagons and 8 triangles, 8^{6}3^8) and truncated octahedron (8 hexagons and 6 squares, 6^{8}4^6), the latter known as the Kelvin cell.

Among the naturally occurring hydrates, methane hydrate is by far the more abundant in ocean-floor sediments and at a number of continental sites in high latitudes, including permafrost. Hydrates decompose above some temperature or below some pressure, liberating their gas, such as carbon dioxide or methane, both of which are greenhouse gases. Decomposition of one mol of hydrate $CO_2 \cdot 5.75H_2O$, of density about 1.1 g/cm^3, would release one mol of CO_2, the volume of which in the atmosphere at 15°C would be about 180 times the volume of the hydrate. With respect to the modern environmental problem of global warming (Chapter 11), there is some concern that as the climate warms, some methane gas hydrates could be destabilized and lead to a release or rapid burst of CH_4 to the atmosphere. The regions of stability of CO_2 hydrates in the ocean and continental crust are discussed in the next section.

5.2 CO_2 in Oceanic Water Column and Continental Crust

The environmental interest in the hydrate and liquid carbon dioxide is focused on two main issues: one is a potential release of carbon dioxide and methane from their hydrates occurring in oceanic sediments in case of a rise in ocean-water temperature due to global warming (e.g., Brewer *et al.*, 1999); the other is the potential disposal of the CO_2 industrial emissions in a liquid or hydrate form in the deep ocean (e.g., Murray

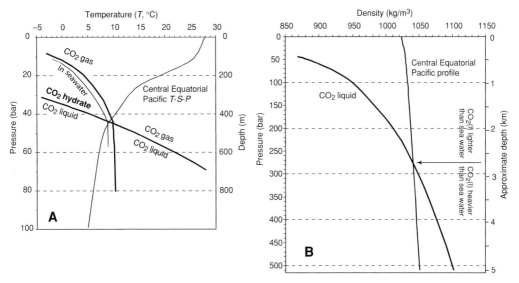

Figure 5.9   A. Temperature and pressure profile in the Central Equatorial Pacific, W180°, N0° (from Chapman, 2003; Gordon, 2000). CO_2 gas-liquid phase boundary (from Lemmon *et al.*, 2003) and hydrate-gas boundary (from Brewer *et al.*, 1999; E. T. Peltzer, *personal communication*, 2005; Murray *et al.*, 1999; Ohgaki *et al.*, 1993; Sloan, 1998). B. Ocean water density calculated from temperature and salinity data (Chapman, 2003; Gordon, 2000). CO_2 liquid density, based on the adiabatic volume compressibility coefficient (Lemmon *et al.*, 2003; Span and Wagner, 1996) is greater than ocean-water density below 2800 m depth. Details in the text.

et al., 1999; Ormerod *et al.*, 2002) and CO_2 injection or its formation in deep wells used for disposal of industrial acidic wastes where acid may react with limestone. At the present temperature and pressure distribution with depth in the oceanic water column (Fig. 5.9A), exemplified by the Central Equatorial Pacific, CO_2 hydrate is stable below about 400 m, although in a colder water of about 1°C it may be stable at shallower depths of about 120 m. These depths include the surface ocean water, the continental shelf, and the upper part of the continental slope (Chapter 9) at the higher latitudes that are particularly sensitive to a global temperature rise that could cause decomposition of CO_2 hydrate and release of CO_2 from the sediments. There are reports of liquid CO_2 reacting with seawater and forming a protective coating of CO_2 hydrate that retards dissolution and, on the other hand, reports of dissolution of hydrate particles in seawater. At different temperatures and depths between 800 and 3600 m, dissolution rates of liquid CO_2 have been reported between 1.7×10^{-6} and $3 \times 10^{-6} \, \mathrm{mol \, cm^{-2} \, s^{-1}}$ (Fer and Haugan, 2003; Peltzer *et al.*, 2000). At these rates, a 100 m thick layer of liquid CO_2, of density 1060 kg/m³ (Fig. 5.9B), would dissolve in 3 to 4 years.

An important consideration in the carbon dioxide sequestration problem in the deep ocean is the density of CO_2 hydrate and liquid. The hydrate density is between 1049 to 1120 kg/m³, which is similar to, or heavier than, the seawater density at about

5000 m and also heavier than liquid CO_2 (Aya $et\ al.$, 1997; Teng $et\ al.$, 1996; Udachin $et\ al.$, 2001). Because the compressibility of liquid CO_2 is greater than that of normal seawater (the coefficient of adiabatic volume compressibility; Lemmon $et\ al.$, 2003; Millero, 2001; Span and Wagner, 1996), the density of liquid CO_2 increases faster with depth and becomes greater than that of seawater at about 2800 m, in the sample density profile for the Central Equatorial Pacific shown in Fig. 5.9B. If liquid CO_2 is injected into a deep ocean or fills depressions on the deep ocean floor, different model calculations give the times of dissolution and CO_2 return to the atmosphere of the order of 10^2 to 10^3 years, depending on the mass of liquid CO_2 sequestered and the assumed chemical kinetic and physical transport parameters in the disposal area (Bacastow and Dewey, 1996; Fer and Haugan, 2003). These projections depend on the mass of the injected liquid CO_2, the volume of ocean water that initially receives the disposed CO_2, and the dynamics of water exchange between the deep and surface ocean. One may also expect changes in the carbonate chemistry of ocean water if liquid CO_2 dissolved, adding CO_2 to solution and thereby lowering its pH and affecting the solubility of calcium carbonate minerals in the water column and on the ocean floor. The effect on the benthic biota has been little studied.

As far as the issue of the potential disposal of industrial CO_2 in the continental crust is concerned, there are sections where temperature and pressure are within the stability field of liquid CO_2, suggesting that the gas may be liquefied there or remain as a liquid phase (Lerman $et\ al.$, 1996; Pruess, 2004; Van der Meer, 1993). These conditions should be below the CO_2 critical temperature of 31°C and the critical pressure of 74 bar (Fig. 5.7, Table 2.4). From the mean Earth's surface temperature of 15°C and the geothermal gradient in the continental lithosphere of 2 to 3°C/100 m, a temperature of 30°C may occur at a depth between 500 and 750 m. From about 580 m depth and greater, a hydrostatic pressure of a saline brine column of density 1300 kg/m^3 would meet the pressure requirement for the existence of liquid CO_2. However, under a lithostatic pressure gradient of 27 bar/100 m due to a higher rock density of 2700 kg/m^3, the pressure in the 500 to 750 m depth interval would be greater and well within the stability field of liquid CO_2. Thus the regions of stability of liquid CO_2 in the continental sediments and crust are primarily limited by the temperatures at depth.

Another clathrate compound, methane hydrate ($CH_4 \cdot nH_2O$), occurs in oceanic sediments and in the permafrost zone of the Eastern and Western Hemispheres. It is much more abundant than the CO_2 hydrate (Fig. 1.5), and its occurrences in continental slope sediments as well as numerous methane leaks from the ocean floor around parts of the continents are a subject of continuing intensive research because methane gas is a fuel and an energy source and it is also a strong greenhouse gas that may increase the warming of the atmosphere. Buffett (2000) provided an extensive review of the properties and geological occurrences of CH_4 hydrate, its formation in oceanic sediments, and the possible mechanisms of its release to the water and, subsequently, to the atmosphere. The source of methane is the organic matter in sediments that can be converted to CH_4 by bacteria or by thermal reactions at greater depths. H_2O that goes into the formation of the hydrate may increase the salinity of the pore water and, conversely, when the hydrate de-

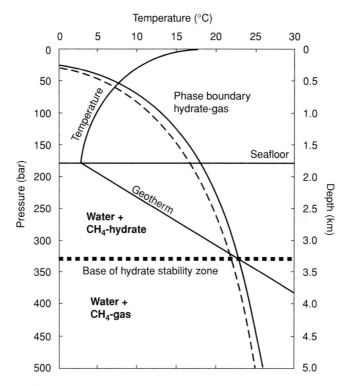

Figure 5.10   Methane hydrate-gas phase boundary and a generalized temperature-depth profile in ocean water and sediment below the sea floor at 1,800 m depth. Solid phase-boundary line is for pure water, dashed line is for seawater (after Buffett and Archer, 2004, by permission of the authors).

composes, pore-water salinity may decrease. The phase boundary between CH_4 hydrate and CH_4 gas is shown in Fig. 5.10 within a framework of ocean water and sediments.

The phase boundary between the hydrate and gas in seawater for the CH_4 hydrate lies slightly below the temperatures in pure water. However, at depths shallower than about 450 m, the CO_2-hydrate stability field (Fig. 5.9A) is slightly bigger than that of the CH_4 hydrate (Fig. 5.10): the CO_2-hydrate field extends to a shallower depth (lower pressure) and slightly higher temperature. This suggests that in colder shallower water, the conditions may be suitable for the occurence of CO_2 hydrates, but not of CH_4 hydrates. Without considering the hydrates, methane is much less soluble in water than carbon dioxide. At the pressure of $P = 1$ bar, the solubility of CO_2 in seawater between $0°$ and $30°C$ is 0.062 to 0.025 mol/kg (Box 5.1). For CH_4 in pure water, in the same temperature range, the solubility is 0.0023 to 0.0012 mol/kg and it decreases by about 22% in a solution of 1 molal NaCl, to 0.0018 and 0.0009 mol/kg solution (Duan et al., 1992a). The low solubility of methane in seawater and its low atmospheric concentration are the conditions that favor transport from seawater to the atmosphere, where, in the absence of a mechanism of faster oxidation, the CH_4 concentration may increase. At present, the main mechanism of the methane oxidation in the atmosphere is a reaction with the hydroxyl radical ($OH^\bullet$), that makes an average residence time of methane between 8 and 9 years (Jacob, 2003).

6   Air-Sea CO_2 Exchange Due to Carbonate and Organic Carbon Formation

6.1   Background

Precipitation of $CaCO_3$ minerals from supersaturated seawater is accompanied by a decrease in total alkalinity and an increase of dissolved CO_2 and H^+-ion concentrations or a lowering of the solution pH (e.g., Schoonmaker, 1981; Tribble and Mackenzie, 1998; Wollast et al., 1980). Mineral precipitation and biological production of organic matter in the surface water of the coastal zone and open ocean results in removal of dissolved inorganic carbon (DIC), redistribution of the concentrations of dissolved carbonate species, and changes in the relative concentration of CO_2 in solution. Such changes affect the CO_2 flux across the air-sea interface and may result in changes in the CO_2 concentrations in other reservoirs. In a global ocean, release of CO_2 due to precipitation of inorganic or biogenic $CaCO_3$ must be compensated by restoration of dissolved inorganic carbon and calcium in water, although there probably were variations in the carbon content and alkalinity of ocean water through geologic time (e.g., Broecker, 2002; Hardie, 1996; Locklair and Lerman, 2005; Milliman, 1993; Tyrrell and Zeebe, 2004). However, CO_2 exchange between the atmosphere and the coastal and open surface ocean is only in part controlled by carbonate mineral precipitation. Other major mechanisms in the system are production, respiration, remineralization, and net sedimentary storage of organic matter in the ocean; storage and release of carbon from the biomass and humus on land that affect atmospheric CO_2 levels and hence CO_2 exchange across the air-sea interface; physical processes, for example oceanic circulation that generally removes CO_2 from the atmosphere in the downwelling zones and returns it in the upwelling regions; and the cycle of carbonate minerals settling from the euphotic zone and their dissolution, deposition, and accumulation in deep-sea and shallow-water sediments. These processes have been discussed in the literature by many authors (e.g., Broecker and Henderson, 1998; Broecker and Peng, 1982; Maier-Reimer, 1993; Maier-Reimer and Hasselmann, 1987; Milliman et al., 1999; Morse and Mackenzie, 1990; Sigman and Boyle, 2000; Woodwell, 1995).

Theoretical analysis of the CO_2 production in surface ocean water and its exchange with the atmosphere has been advanced by a number of investigators using different approaches. Wanninkhof (1992) related the seawater-air CO_2 flux to wind velocity. Smith (1985) analyzed the CO_2 exchange flux at the air-water interface as a gas diffusional process across a boundary layer and his results show an increase in dissolved CO_2 during $CaCO_3$ precipitation in a closed system and subsequent transfer of dissolved inorganic carbon as CO_2 to the atmosphere. Smith's (1985) model analysis was for ocean water of normal salinity at 25°C, a water layer about 1 m thick, ambient atmospheric P_{CO_2} of 340 ppmv, and $CaCO_3$ deposition at a rate equivalent to global precipitation of carbonate at about 1.3×10^{12} mol C/yr, which is significantly lower than $CaCO_3$ production or net storage rates in the ocean as a whole. Frankignoulle *et al.* (1994) estimated the release of CO_2 from ocean water in response to $CaCO_3$ precipitation by means of an analytic function $\Psi =$ (released CO_2)/(precipitated CO_3^{2-}). Their results show an increase in the CO_2 fraction released with decreasing temperature from 25° to 5°C and with increasing atmospheric P_{CO_2} in the range from 290 to 1000 ppmv, which they indentified as a positive feedback of atmospheric CO_2 on CO_2 emissions from surface water. The study of Frankignoulle *et al.* (1994) deals with an atmosphere of constant CO_2 and gives no data on the mass or rate of $CaCO_3$ precipitation. To make the analysis of the relationships between CO_2 release and $CaCO_3$ and C_{org} deposition in this chapter clear and reproducible, the essential computational steps leading to the results are outlined below.

The precipitation of $CaCO_3$ removes 1 mol C from solution and produces 1 mol CO_2, while dissolution consumes CO_2 and produces HCO_3^-:

$$\overset{\text{Precipitation}}{Ca^{2+} + 2HCO_3^- \underset{\text{Dissolution}}{\rightleftharpoons} CaCO_3 + H_2O + CO_2} \tag{5.21}$$

Primary production or net primary production ($NPP = GPP - R_{auto} \geq 0$) consumes CO_2 and respiration or remineralization of organic matter produces CO_2:

$$\overset{\text{Gross Primary Production (GPP)}}{CO_2 + H_2O \underset{\text{Autorespiration } (R_{auto})}{\rightleftharpoons} CH_2O + O_2} \tag{5.22}$$

Net primary production in the euphotic zone, removing dissolved CO_2, essentially competes with the carbonate precipitation process generating CO_2 that can be emitted to the atmosphere, as follows from the sum of reactions (5.21) and (5.22):

$$Ca^{2+} + 2HCO_3^- \rightleftharpoons CaCO_3 + CH_2O + O_2 \tag{5.23}$$

Some of the calcium carbonate produced in surface coastal and open ocean water, perhaps as much as 60%, dissolves in the water column and in sediments driven to a significant extent by remineralization of organic matter sinking in the ocean or deposited in sediments that produces CO_2 (Archer and Maier-Reimer, 1994; Emerson and Bender, 1981; Mackenzie *et al.*, 2004; Milliman, 1993; Milliman *et al.*, 1999; Morse and Mackenzie, 1990; Moulin *et al.*, 1985; Wollast, 1994). These processes result in return of some of the carbon taken up in $CaCO_3$ and organic matter production back to ocean water. Most of the calcium carbonate produced in the more recent geological past is primarily biogenic in origin, and lesser quantities of $CaCO_3$ in the modern oceans and for part of the geologic past have been produced as cements in sediments and as whitings and ooids (e.g., Bathurst, 1974; Morse and Mackenzie, 1990).

Ties between biological calcification and primary production in various ecosystems, and their effects on the carbon fluxes, are a subject of intensive research (e.g., Andersson and Mackenzie, 2004; Gattuso *et al.*, 1995; Gattuso and Buddemeier, 2000; Suzuki *et al.*, 1998). In a strongly autotrophic ecosystem, CO_2 production by carbonate pre-cipitation may be counteracted by uptake of the generated CO_2 by organic productivity, resulting in a lower transfer of CO_2 from water to the atmosphere or in a transfer in the opposite direction, as shown in reactions (5.21)–(5.23). Removal of dissolved inorganic carbon (DIC) by primary production also lowers the degree of saturation of surface ocean water with respect to carbonate minerals, which may have a kinetic effect and lead to a slower rate of carbonate precipitation. The rise in atmospheric CO_2 in the Industrial Era and into the future may result in lower production of $CaCO_3$ by calcifying organisms because of the consequent lowering of the saturation state of surface ocean waters, both in the shallow coastal and open ocean euphotic zone where calcareous organisms live. This overall process, termed ocean acidification, leads to a weak negative feedback to the release of CO_2 to the atmosphere from $CaCO_3$ precipitation that increases with increasing atmospheric CO_2. The subject is addressed in more detail in Chapter 9.

The main site of marine calcification is the euphotic zone of the ocean where the major groups of organisms producing $CaCO_3$ as their skeletal material thrive: phyto-plankton (Coccolithophoridae, of the phylum Haptophyta, Table 4.2) and phytobenthos (calcareous algae), as well as zooplankton (e.g., foraminifera, pteropods) and zooben-thos (e.g., foraminifera, molluscs, corals). The euphotic zone is a surface ocean layer of 50 m average thickness, where incident solar radiation declines to 0.1–1% of its surface irradiance value, although this zone varies in thickness and in some areas of the ocean may be as thick as 150–200 m (e.g., Ketchum, 1969; Krom *et al.*, 2003; Yentsch, 1966). It is important to bear in mind that net storage of $CaCO_3$ and organic carbon (C_{org}) in sediments is the difference between their production rates and dissolution or remineralization rates, and sequestration in sediments represents net removal of car-bonate and organic carbon from the euphotic zone. If the net removal of carbon into sediment storage is not matched by its input from land and oceanic crustal sources, then the oceanic reservoir is out of balance, and its inorganic and organic carbon content changes with time.

A measure of the net $CaCO_3$ and organic carbon production is often taken as their proportions in sediments and sedimentary rocks. In the geologically long-term sed-imentary record, the ratio of carbon in carbonate rocks to carbon in organic matter, $CaCO_3/C_{org}$, ranges from about 5/1 to 2.5/1 (Fig. 1.4; Holland, 1978, p. 215, with a review of older literature; Hayes *et al.*, 1999; Li, 2000; Turekian, 1996). It is about 2/1 in Recent sediments that include C_{org} brought in from land or 4/1 if only the *in situ* pro-duced organic matter is counted (Mackenzie *et al.*, 2004; Wollast and Mackenzie, 1989). Significantly, riverine input to the ocean in pre-industrial time supplied only about 25% more inorganic than organic carbon (ratio 1.23/1; Mackenzie *et al.*, 1993; Ver *et al.*, 1999): 32×10^{12} mol/yr of dissolved inorganic carbon and 26×10^{12} mol/yr of reactive organic carbon, consisting of dissolved organic carbon (DOC, 18×10^{12} mol/yr) and a 50%-fraction or 8×10^{12} mol/yr of total particulate carbon input (Smith and Hollibaugh, 1993). The remaining fraction of particulate organic carbon (POC), 8×10^{12} mol/yr, is considered refractory and stored in coastal zone and continen-tal slope sediments or transported to depth in the open ocean where it may be in part remineralized. The preceding numbers on the input and storage ratios in sediments show that a large fraction of reactive organic carbon brought to the ocean undergoes remineralization.

6.2   CO$_2$ Emission: General Case

The mass of CO$_2$ that can be released from a surface ocean layer depends not only on the dissolved carbonate concentration, the acid-base balance, and temperature, but also on the masses of precipitated CaCO$_3$ and photosynthetically produced C$_{org}$, inputs of carbon from land, and on the state of atmospheric CO$_2$ that may be either constant or rising due to emissions from the surface ocean. Computations of the CO$_2$ transfer in the present time have been based on gas transfer velocity equations (e.g., Nightingale *et al.*, 2000; Smith, 1985; Takahashi *et al.*, 2002) that depend on such parameters as the degree of mixing of the water layer, properties of the diffusive boundary layer at the air-water interface, and wind velocity. To reduce some of the uncertainties associated with the past environmental conditions, a different approach is adopted here that is based on material balances and gas-solution partitioning of the generated CO$_2$. A general diagram of a model euphotic zone and the main carbon fluxes is shown in Fig. 5.11.

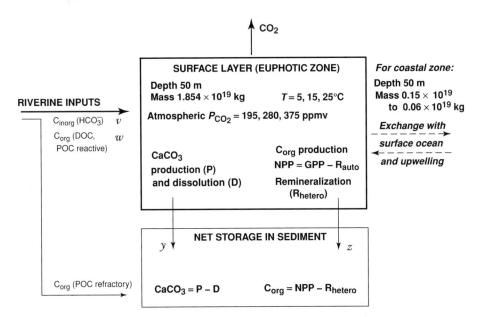

Figure 5.11   Schematic representation of the surface ocean water layer (euphotic zone) with inputs of inorganic and organic carbon, and carbonate and organic carbon storage in sediments. For the euphotic layer in the shallow-ocean coastal zone, exchange with the open ocean and coastal upwelling are shown. Water masses in the coastal zone correspond to the present and LGM time when the coastal zone area was smaller.

Table 5.4   Upper part: initial conditions of atmospheric P_{CO_2}, water temperature, total alkalinity (A_T), pH, and dissolved inorganic carbon (DIC) used in CO_2 release calculations. From pH = 8.35 (Sanyal et al., 1995) and atmospheric P_{CO_2} = 195 ppmv, A_T is given at the three temperatures; these A_T values are used to compute the pH and DIC at the higher atmospheric CO_2 concentration. Lower part: an alternative set of initial conditions, based on constant A_T, that shows small differences in the pH and DIC values from the preceding.

CO_2	195 ppmv			280 ppmv			375 ppmv		
Temperature	5°C	15°C	25°C	5°C	15°C	25°C	5°C	15°C	25°C
A_T (10^{-3} mol-equivalent/kg)	2.582	2.613	2.808	2.582	2.613	2.808	2.582	2.613	2.808
pH	8.35	8.35	8.35	8.22	8.23	8.23	8.11	8.12	8.14
DIC (10^{-3} mol/kg)	2.255	2.164	2.190	2.332	2.253	2.293	2.388	2.319	2.373
ALTERNATIVE:									
A_T (10^{-3} mol-equivalent/kg)	2.582	2.582	2.582	2.582	2.582	2.582	2.582	2.582	2.582
pH	8.350	8.346	8.322	8.219	8.222	8.205	8.109	8.118	8.106
DIC (10^{-3} mol/kg)	2.255	2.139	2.027	2.322	2.227	2.122	2.388	2.294	2.191

The sea-air exchange of CO_2 will be discussed for three representative temperatures of 5°, 15°, and 25°C, and three values of atmospheric P_{CO_2}: 195 ppmv, representative of the Last Glacial Maximum (LGM) 18,000 years ago (Petit et al., 1999; Wahlen, 2002), 280 ppmv for the end of pre-industrial time near the year 1700 (Siegenthaler and Oeschger, 1987, and Fig. 6.8), and 375 ppmv for the near present (Keeling and Whorf, 2003). Besides the temperature and atmospheric CO_2 concentration that represent a range of conditions since the LGM to the present, there have been other changes in the chemical composition of seawater. At the LGM the sea level was about 120 m lower (Fairbanks, 1989) and the salinity and individual dissolved species concentrations in seawater were correspondingly 3% higher than the modern values. These differences are not considered in the calculation that is done for a constant salinity of 35. Also, the volume of the euphotic zone is taken as constant: the ocean surface area covered by ice at the LGM might have been 10 to 15% of the total area; such a change in the volume of the model surface ocean layer would be compensated by a change equivalent to 5 to 10 m in its thickness. Two sets of the initial conditions for the CO_2-transfer calculations are given in Table 5.4. One set is based on the pH value of about 8.35 for surface ocean water in the LGM, reported by Sanyal et al. (1995), and the temperatures and atmospheric CO_2 concentrations that give the values of total alkalinity and DIC. These alkalinity values are conserved at the higher atmospheric CO_2 concentrations. In an alternative set of conditions in Table 5.4, total alkalinity is calculated for P_{CO_2} = 195 ppmv and 5°C, and this value is taken as constant at other temperatures and CO_2 partial pressures. The differences between the two sets of conditions are not large. In addition to the surface ocean water pH ≈ 8.35 near the LGM, Sanyal et al. (1995) also reported a higher pH for the deep ocean. However, Anderson and Archer (2002) found no increase in the deeper-ocean water pH between the depths of 1600 and 4200 m on the basis of restored carbonate-ion concentrations. A somewhat lower pH of 8.17 to 8.30 of the surface water in the Equatorial Pacific between about 20 and 16 ka before present was reported by Palmer and Pearson (2003).

The present-day total alkalinity and DIC of the surface ocean are lower than the values in Table 5.4; $A_T = 2.29$ to 2.38×10^{-3} mol-equivalent/kg and DIC $= 1.98$ to 2.13×10^{-3} mol/kg (Takahashi, 1989). If the present-day surface ocean water were at equilibrium with atmospheric CO_2 of about 340 ppmv (1975–1985 range is about 330 to 345 ppmv; Keeling and Whorf, 2003), its total alkalinity would correspond to the pH of 8.10 to 8.13, about 0.2 to 0.1 pH units lower than the values reported for near the Last Glacial Maximum.

Equation (5.24) shows that on an annual basis, riverine input adds per 1 kg of surface water layer v mol HCO_3^- and w mol C_{org}. At this stage, DIC and total alkalinity (A_T) are increased by v mol/kg. Reactive organic carbon is remineralized and respired, producing CO_2 according to reaction (5.22) and adding w mol CO_2 to DIC. Precipitation of $CaCO_3$ is a net removal of y mol $CaCO_3$ from surface ocean water and z mol C are removed as CO_2 in primary production by the formation and storage of organic matter. The changes in the initial values of DIC, Ca^{2+}, and A_T are:

$$DIC = DIC_0 + v + w - y - z \tag{5.24}$$

$$[Ca^{2+}] = [Ca^{2+}]_0 - y \tag{5.25}$$

$$A_T = A_{T,0} + v - 2y \tag{5.26}$$

The sum of the individual input, precipitation, and respiration processes usually increases the initial dissolved CO_2 concentration because of the net removal of $CaCO_3$ (y) and oxidation of organic matter (w). When the changed DIC content of the water that contains newly produced CO_2 equilibrates with atmospheric CO_2, transfer occurs between ocean water and the atmosphere. The DIC concentration at equilibrium with atmospheric CO_2 (DIC_{eq}) is usually smaller than DIC and the difference between them is the mass of inorganic carbon transferred to the atmosphere as CO_2. The change in DIC due to this CO_2 transfer is, from the preceding:

$$\Delta[DIC] = [DIC] - [DIC]_{eq} \quad (mol\, C/kg) \tag{5.27}$$

Quotient θ is a measure of the CO_2 emission per unit of $CaCO_3$ mass precipitated, y mol $CaCO_3$/kg:

$$\theta = \Delta[DIC]/y \quad (mol/mol) \tag{5.28}$$

Before discussing the sea-air transfer of CO_2, the meaning of the quotient θ should be clarified. In seawater that receives no external inputs of DIC and organic carbon—that is, $v = 0$ and $w = 0$ in equation (5.24)—the release of CO_2 due to production and net storage of $CaCO_3$ in sediments (y) is counteracted by the uptake of CO_2 in primary production and its net sequestration in sediments (z). For $CaCO_3$ precipitation alone, θ is a fraction that depends on water temperature, mass of $CaCO_3$ removed from seawater, and the atmospheric P_{CO_2} at equilibrium with the surface water. Under these conditions, θ has values between 0.44 and 0.79, for a temperature range from 5 to 25°C and P_{CO_2} from 195 to 375 ppmv. These values are shown in Table 5.5 and their derivation is explained in the next section.

6.3   Carbonate Precipitation at a Fixed Atmospheric CO_2

In any model of CO_2 emissions to an atmosphere of a constant P_{CO_2}, the emissions do not increase the atmospheric content, as if they were either removed from the carbon balance or taken up by other reservoirs. In a system with no external inputs and no biological production of C_{org}, only $CaCO_3$ precipitation produces CO_2, and equations (5.24)–(5.26) are then simplified to the following form:

$$DIC = DIC_0 - y \tag{5.29}$$

$$[Ca^{2+}] = [Ca^{2+}]_0 - y \tag{5.30}$$

$$A_T = A_{T,0} - 2y \tag{5.31}$$

At a constant atmospheric P_{CO_2}, dissolved $[CO_2]$ is also constant and in this case equations (5.27) and (5.28) become:

$$\Delta[DIC] = [DIC_0 - y] - [DIC]_{eq} \quad (mol\ C/kg) \tag{5.32}$$

and

$$\theta = \frac{\Delta[DIC]}{y} = \frac{(1 - f)[DIC_0] - [DIC]_{eq}}{f[DIC_0]} \tag{5.33}$$

where f is a fraction of the initial DIC concentration removed as $CaCO_3$. The losses of CO_2 from a surface water layer, taken as a 50-m-thick euphotic zone, due to $CaCO_3$ precipitation are plotted in Fig. 5.12 as the ratio θ for three temperatures and atmospheric CO_2 partial pressures. In this case, the CO_2 transfer from ocean water to the atmosphere, calculated as explained in Box 5.2, depends on the initial DIC content of the euphotic zone, mass of $CaCO_3$ precipitated, the atmospheric CO_2 concentration, and temperature. For a comparison with average oceanic conditions, the net storage of $CaCO_3$ is about 32×10^{12} mol C/yr (Chapter 10, Table 10.3), which corresponds to an annual net removal rate of 1.7×10^{-6} mol C/kg. Multiplied by $\theta \approx 0.65$, the preceding $CaCO_3$ storage rate gives a CO_2 flux of 21×10^{12} mol C/yr, which is slightly smaller than the values that take into account the input of inorganic and reactive organic carbon from land, as is shown in a later section.

Figure 5.12 shows that the CO_2 emission increases with increasing $CaCO_3$ precipitation that corresponds to a larger fraction of DIC and A_T removed, and it increases with an increasing atmospheric CO_2. At a higher atmospheric P_{CO_2} at equilibrium with ocean water, there is more dissolved CO_2 and its internal increase due to carbonate precipitation is greater than that at lower CO_2 atmospheric pressures. This accounts for a greater release to the atmosphere and, consequently, a higher value of θ. The increase in the CO_2 emission at a lower temperature is due to the distribution of the

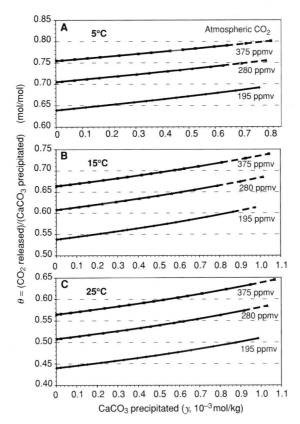

Figure 5.12 CO_2 release due to $CaCO_3$ precipitation (θ) from surface ocean layer into an atmosphere of a constant CO_2 of 195, 280, and 375 ppmv, at 5°, 15°, and 25°C (Table 5.5). Initial ocean-water conditions given in Table 5.4. Lowest values of y are 2.2 to 2.4 × 10^{-6} mol C/kg. Note the differences in the horizontal and vertical scales. Dashed lines show regions of calcite undersaturation caused by precipitation of larger fractions of dissolved inorganic carbon (DIC).

dissolved carbonate species: at a lower temperature, the relative abundance of $[CO_2]$ increases, such that there is more dissolved CO_2 at the start. In a normal ocean water of a constant DIC, within the pH range from 7.7 to 8.4, dissolved CO_2 concentration at 5°C is 60 to 80% greater than at 25°C: $[CO_2]_{5°}/[CO_2]_{25°} \approx 1.6$ to 1.8. The $CaCO_3$ mass precipitated that is shown in Fig. 5.12 corresponds to increasing fractions of initial DIC, from ≤ 0.1 to 45%, constrained by the condition that the final supersaturation with respect to calcite does not fall below 1, $\Omega \geq 1$ (Box 5.2).

The results shown in Fig. 5.12 are in general agreement with those of Frankignoulle *et al.* (1994) as to the direction of change of θ: it increases with increasing atmospheric CO_2 and with decreasing temperature. The values of θ between 280 and 375 ppmv

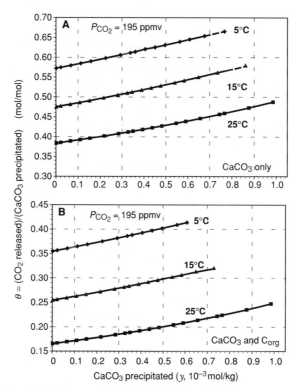

Figure 5.13 CO_2 release due to carbonate and organic carbon deposition that increases atmospheric CO_2, starting at 195 ppmv. Lowest value of y is 2.2×10^{-6} mol C/kg. A. Release of CO_2 due to $CaCO_3$ storage at 25°, 15°, and 5°C in surface ocean layer (a 50-m-thick euphotic zone, of water mass $M_w = 1.854 \times 10^{19}$ kg). Dashed lines indicate regions of calcite undersaturation. Compare with the curves in Figure 5.12. B. Release of CO_2 from ocean water where both $CaCO_3$ and organic carbon are stored (molar ratio of $CaCO_3/C_{org} = 4/1$) at the same temperatures as in A. Note that θ is smaller when dissolved CO_2 is removed into organic carbon storage than in the case of $CaCO_3$ precipitation only (Table 5.5).

CO_2 near the low end of the precipitated carbonate mass are also close to those of Frankignoulle *et al.*, between 290 and 400 ppmv CO_2, at the three temperatures of 5°, 15°, and 25°C.

Primary production removing CO_2 while $CaCO_3$ precipitates also reduces the CO_2 amount releasable to the atmosphere and lowers the fraction released θ to 0.17 to 0.35, at the conditions as shown in Fig. 5.13:

$$DIC = DIC_0 - y - z \qquad (5.34)$$

$$A_T = A_{T,0} - 2y \qquad (5.35)$$

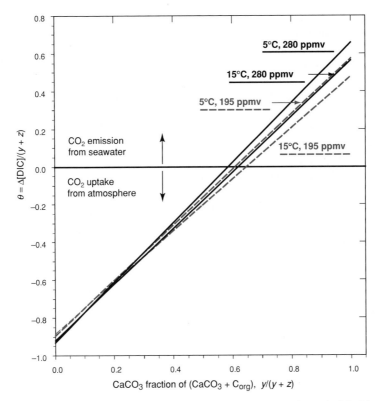

Figure 5.14 CO_2 transfer across the air-sea interface as a function of $CaCO_3/(CaCO_3 + C_{org})$ ratio of total carbon removed from ocean water. Initial conditions from Table 5.4 (Alternative): $A_T = 2.582 \times 10^{-3}$ mol-equivalent/kg at 5°C and 15°C, and $P_{CO_2} = 195$ and 280 ppmv; carbon removal is 1% of DIC, $y + z = 0.01 \times [DIC_0]$. Values of $\theta < 0$ indicate CO_2 transfer from the atmosphere to ocean water, $\theta > 0$ is flux to the atmosphere. Model surface ocean layer 50-m thick is a CO_2 sink as long as the $CaCO_3$ fraction is less than about 57 to 65% of total carbon stored.

where the ratio of carbonate to organic carbon removal is $y/z = 4/1$, similar to this ratio in sediments. More generally, the mutually competing roles of the formation of calcium carbonate and organic carbon on the direction of CO_2 flux across the air-sea interface are shown in Fig. 5.14. Deposition of only C_{org} makes the surface water a sink of atmospheric $CO_2(\theta < 0)$. As the amount of $CaCO_3$ increases relative to C_{org}, surface water becomes a source of atmospheric CO_2, as reflected in the values of $\theta > 0$. The break-even point of the CO_2 uptake or emission ($\theta = 0$) at the temperatures and initial atmospheric CO_2 concentrations shown in Fig. 5.14 is a carbonate fraction of about 57 to 65% of total stored carbon. This relationship represents a known observation (Chapter 9) of a net carbon storage in an autotrophic ecosystem where primary production exceeds autorespiration and remineralization of organic matter. Conversely, emission of CO_2 corresponds to a case where relatively little organic carbon is stored either due to a low level of primary production or strong remineralization, as in a case of a heterotrophic ecosystem.

6.4   Rising Atmospheric CO_2 due to Carbonate Precipitation

A case where the CO_2 released from seawater increases the CO_2 content of the atmosphere is different from the one discussed in the preceding section. A balance between the CO_2 released from surface ocean water and added to the atmosphere requires knowledge of the water volume from which the release takes place. As in the preceding section, an average 50-m-thick euphotic zone first is considered that releases CO_2 into an atmosphere of an initial concentration 195 ppmv. This euphotic-zone water mass is 1.854×10^{19} kg, of mean density 1027 kg/m^3, over the global ocean surface area of 3.61×10^{14} m^2. The density difference of about 4 kg/m^3 between 5° and 25°C is disregarded. The increase in atmospheric CO_2 due to carbonate and organic carbon storage was computed by a numerical iteration routine because neither the final atmospheric CO_2 concentration nor the final DIC in surface ocean water is known. The computational procedure is detailed in Box 5.2 at the end of this chapter. The release of CO_2 is shown in Fig. 5.13A and Table 5.5. If the surface layer volume is doubled, the precipitated carbonate mass also doubles, but the CO_2 mass emitted to the atmosphere is somewhat less than double because the rising CO_2 exerts back-pressure on ocean water, thereby increasing its DIC_{eq}. This non-linearity is reflected in the slightly lower values of θ for the thicker model surface layers of 50, 100, and 200 m (Table 5.5).

Table 5.5   Factor θ defining CO_2 release to the atmosphere as a function of $CaCO_3$ formation in a surface ocean layer at different temperatures and constant atmospheric CO_2 concentrations. θ for rising atmospheric CO_2 from its initial 195 ppmv: carbonate precipitation in a surface layer 50, 100, and 200 m thick (mass of a 50 m layer $M_w = 1.854 \times 10^{19}$ kg; Fig. 5.11); $CaCO_3$ and C_{org} storage in a surface layer, Figs. 5.12–5.14. The range of θ within each cell is for increasing amounts of precipitated $CaCO_3$, from ≤ 0.1 to up to 45% of initial DIC concentration, limited by $\Omega \leq 1.1$ to 1.6.

		Temperature		
CO_2 release process		*5°C*	*15°C*	*25°C*
θ for $CaCO_3$ formation at	$P_{CO_2} = 195$ ppmv	0.64–0.69	0.54–0.60	0.44–0.51
constant atmospheric CO_2	$P_{CO_2} = 280$ ppmv	0.70–0.74	0.61–0.66	0.51–0.57
	$P_{CO_2} = 375$ ppmv	0.75–0.79	0.66–0.71	0.56–0.63
θ for $CaCO_3$ formation	50 m layer	0.57–0.65	0.48–0.56	0.38–0.49
and rising atmospheric CO_2	100 m layer	0.52–0.63	0.43–0.54	0.35–0.47
Initial $P_{CO_2} = 195$ ppmv	200 m layer	0.48–0.57	0.40–0.51	0.33–0.43
θ for $CaCO_3$ and C_{org} formation (4:1)		0.35–0.42	0.25–0.34	0.17–0.25
and rising atm. CO_2 (50 m layer)				
Initial $P_{CO_2} = 195$ ppmv				

6.5 Controls on the Sea-Air CO_2 Flux

As pointed out earlier, the $CaCO_3$ production and removal rates in the surface ocean layer account for only small fractions of the DIC mass in surface water: at the DIC concentration of about 2.2×10^{-3} mol/kg, the annual $CaCO_3$ production in a 50-m-thick euphotic zone is a fraction 0.13 to 0.23% of the total DIC mass and the net $CaCO_3$ storage in sediments is even a smaller fraction. A potential rise in atmospheric CO_2 due to calcite precipitation from a surface ocean layer is a function not only of the DIC fraction removed as $CaCO_3$ but also of temperature and the surface layer thickness, as more CO_2 is emitted from a thicker water layer. The mass of CO_2 needed for an increase of 85 ppmv in atmospheric CO_2 since the LGM to the end of pre-industrial time is equivalent to the mass that could be released from a 100-m-thick surface ocean layer at different combinations of temperature and net $CaCO_3$ storage. Figure 5.15 shows that about 30% of the initial DIC precipitated as $CaCO_3$ at 5°C results in the same increase as 45% precipitated at 25°C. A 50-m-thick model layer does not provide enough CO_2, adding not more than 45 ppmv CO_2 or about one-half of the CO_2 comparable to the rise from LGM to pre-industrial time. It should be noted that the $CaCO_3$ values that are shown on the x-axis of Fig. 5.15 are one-step precipitation without replenishment of carbon and alkalinity lost, and no time length is assigned to each precipitation step. Near the high end, the precipitated amounts of 0.65 to 1×10^{-3} mol $CaCO_3$/kg correspond to fractions of 30 to 45% of the DIC content of the surface ocean layer, depending on the temperature, where the $CaCO_3$ precipitation reduces the ocean water saturation with respect to calcite to $\Omega \approx 1.1$ to 1.5. An increase of 85 ppmv over a period of 18 ka translates into a mean increase rate of 0.84×10^{12} mol C/yr or 0.01 Gt C/yr, although the ice-core record shows that the increase was neither smooth nor linear with time (Petit et al., 1999; Wahlen, 2002). As is discussed in more detail in Chapter 10, it is difficult to incorporate such a low rate of atmospheric CO_2 increase in a model of the global carbon cycle because the interreservoir fluxes are not known with a sufficient accuracy at 10^2 to 10^4-year time scales that would give a reliable net rate of change in the atmosphere of the order of 0.01 Gt C/yr or 0.84×10^{12} mol C/yr. It is further discussed in Chapters 9 and 10 that the CO_2 sea-air fluxes are much greater than that, whereas the results in Fig. 5.15 only indicate the mass relationships between the carbonate sequestered from a surface ocean layer and the resultant production of the CO_2 equivalent to the net increase of 1.5×10^{16} mol C or 180 Gt C in the atmosphere since the Last Glacial Maximum to the end of pre-industrial time.

Although the values of the quotient θ that are derived in the preceding section may give the CO_2 flux out of, or into, the surface water layer if the net removal rates of $CaCO_3$ and C_{org} are known, θ alone does not, in general, define the sea-air CO_2 transfer because the latter also depends on the input of dissolved inorganic and reactive organic carbon from land, as given in equation (5.24). These inputs affect the DIC and total alkalinity balance of seawater and, in particular, input of reactive organic carbon that is remineralized in seawater and may produce additional CO_2 that would be a significant component of the sea-air flux. The pre-industrial inputs from land and sedimentary storage rates are given in Chapter 10 (Tables 10.2 and 10.3), and by the year 1980,

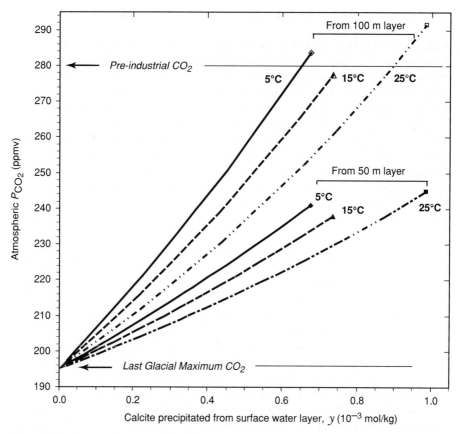

Figure 5.15   Increase in atmospheric CO_2 due to carbonate storage from a surface ocean layer of 50 m and 100 m thickness in comparison to the 85 ppmv CO_2 increase from the Last Glacial Maximum to the end of pre-industrial time. Results for 5°, 15°, and 25°C, without replenishment of the dissolved carbon and calcium lost to precipitation. At the higher end of the abscissa, precipitated $CaCO_3$ amounts to 30 to 45% of the initial DIC concentration (2.19×10^{-3} mol C/kg) where the degree of calcite saturation has decreased to $\Omega \approx 1.1$ to 1.5 (Box 5.2).

the land inputs of inorganic and organic carbon to the ocean increased more than 20% since the end of pre-industrial time and the atmospheric CO_2 concentration was about 340 ppmv.

In modern, industrial time, CO_2 is emitted from the tropical ocean to the atmosphere from an area between the latitudes of S15° and N15° that accounts for about 28% of total ocean surface area. The outgoing CO_2 flux was reported as 135×10^{12} mol/yr (1.62 Gt C/yr; Tans et al., 1990) and 89×10^{12} mol/yr (1.07 Gt C/yr; Takahashi et al., 2002). It should, however, be borne in mind that the *global net* CO_2 flux is from the atmosphere to the ocean because the higher latitudes are a bigger sink of atmospheric CO_2 than the source in the tropical ocean. The magnitudes and directions of the CO_2

air-sea fluxes vary greatly between different areas. In the present-day, the equatorial ocean emits about 1.1 mol C m^{-2} yr^{-1}. Shallower near-coastal sections, such as in the Arabian Sea, Bermudian reefs, South Atlantic Bight, and other locations emit between 0.5 and 2.5 mol C m^{-2} yr^{-1}, and much larger fluxes, up to about 300 mol C m^{-2} yr^{-1} have been reported on the basis of short-term measurements in European estuaries, where they are likely to be driven by the remineralization of organic matter (Table 9.2; Borges et al., 2004).

The various aspects of the natural carbon system of gas-solution-mineral that are treated in this chapter underlie in part the explanations of the processes in the inorganic and organic world that are discussed in Chapters 6 and 7, the role of CO_2 in mineral weathering and transport of the products of weathering from land to the ocean (Chapter 8), the biogeochemical processes that control ecosystem dynamics and sediment-pore water reactions and sediment composition in the shallow oceanic coastal zone (Chapter 9), and the more recent history of the Earth's surface and near-future projections for the environment, as developed in Chapters 10 and 11.

Box 5.2   Calculation of CO_2 transfer between seawater and atmosphere

1. From known values of total alkalinity (A_T) and total dissolved inorganic carbon (DIC), given in equations (5.12) and (5.5), the H^+-ion concentration or the pH of seawater can be computed as a root of the 5th-degree polynomial:

$$a_5[H^+]^5 + a_4[H^+]^4 + a_3[H^+]^3 + a_2[H^+]^2 + a_1[H^+] + a_0 = 0 \quad (5.2.1)$$

The coefficients a_i are algebraic expressions that include the individual apparent dissociation constants K_i' (Box 5.1), and the A_T and DIC values. The coefficients are given in Bacastow and Keeling (1973, pp. 130–133) and Zeebe and Wolf-Gladrow (2001, pp. 271–277). When total alkalinity and either dissolved CO_2 concentration or the partial pressure of atmospheric CO_2 at equilibrium with ocean water are known (A_T and [CO_2] or P_{CO_2}), then [H^+] is a root of the 4th-degree polynomial:

$$a_4[H^+]^4 + a_3[H^+]^3 + a_2[H^+]^2 + a_1[H^+] + a_0 = 0 \qquad (5.2.2)$$

where the coefficients a_i are obtainable from Zeebe and Wolf-Gladrow (2001, loc. cit.).

2. Calculation of θ

Starting with initial DIC and Ca^{2+}-ion concentrations and total alkalinity (A_T), removal of y mol CaCO$_3$/kg results in the following new concentrations:

$$DIC = DIC_0 - y \qquad (5.2.3)$$

$$[Ca^{2+}] = [Ca^{2+}]_0 - y \qquad (5.2.4)$$

$$A_T = A_{T,0} - 2y \qquad (5.2.5)$$

2.1 Release into Atmosphere of Constant CO_2

At a constant atmospheric P_{CO_2}, dissolved CO_2 concentration is also constant and changes in DIC due to CO_2 transfer from ocean water to the atmosphere do not affect the total alkalinity value after the carbonate precipitation. From the latter value of A_T and P_{CO_2}, new values of $[H^+]$ and $[DIC]_{eq}$ are obtained from equation (5.2.2). The DIC mass transferred to the atmosphere is

$$\Delta[DIC] = [DIC_0 - y] - [DIC]_{eq} \quad (mol/kg) \qquad (5.2.6)$$

where $[DIC_{eq}]$ is smaller than DIC after carbonate precipitation, $[DIC_0 - y]$.

Parameter θ is the loss of DIC as CO_2 from surface ocean water per mol of $CaCO_3$ removed from ocean water by precipitation and storage in sediment:

$$\theta = \frac{\Delta[DIC]}{y} \quad (mol/mol) \qquad (5.2.7)$$

In this model, the emitted CO_2 is not added to the atmospheric CO_2 mass that remains constant.

2.2 CO_2 Emission that Increases Atmospheric P_{CO_2}

An increase in atmospheric P_{CO_2} by CO_2 emission from surface ocean water is calculated by an iterative routine because neither the P_{CO_2} nor DIC is known after the ocean water, from which carbonate precipitated, reequilibrated with the new atmospheric CO_2.

First, precipitation of $CaCO_3$ takes place in a temporarily closed system where DIC, $[Ca^{2+}]$, and total alkalinity decrease, as given in equations (5.2.3–5). Dissolved CO_2 and $[H^+]$ concentrations in this closed system increase strongly (Fig. 5.2.1). In the next step, the ocean-water volume is opened to the atmosphere and the CO_2 mass transferred from water to the atmosphere (x moles) is approximated by an equilibrium partition between the dissolved CO_2 mass at the end of the precipitation stage and the atmospheric mass:

$$\frac{n_w - x}{n_a^\circ + x} = \frac{K_0' RTM_w}{V} \qquad (5.2.8)$$

where M_w is the water mass of a 50-m-thick euphotic zone (Table 5.2.1), n_w is the mass of dissolved CO_2 in the closed system after the precipitation of $CaCO_3$ ($n_w = [CO_2] \times M_w$), and $n_a^\circ = 3.487 \times 10^{16}$ mol is the atmospheric CO_2 mass at 195 ppmv CO_2. Other parameters in equation (5.2.8) are: K_0' is the CO_2 solubility coefficient given in equation (5.1.1), the gas constant $R = 8.315 \times 10^{-5}$ bar m³ K⁻¹ mol⁻¹, atmospheric temperature $T = 288.15$ K, and atmospheric volume $V = 4.284 \times 10^{18}$ m³ (thickness 8,400 m over the Earth's surface area of 5.1×10^{14} m²). Equation (5.2.8) follows from the CO_2 solubility relationship

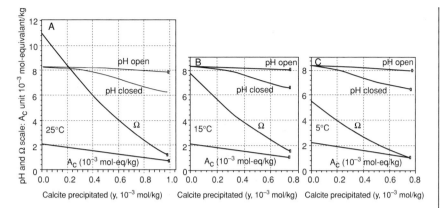

Figure 5.2.1. Changes in the pH, carbonate alkalinity (A_C), and degree of saturation with respect to calcite (Ω) due to $CaCO_3$ precipitation at three temperatures, at starting atmospheric CO_2 195 ppmv and pH = 8.35 (Table 5.2.1). pH of the closed system is after calcite precipitation and pH of the open system is after reequilibration with the atmosphere and A_C values are for an open system. Note that the scale for pH and A_C are the numbers on the vertical axis; for carbonate alkalinity (A_C) the same scale is in units of 10^{-3} mol-equivalent/kg.

(Table 5.1.1, Box 5.1):

$$K_0' = \frac{[CO_2]}{P_{CO_2}} \quad (\text{mol C kg}^{-1} \text{ bar}^{-1}) \tag{5.2.9}$$

where the terms $[CO_2]$ and P_{CO_2}, taking into account the transfer of x moles carbon from the ocean to the atmosphere, can be written as:

$$[CO_2] = \frac{n_w - x}{M_w} \quad (\text{mol C/kg}) \tag{5.2.10}$$

and

$$P_{CO_2} = \frac{(n_a^\circ + x)RT}{V} \quad (\text{bar}) \tag{5.2.11}$$

Using the new $[CO_2]$ value from equation (5.2.10) and A_T from equation (5.2.5), a new $[H^+]$ is computed from equation (5.2.2) and a first approximation to $[DIC]_{eq}$ is calculated. Then the first approximations of the amount transferred $\Delta[DIC]$ and new $[CO_2]$ are computed. The latter value of $[CO_2]$ and A_T are used to repeat the steps, three or four iterations, until stable values of DIC loss converging within $\leq 0.1\%$ are obtained.

The initial data, intermediate results, and final values at a number of precipitation steps for a case of CO_2 rise from 195 ppmv at 25°C are shown in Fig. 5.2.1 and Table 5.2.1.

Table 5.2.1   Release of CO_2 from $CaCO_3$ formation and atmospheric CO_2 increase from 50-meter-thick euphotic zone. Salinity $S = 35$, 25°C, initial $P_{CO_2} = 195$ ppmv, water mass in euphotic zone $M_w = 1.854 \times 10^{19}$ kg. Figure 5.13A.

Precipitation of $CaCO_3$ and increase of $[CO_2]$

Fraction of DIC_0 removed f	Mass C or Ca removed y 10^{-3} mol/kg	Mass calcite removed yM_w 10^{12} mol	Change in DIC $DIC_0 - y$ 10^{-3} mol C/kg	Change in total alk. $A_{T,0} - 2y$ 10^{-3} mol/kg	pH eq. (5.2.1)	Change in $[Ca^{2+}]$ $[Ca^{2+}]_0 - y$ 10^{-2} mol/kg	$[CO_2]$ in solution mol/kg	Internal P_{CO_2} ppmv	$[CO_2] \times M_w$ n_w 10^{12} mol
0.000	0	0	2.190	2.808	8.350	1.028	5.537E-06	195	102.6
0.001	0.0022	40.6	2.188	2.803	8.348	1.028	5.555E-06	196	103.0
0.010	0.0219	406	2.168	2.764	8.334	1.026	5.729E-06	202	106.2
0.100	0.219	4,060	1.971	2.370	8.159	1.006	8.353E-06	294	154.8
0.200	0.438	8,120	1.752	1.932	7.836	0.984	1.682E-05	592	311.7
0.450	0.986	18,270	1.205	0.837	6.207	0.929	3.708E-04	13061	6874.5

Equilibrating open system

Fraction of DIC_0 removed f	1st approximation of CO_2 transfer							Final results			DIC released as CO_2 eq.(5.2.6)	
	$[CO_2]$ mass transferred x eq. (5.2.8) mol	CO_2 in ocean water $n_w - x$ mol	CO_2 in atmosphere $n_a^\circ + x$ mol	DIC 10^{-3} mol/kg	pH eq. (5.2.2)	$[CO_3^{2-}]$ 10^{-3} mol/kg	Calcite saturation Ω	$[CO_2]$ mol/kg	Atmosphere P_{CO_2} ppmv	$\Delta[DIC]$ 10^{-3} mol/kg	θ mol/mol
---	---	---	---	---	---	---	---	---	---	---	---
0.000	0	1.026E+14	3.487E+16	2.190	8.350	0.459	11.05	5.537E-06	195.0	0	
0.001	3.454E+11	1.026E+14	3.487E+16	2.187	8.349	0.458	11.02	5.539E-06	195.1	0.0008	0.383
0.010	3.564E+12	1.026E+14	3.487E+16	2.160	8.343	0.447	10.74	5.561E-06	195.9	0.0084	0.385
0.100	5.206E+13	1.028E+14	3.492E+16	1.883	8.279	0.346	8.15	5.796E-06	204.1	0.0880	0.402
0.200	2.085E+14	1.032E+14	3.507E+16	1.567	8.192	0.244	5.62	6.082E-06	214.2	0.1852	0.423
0.450	6.752E+14	1.225E+14	4.162E+16	0.724	7.836	0.054	1.17	6.951E-06	244.8	0.4807	0.488

Chapter 6

Isotopic Fractionation of Carbon: Inorganic and Biological Processes

Stable and radioactive isotopes are extensively used as tracers of numerous processes in the planetary and terrestrial environment. The relative abundances of isotopic species measured by their ratios provide indications of the origin of various materials and differences in the abundance ratios that develop in different processes make it possible to identify the mechanisms behind a variety of phenomena in extraterrestrial space, within the solid Earth, on its surface, and in the biosphere. The improvements in the sensitivity and precision of mass spectrometers used for the determination of isotope abundances are continually expanding the number of isotopes that can be identified in natural materials as well as the understanding of the mechanisms that drive many parts of the Earth's system. The involvement of carbon dioxide in many geochemical inorganic as well as biogeochemical processes focused long ago attention on the behavior of the different isotopic species of CO_2 and made possible many new interpretations of processes in the atmosphere, on land, in the oceanic and continental waters, and within the biosphere. The goal of this Chapter is to review the essentials of the isotope geochemistry of carbon dioxide and the mechanisms of its isotopic fractionation, and to discuss the broader aspects of the global carbon cycle that are based on the carbon-isotope geochemistry.

1   Isotopic Species and Their Abundance

There are two stable isotopes of carbon, ^{12}C and ^{13}C. The lighter and even-numbered isotope is more abundant than the heavier and odd-numbered, as is the general rule for

Table 6.1 CO_2 stable isotopic species and their relative abundances (atom percent)

C and O nuclides (atom %)		CO₂ isotopic species (atom %)			
^{12}C	98.93	$^{12}C^{16}O_2$	98.450	$^{13}C^{16}O_2$	1.065
^{13}C	1.07	$^{12}C^{16}O^{17}O$	0.075	$^{13}C^{16}O^{17}O$	—
^{16}O	99.757	$^{12}C^{16}O^{18}O$	0.405	$^{13}C^{16}O^{18}O$	0.0044
^{17}O	0.038	$^{12}C^{17}O^{18}O$	—	$^{13}C^{17}O^{18}O$	—
^{18}O	0.205	$^{12}C^{17}O_2$	—	$^{13}C^{17}O_2$	—
		$^{12}C^{18}O_2$	—	$^{13}C^{18}O_2$	—

Note: Abundances of individual C and O isotopes from Holden (2002). Dashes indicate very low computed abundance, <0.001 atomic %.

isotopic abundances, with a few notable exceptions in the periodic table. On Earth, ^{13}C accounts for slightly more than 1 atom % of carbon (Table 6.1), making the atomic ratio $^{12}C/^{13}C$ about 92. A higher abundance of ^{13}C has been reported in interstellar space, with $^{12}C/^{13}C$ ratios of 60 ± 7 and 67 ± 19, and organic matter of carbonaceous chondrite meteorites is strongly enriched in ^{13}C, where the ratio $^{12}C/^{13}C$ is about 42 (Hoefs, 1987). The enrichment has been variably attributed to the formation of organic matter by different processes at high temperatures and to its subsequent alteration.

The two stable isotopes of carbon combined with the three stable isotopes of oxygen (^{16}O, ^{17}O, and ^{18}O) make a total of 12 isotopically different species of CO_2. The general formula for the number of distinguishable isotopic species of n isotopes of an element, where each species has m sites that are occupied by different combinations of the n isotopes, is:

$$N = \frac{(n + m - 1)!}{(n - 1)!m!} \tag{6.1}$$

Of the three oxygen isotopes ($n = 3$), any two make one isotopically distinct O_2 molecule (two sites, $m = 2$), which results in a total of six distinguishable species: $^{16}O_2$, $^{16}O^{17}O$, $^{16}O^{18}O$, $^{17}O^{18}O$, $^{17}O_2$, and $^{18}O_2$. The total number of species is larger, n^m, but such isotopomers as $^{16}O^{18}O$ and $^{18}O^{16}O$ may be indistinguishable one from the other if they do not differ in their chemical properties. Their relative abundances are products of the individual isotope abundances multiplied by 2. Each of the six O_2 species combines with either ^{12}C or ^{13}C, making a total of 12 CO_2 species that are shown in Table 6.1 with their relative abundances.

The three most abundant CO_2 species are $^{12}C^{16}O_2$, $^{13}C^{16}O_2$, and $^{12}C^{16}O^{18}O$. The isotopic variety of the aqueous carbonate and bicarbonate species is even greater than that of CO_2. As each has three oxygen sites, the total number of species rises to 20 for CO_3^{2-} and to 40 for HCO_3^-, because of the occurrence of the two stable hydrogen isotopes, ^1H and ^2H (deuterium or D). However, the abundance of most of the species among those of the carbonate and bicarbonate ions is also very low, and studies of the isotopic processes in the Earth's surface environment often deal mainly with the ^{13}C and ^{12}C species, irrespective of their oxygen isotopic composition, or with the ^{16}O and ^{18}O species in the CO_2-water natural system, without distinguishing between their carbon isotopic composition.

Because of the differences in the chemical and physical properties of the isotopic species of an element, there are differences in the behavior of $^{12}CO_2$ and $^{13}CO_2$. In general, textbooks of physical chemistry explain that the heavier isotopes form stronger bonds, making their compounds also slower to react than those containing the lighter isotope. The differences between the behavior of the light and heavy CO_2 species fall into two main classes: (1) in inorganic atmosphere-water-mineral systems, the distribution of the two CO_2 isotopic species is controlled by their chemical thermodynamic and kinetic properties; (2) in biologically driven systems of chemosynthesis or photosynthesis and respiration or decomposition of organic matter, the differences between the ^{12}C and ^{13}C species come primarily into play because of the general preference of organisms to use the lighter isotope, ^{12}C, in the building of organic matter. This is a biological kinetic effect that fractionates ^{12}C from ^{13}C in such processes as transfer of CO_2 from the atmosphere to the land or aquatic biosphere. Separation of the pathways of ^{12}C from ^{13}C that results in their fractionation is usually much stronger in biological systems than in the geochemical inorganic processes that occur at Earth's surface temperatures.

2   Isotopic Concentration Units and Mixing

Isotope abundances are commonly expressed not in conventional concentration units of mass per unit of volume or per unit of mass, but in units of the abundance ratio relative to a standard, denoted δ. For ^{13}C,

$$\delta^{13}C = \frac{(^{13}C/^{12}C)_{\text{sample}} - (^{13}C/^{12}C)_{\text{standard}}}{(^{13}C/^{12}C)_{\text{standard}}} \times 1000 \quad (\permil) \qquad (6.2)$$

where $^{13}C/^{12}C$ are atomic abundance ratios of the heavy to the light isotope in the sample and a standard. This convention follows the mass spectrometric practice where relative abundances of the isotopes are determined in a sample and then compared to a known standard. The values of δ can be either positive, indicating enrichment in ^{13}C relative to the standard, or negative, indicating depletion of the heavier isotope (Box 6.1).

Box 6.1 δ values and different isotopic standards

The definition of δ as a fraction in equation (6.2), for an isotope X relative to a standard A that has the isotope ratio R_A, is:

$$\delta X_A = \frac{R_{\text{sample}}}{R_A} - 1 \qquad (6.1.1)$$

The value of δX_B of the same sample, relative to another standard B, is:

$$\delta X_B = \frac{R_{\text{sample}}}{R_B} - 1 \qquad (6.1.2)$$

It follows from the preceding two relationships that conversion depends on the isotope abundance ratios in the two standards:

$$1 + \delta X_B = (1 + \delta X_A)\frac{R_A}{R_B} \qquad (6.1.3)$$

The relationships between δX_B and δX_A can be written with explicit numerical values of the abundance ratios in the two standards, R_A / R_B:

$$\delta X_B = \delta X_A \frac{R_A}{R_B} + \frac{R_A}{R_B} - 1 \qquad (6.1.4)$$

Alternatively, δX_B can be expressed in terms of the old δX_A and the value of the new standard relative to the old, $\delta R_{(B-A)}$:

$$\delta X_B = \frac{\delta X_A - \delta R_{(B-A)}}{1 + \delta R_{(B-A)}} \qquad (6.1.5)$$

where the δ value of the new standard relative to the old is, as a fraction:

$$\delta R_{(B-A)} = \frac{R_B}{R_A} - 1 \qquad (6.1.6)$$

Values of the carbon and oxygen isotope atomic ratios in some of the commonly used standards are given below.

No.	Isotope	Abundance ratio		Standard name	References
1	$^{13}C/^{12}C$	0.011237	PDB	Pee Dee Formation, belemnite	Craig (1957)
2	$^3C/^{12}C$	0.011056	VPDB	Vienna-PDB	Coplen *et al.* (2002), Coplen (*pers. comm.*, 2003)
3	$^3C/^{12}C$	0.011078	NBS19	CaCO₃ marble $\delta^{13}C(VPDB) =$ +1.95 ‰	NIST (1992)
4	$^{13}C/^{12}C$	0.011180	VPDB	Vienna-PDB	NIST (2002)
5	$^3C/^{12}C$	0.011202	NBS19	CaCO₃ marble $\delta^{13}C(VPDB)$ = +1.95 ‰	NIST (2002), Coplen (*pers. comm.*, 2003)
6	$^{18}O/^{16}O$	0.0020672	VPDB	Vienna-PDB	NIST (2002), Clark and Fritz (1997)
7	$^{18}O/^{16}O$	0.0019934	SMOW	Standard Mean Ocean Water	Craig (1961), Fritz and Fontes (1980)
8	$^{18}O/^{16}O$	0.0020052	VSMOW	Vienna-SMOW	Baertschi (1976), Clark and Fritz (1997)

Many of the $\delta^{18}O$ values of sedimentary carbonates have been reported on the PDP or VPDB standard, and their range often is $\pm X$ ‰ in the single digits. On the SMOW or VSMOW standard, these $\delta^{18}O$ values would be about 30 ‰ heavier. $\delta^{18}O$ for water and atmosphere are usually based on the SMOW or VSMOW standard. Conversion of $\delta^{18}O$ between VPDB and VSMOW standards, based on abundance ratios in Nos. 6 and 8 tabulated above and equation (6.1.4), is:

$$\delta^{18}O_{VSMOW} = 1.03092\ \delta^{18}O_{VPDB} + 30.92\ ‰$$

$$\delta^{18}O_{VPDB} = 0.9700\ \delta^{18}O_{VSMOW} - 29.99\ ‰$$

The $^{13}C/^{12}C$ abundance ratio in the VPDB standard can be computed from the ratio value in marble NBS19 (No. 3 in the table above) and its $\delta^{13}C = 1.95$ ‰ relative to VPDB, using equation (6.1.6):

$$R_A = \frac{R_B}{1 + \delta R_{(B-A)}} = \frac{0.011078}{1.00195} = 0.011056$$

The difference between the $^{13}C/^{12}C$ ratios in the old PDB and the newer Vienna-PDB standards is not interpreted as affecting the measured $\delta^{13}C$ values and no correction is made between the $\delta^{13}C$ values based either on the PDB or the VPDB standard, because the reference material for mass-spectrometric measurements is the NIST standard, marble called NBS19 (B. N. Popp, *pers. comm.*, 2003; T. B. Coplen, *pers. comm.*, 2003; Coplen *et al.*, 1983; Hoefs, 1987).

In the concentration units of permil (‰), the values usually vary from single digits to tens permil for the isotopic ratios of such elements as carbon, nitrogen, and oxygen in the surface environment. They may extend into the hundreds for the lighter isotopes of hydrogen[1] or in stronger fractionation processes of other isotopes. In cases where one of the isotopes is much less abundant than the other, such as ^{13}C at about 1 atom % and ^{12}C at 99 atom %, the ratio of the isotopic abundances is very nearly equal to the atomic fraction of the heavy isotope:

$$R_{sample} = \left(\frac{^{13}C}{^{12}C}\right)_{sample} = \frac{n_{13}}{n_{12}} \approx \frac{n_{13}}{n_{12} + n_{13}} = \frac{R_{sample}}{1 + R_{sample}} = x_{13} \qquad (6.3)$$

where n_{13} and n_{12} are the numbers of moles of ^{13}C and ^{12}C in the sample, and x_{13} is the mol or atomic fraction of ^{13}C. In general, $n_{13} \ll n_{12}$ and $R_{sample} \ll 1$.

The δ scale requires a standard, and the present standard for the $^{13}C/^{12}C$ ratio is limestone material, called Vienna-PDB, that replaced after intercalibration the older, no longer available standard of calcite of the shell of a Cretaceous cephalopod belemnite from the Pee Dee formation in South Carolina, and accordingly called PDB. The isotopic

---

[1] Writing δ values as fractions rather than in permil circumvents the need of inserting 1000 into equations. In this chapter, it is always indicated whether δ values are fractions or ‰.

ratio values in these and some other standards, and conversion of δ values from the scale of one standard to another are given in Box 6.1.

The atomic fraction of ^{13}C, x_{13}, is obtainable from the measured value of δ^{13}C, as a fraction, and the isotope abundance ratio in the standard:

$$x_{13} = (1 + \delta^{13}C)R_{\text{standard}} \tag{6.4}$$

Values of δ can often be used as mol fractions of ^{13}C in a mixture of carbon species to estimate the mean δ^{13}C of the whole. Dissolved inorganic carbon (DIC), the definition of which is given in equation (5.4), will have a mean δ that follows from the δ^{13}C values of the three species and their mol fractions:

$$\delta^{13}_{\text{DIC}}[\text{DIC}] = \delta^{13}_{\text{CO}_2}[\text{CO}_2] + \delta^{13}_{\text{HCO}_3^-}[\text{HCO}_3^-] + \delta^{13}_{\text{CO}_3^{2-}}[\text{CO}_3^{2-}] \tag{6.5}$$

Using the fractions of DIC for each of the three carbonate species, as shown in equation (5.7) and Fig. 5.1, total DIC has a mean δ^{13}C value of:

$$\delta^{13}_{\text{DIC}} = \delta^{13}_{\text{CO}_2}\,\alpha_0 + \delta^{13}_{\text{HCO}_3^-}\,\alpha_1 + \delta^{13}_{\text{CO}_3^{2-}}\,\alpha_2 \tag{6.6}$$

The mean δ^{13}C of the Earth's surface environment is the weighted sum of the δ^{13}C values of the carbon masses in the major reservoirs (Fig. 1.4): in order of decreasing reservoir mass, these are sedimentary carbonates, sedimentary organic matter, dissolved carbon in the oceans, the atmosphere, the land biomass, and the aquatic biomass. For the Earth's surface, the mean is heavily weighted by the two largest sedimentary reservoirs of carbonates and organic matter, variably estimated as a δ^{13}C of −7 to −5 ‰ (see p. 184). Values of δ^{13}C for some inorganic and organic materials and reservoirs are shown in Fig. 6.1.

Notice that there is a large overlap of the δ^{13}C ranges of the various constituents of terrestrial and marine organic matter, including fossil fuels. Interpretation of these data is to a large extent based on our understanding of the physical, chemical, and biological processes that are reviewed in the subsequent sections.

3   Fractionation in Inorganic Systems

An isotopic exchange reaction between heavy and light CO_2 species in the gas phase and in an aqueous solution can be written as:

$$^{12}CO_2(\text{aq}) + {}^{13}CO_2(\text{g}) = {}^{12}CO_2(\text{g}) + {}^{13}CO_2(\text{aq}) \tag{6.7}$$

An equilibrium fractionation factor α, analogous to an equilibrium constant K, for the preceding reaction is:

$$\alpha_{CO_2(\text{aq})-CO_2(\text{g})} = \frac{[^{13}CO_2(\text{aq})]/[^{12}CO_2(\text{aq})]}{[^{13}CO_2(\text{g})]/[^{12}CO_2(\text{g})]} = \frac{(^{13}C/^{12}C)_{\text{aq}}}{(^{13}C/^{12}C)_{\text{g}}} \tag{6.8}$$

which is a quotient of the isotopic abundance ratio ^{13}C/^{12}C in solution to the abundance ratio in the gas phase. Isotopic fractionation factors for many reactions in the Earth's

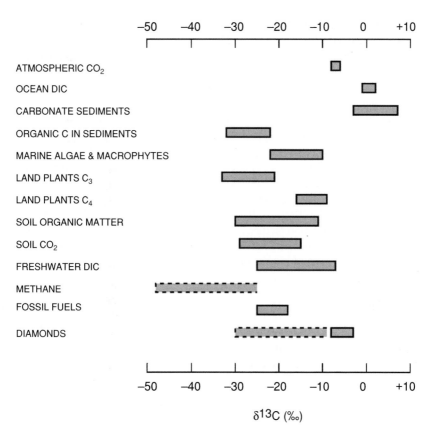

Figure 6.1   Carbon isotopic composition of some Earth surface and mantle materials. $\delta^{13}C$ values on the PDB or VPDB scale (from numerous sources; see Box 6.1).

surface environment that involve hydrogen, oxygen or carbon in water, carbon dioxide, and carbonate minerals have values of α close to 1 or ln α close to zero. The change in the Gibbs free energy for such reactions in the standard state (ΔG_R^o) is of an order of magnitude of 10^0-10^1 joule per mol, in comparison to 10^0-10^2 kilojoule per mol for many mineral-solution-gas reactions of environmental geochemical interest (O'Neil, 1986).

The order of the species and phases in the subscript of α in equation (6.8) should be noted in relation to the abundance ratios in the numerator and denominator. In general, for an equilibrium exchange between isotopic species X_{heavy} and X_{light} in phase 1 and phase 2, the notation is:

$$\alpha_{X_1-X_2} = \frac{(X_{heavy}/X_{light})_1}{(X_{heavy}/X_{light})_2} \qquad (6.9)$$

An inverse order of the subscripts in α for the same fractionation as in equations (6.8) and (6.9) is also used in some publications (Box 6.1).

Theoretical derivation of equilibrium fractionation between isotopic species was originally developed by Urey (1947). The methodology is based on calculation of the partition functions of diatomic or multiatomic molecules in an isotopic exchange equilibrium, as in reaction (6.7). Its applications to systems of geochemical interest and sources of errors in the calculation have been discussed and demonstrated by Bottinga (1968), Richet *et al.* (1977), O'Neil (1986), Chacko *et al.* (2001), and others cited in those references. The dependence of isotopic exchange on temperature is such that fractionation factor α tends to 1 at high temperature, varying in proportion to $1/T^2$, but the manner in which this change occurs varies greatly from one system to another. Also, the carbon isotopic difference between calcite and organic matter is greatly reduced as sediments undergo progressive metamorphism through greenschist ($300°$-$450°C$), amphibolite ($450°$-$650°C$), and granulite facies ($\geq 650°C$) with the conversion of organic carbon to graphite and isotopic exchange with the carbonate (Des Marais, 2001).

Equilibrium fractionation factors, as in equations (6.8) and (6.9), are usually written in the δ notation of the species concentrations. For the exchange between gaseous and aqueous CO_2, the fractionation factor is, with the δ values taken as fractions, where $\delta = (\delta \text{ ‰})/1000$:

$$\alpha_{CO_2(aq)-CO_2(g)} = \frac{1 + \delta^{13}C_{aq}}{1 + \delta^{13}C_g} \tag{6.10}$$

As long as δ is much smaller than 1 and α is close to 1, two approximations can be used in (6.10): $\ln(1 + \delta) \approx \delta$ and $\ln \alpha \approx \alpha - 1$. These give logarithmic and linear forms of α:

$$\ln \alpha_{CO_2(aq)-CO_2(g)} = \delta^{13}C_{aq} - \delta^{13}C_g \tag{6.11}$$

and

$$\alpha_{CO_2(aq)-CO_2(g)} - 1 = \delta^{13}C_{aq} - \delta^{13}C_g \tag{6.12}$$

From the preceding it follows that, for practical purposes, the two fractionation relationships may give essentially the same result that is also denoted ε:

$$\ln \alpha_{12} \approx \alpha_{12} - 1 = \varepsilon_{12} \tag{6.13}$$

For δ values in permil (‰), $\ln \alpha$, $(\alpha - 1)$, and ε are multiplied by 1000. The convenience of relationship (6.12) is that it gives the enrichment or depletion of one phase relative to the other in δ units. It is also consistent with the notation of the subscript in α that stands for the difference between the δ values in phases 1 and 2. Equilibrium fractionation factors as functions of temperature for some of the exchange reactions in the system CO_2-H_2O and for calcite are listed in Table 6.2. As explained in the Table footnotes, some of the fractionation factors have been computed from theoretical calculations using partition functions, some have been measured experimentally, and some have been evaluated using the additivity rule of chemical reactions. Positive values of $\ln \alpha$ or $\alpha - 1$ indicate that phase 1 (on the left-hand side of the pair) is enriched in ^{13}C relative to phase 2. It may be noted, that dissolved CO_2 is slightly depleted relative to gaseous CO_2 (Table 6.2, Nos. 1, 2). The enrichment of HCO_3^-, usually the most

Table 6.2 $^{13}C/^{12}C$ equilibrium fractionation between species shown as 1 and 2 in gaseous phase (g), aqueous solution (aq), calcite (cal), and aragonite (arag)

	Species and phase				$(\alpha_{12} - 1) \times 10^3$(‰)	
No.	1	2	ln α_{12} or $\alpha_{12}-1$, as shown	t range ($^\circ C$)	$5\,^\circ C$	$25\,^\circ C$
1	CO_2(aq)	CO_2(g)	$\alpha_{12} - 1 = 4.1 \times 10^{-6} \times t^\circ C + 0.00118^{(1)}$	0–60	−1.2	−1.1
2	CO_2(aq)	CO_2(g)	$\alpha_{12} - 1 = -0.373/T + 0.00019^{(2)}$	0–60	−1.2	−1.1
3	HCO_3^-(aq)	CO_2(g)	ln $\alpha_{12} = 1009.9/T^2 0.00454^{(3)}$	—	8.5	6.8
4	HCO_3^-(aq)	CO_2(g)	$\alpha_{12} - 1 = 9.552/T - 0.02410^{(4)}$	5–125	10.2	7.9
5	HCO_3^-(aq)	CO_2(aq)	$\alpha_{12} - 1 = 9.866/T - 0.02412^{(5)}$	5–125	11.3	9.0
6	CO_3^{2-}(aq)	CO_2(g)	ln $\alpha_{12} = 870/T^2 - 0.0034^{(6)}$	—	7.8	6.4
7	CO_3^{2-}(aq)	CO_2(g)	ln $\alpha_{12} = 9.037/T - 0.02273^{(7)}$	0–100	9.8	7.6
8	CO_3^{2-}(aq)	CO_2(g)	ln $\alpha_{12} = 8.886/T - 0.02225^{(8)}$	0–100	9.7	7.6
9	CO_3^{2-}(aq)	HCO_3^-(aq)	ln $\alpha_{12} = -0.867/T + 0.00252^{(9)}$	0–35	−0.6	−0.4
10	CO_3^{2-}(aq)	HCO_3^-(aq)	ln $\alpha_{12} = -0.666/T + 0.00185^{(10)}$	5–125	−0.5	−0.4
11	$CaCO_3$(cal)	CO_2(g)	ln $\alpha_{12} = 1194/T^2 - 0.00363^{(11)}$	—	11.8	9.8
12	$CaCO_3$(cal)	CO_2(g)	ln $\alpha_{12} = 2988.0/T^2 - 7.6663/T + 0.0024612^{(12)}$	0–600	13.5	10.4
13	$CaCO_3$(cal)	HCO_3^-(aq)	ln $\alpha_{12} = -4.232/T + 0.0151^{(13)}$	0–35	−0.1	0.9
14	$CaCO_3$(cal)	HCO_3^-(aq)	ln $\alpha_{12} = 2988.0/T^2 - 17.1883/T + 0.02656^{(14)}$	5–125	3.4	2.5
15	$CaCO_3$(cal)	HCO_3^-(aq)	$^{(15)}$	25		0.9 ± 0.2
16	$CaCO_3$(arag)	HCO_3^-(aq)	$^{(15)}$	25		1.8 ± 0.2

Notation of the fractionation factor α_{12} as in text equations (6.8) and (6.9). In α_{12} and $\alpha_{12} - 1$ defined in (6.11)–(6.13). Temperature T is in kelvin, t in degrees Celsius.

[1] Vogel et al. (1970).

[2] Equation fitted by Mook et al. (1974) to data of Vogel et al. (1970).

[3] Deines et al. (1974).

[4] Mook et al. (1974), as recalculated by Friedman and O'Neil (1977, Fig. 27).

[5] Mook et al. (1974).

[6] Deines et al. (1974).

[7] Equation attributed to data of Thode et al. (1965) (Clark and Fritz, 1997, inside front cover).

[8] Regression fit to data of Thode et al. (1965, Table III).

[9] Equation based on data in Salomons and Mook (1986) (Clark, pers. comm., 2003; Clark and Fritz, 1997, inside front cover); it differs from the ratio of α's of reaction No. 7 to No. 4, that is α of reaction No. 9.

[10] Ratio of α's of reaction No. 7 to No. 4, as difference of their ln α_{12}.

[11] Deines et al. (1974).

[12] Bottinga (1968).

[13] Equation based on data of Salomons and Mook (1986, Table 6–2A) (Clark, pers. comm., 2003; Clark and Fritz, 1997, inside front cover).

[14] Fractionation as difference between equation No. 12 and No. 4. Note that it is based on a theoretical calculation of α between calcite and CO_2 gas in reaction No. 12, and it gives a greater fractionation between calcite and HCO_3^- aqueous ion than in No. 13.

[15] Rubinson and Clayton (1969).

abundant DIC species in natural waters, is 7 to 11 ‰ relative to CO_2 gas (Nos. 3–5). This is about the same as the difference between the $\delta^{13}C$ of ocean-water DIC, about $+1$ ‰, and $\delta^{13}C$ of atmospheric CO_2, about -7 ‰ (Fig. 6.1). Fractionation between calcite and the bicarbonate ion is about 1 ‰ or smaller (No. 13). Because of this small fractionation, variations in $\delta^{13}C$ of carbonate rocks in the geologic record are usually interpreted as reflecting similar variations in $\delta^{13}C$ of ocean water, even if the underlying assumption of equilibrium may be questionable in some cases. For example, $^{13}C/^{12}C$ fractionation between CO_2 and CH_4 in geothermal gases below 200°C from Iceland, New Zealand, and Yellowstone National Park is generally lower than the calculated equilibrium fractionation factors (Cole and Chakraborty, 2001).

Outside of the domain of chemical equilibrium, isotope exchange has been exten-sively studied in non-equilibrium reactions in the carbon system between gases, aque-ous species, and solids. However, studies of the ^{13}C and ^{18}O fractionation in calcite or dolomite reacting with CO_2 or H_2O have been mostly done in a temperature range from 300° to 900°C where reaction rates are faster than at the Earth's surface temperatures.

4   Photosynthesis and Plant Physiological Responses to CO_2

Photosynthesis and respiration are important biological processes in the global carbon cycle that fractionate the isotopes of the nutrient elements involved in the production and decomposition of organic matter. This section reviews the main photosynthetic mechanisms before proceeding to a discussion of the carbon isotope fractionation by biological mechanisms in the past and present.

4.1   The Photosynthetic Process

There are few biological mechanisms on Earth that can reduce inorganic carbon to organic carbon. On a global basis, photosynthesis is the most important and is one of the principal achievements of biological evolution. The essentials of photosynthesis as a reaction between carbon dioxide and water producing glucose and free oxygen are discussed briefly in Chapter 2. The major elemental components of living organic matter are C, H, and O, followed by N and P, accounting for 95% of plant matter, and a number of other elements in minor proportions that are important to various physiological functions of the plant (Table 6.3).

The production of organic matter by freshwater and marine planktonic organisms takes carbon, nitrogen, and phosphorus in the atomic ratio C:N:P = 106:16:1, named the Redfield ratio after its principal discoverer (Redfield et al., 1963). This ratio is also used for the phytoplankton and the photosynthetic reaction between CO_2 and aqueous nitrate and phosphate ions:

$$106CO_2 + 16NO_3^- + HPO_4^{2-} + 18H^+ + 122H_2O$$
$$\rightleftharpoons (CH_2O)_{106}(NH_3)_{16}(H_3PO_4) + 138O_2 \qquad (6.14)$$

Table 6.3   The essential elements necessary for plant growth (from Nebel, 1990)

Mass % of dry plant matter	Essential element	Source as a nutrient	Occurs in:	Role in plants, essential in:
95	Carbon, C	Carbon dioxide, CO_2	Air or dissolved in water	Structure of organic molecules
	Hydrogen, H	Water, H_2O	Water	
	Oxygen, O	Carbon dioxide, CO_2	Air or water	
	Nitrogen, N	Nitrate, NO_3^-	Water solution or soil minerals	
		Ammonium, NH_4^+		
		N_2 gas (N fixation)	Air	
	Sulfur, S	Sulfate, SO_4^{2-}		Most proteins
	Phosphorus, P	Phosphate, HPO_4^{2-}		Nucleic acids and energy transfer
5	Potassium, K	Ions in aqueous solution or some soil minerals	Water solution or some soil minerals	Maintenance of water balance, in some enzymes
	Calcium, Ca			Cell membrane functions, cell walls of most plants
	Magnesium, Mg			In chlorophyll molecules
	Iron, Fe			In photosynthesis and energy transfer
	Chlorine, Cl			In photosynthesis
	Manganese, Mn			
	Boron, B			
	Zinc, Zn			
	Copper, Cu			In functions of certain enzymes
	Molybdenum, Mo			
	Cobalt, Co			

In the latter reaction the ratio of oxygen produced to carbon used, $O_2/C = 1.3$, is higher than $O_2/C = 1$ in a simpler photosynthetic reaction (2.33) because of the additional reduction of NO_3^- to NH_3. Respiration or oxidation of organic matter in (6.14), proceeding from the right to the left, produces CO_2, nitrate, phosphate, and water. Under anaerobic conditions, the reaction can lead to the formation of CO_2, methane, and ammonia by heterotrophic bacteria:

$$(CH_2O)_{106}(NH_3)_{16}(H_3PO_4) + 14H_2O \rightarrow 53CO_2 + 53CH_4 + 16NH_4^+$$
$$+ HPO_4^{2-} + 14OH^- \qquad (6.15)$$

In land plants, average C:N:P atomic ratios vary from 510:4:1 (Delwich and Likens, 1977) to 822:9:1 (Deevey, 1973) and 2057:17:1 (Likens et al., 1981). Land-plant photosynthesis produces organic matter with a relatively much higher concentration of carbon than in the aquatic phytomass, in such a reaction as, for example:

$$2057\,CO_2 + 17NO_3^- + HPO_4^{2-} + 19H^+ + 2074\,H_2O$$
$$= (CH_2O)_{2057}(NH_3)_{17}(H_3PO_4) + 2091\,O_2 \qquad (6.16)$$

Photosynthesis in terrestrial and aquatic plants involves net transport of carbon from the surrounding medium into the plant cell. For terrestrial plants, the medium is the atmosphere with its gaseous CO_2; for aquatic plants, the medium is the water with its dissolved inorganic carbon. The site of carbon fixation in any eukaryotic cell is the chloroplast and its smaller morphological units of grana, thylakoids, and stroma surrounding the latter and where CO_2 is stored. Inorganic carbon may cross the membranes simply by diffusion along a concentration gradient from the external medium to the site of carbon fixation, the chloroplast. In this case, because CO_2 is the only major carbon species that can freely cross the chloroplast membranes due to the fact that it is an uncharged molecule, the gas must dissolve in the membrane lipid phases, diffuse across the membranes, and dissolve back into the aqueous phase of the chloroplast. In terrestrial plants with openings in the plant epidermis, known as stomata, the rate of gaseous diffusion to the cell walls of the mesophyll is enhanced through the stomata and is higher than their total area would suggest; hence the concentration of CO_2 in the intercellular spaces of the mesophyll is usually very close to that of the ambient air for most plants.

In aquatic plants the slow kinetic equilibration between CO_2 and H_2CO_3/HCO_3^- is rapidly catalyzed by the enzyme carbonic anhydrase, promoting the conversion of HCO_3^- to CO_2 for photosynthesis. Carbonic anhydrase is a group of related enzymes, all of which use Zn in their active site. It is one of the most catalytically reactive enzymes known, with a not unusual turnover rate of 600,000 s^{-1} (Falkowski and Raven, 1997). The activity of the enzyme is located mainly on the plasmalemma and as a soluble, extracellular enzyme. Although in some unicellular algae and macrophytes, the rates of diffusive fluxes of CO_2 seem adequate to support photosynthetic demands, it appears that in the majority of aquatic plants, there are carbon-concentrating mechanisms that actively transport carbon to the site of fixation. The location and mechanism of this active carbon pump have been difficult to pinpoint; however, the plasmalemma appears to be the only possible location and semi-crystalline arrays of the enzyme Rubisco associated with carbonic anhydrase activity appear to be involved in the carbon-concentrating mechanism.

4.2   Mechanisms of Carbon Incorporation

Photosynthesizing organisms use preferentially the light carbon isotope ^{12}C in making their organic matter, which is thought of as an evolutionary adaptation that uses the lighter-isotope compounds where bond energies are lower and reaction rates faster than in their heavy-isotope equivalents. This fractionation differs from the equilibrium fractionation discussed in Section 3, insofar as the process of organic matter formation from CO_2 and water is not a chemical equilibrium process:

$$\delta^{13}C \text{ of organic matter} = (\delta^{13}C \text{ of } CO_2 \text{ source}) - \varepsilon^{13}C \qquad (6.17)$$

where $\varepsilon^{13}C$ is a fractionation factor in the same units as $\delta^{13}C$. A typical value of biological fractionation is, for example, the difference between $\delta^{13}C$ of atmospheric CO_2 (Fig. 6.1) and that of the C_3 land plants: $\varepsilon^{13}C \approx (-7 \text{ ‰}) - (-27 \text{ ‰}) = 20 \text{ ‰}$. Plants build their tissue from CO_2 that is converted to the basic form of $C_6H_{12}O_6$ by three main photosynthetic pathways. The more wide-spread C_3 pathway (called C_3 because the initial stable product of this photosynthetic pathway is a 3-carbon compound 3-phosphoglycerate) characterizes 85 to 96% of known plant species, the C_4 pathway (where the first stable end product is a 4-carbon compound oxaloacetate) is common to 3 to 14% of plant species, and the crassulacean acid metabolism pathway (CAM) is exhibited by only 1% of plant species. Sage (2004) estimated that about 7500 species of flowering plants or 3% of all the land plant species use C_4 photosynthesis. Although C_4 plants represent only approximately 4% of global plant biomass, they account for 20 to 27% of global primary production (Randerson et al., 2005). The C_3 plants are mainly trees, shrubs, and cool-climate grasses. Forests, shrubland, and prairies are generally C_3 plants that respire up to 50% of the photosynthetically fixed carbon. Figure 6.2 shows the essential features of CO_2 fixation in the C_3 and C_4 pathways. The C_3 photosynthetic mechanism characterizes all the plants, but the C_4 pathway, also known as Krantz or Hatch-Slack pathway, is an addition to some of the plant groups. The smaller group of C_4 plants includes tropical and subtropical grasses, and marsh and wetland plants. They retain most of the photosynthetically produced carbon, respiring only about 1% of it. C_4 photosynthesis in the terrestrial environment probably became widespread about 6 to 8 million year ago with the expansion of grasslands dominated by C_4 species in response to declining atmospheric CO_2 concentrations (Cerling, 1999). The oldest record so far of a C_4 plant is from a 12.5 Ma Late Miocene formation in California (Tidwell and Nambudiri, 1989). C_4 plants have evolved a unique cell architecture called Krantz leaf anatomy and have developed very effective mechanisms for modification of their internal CO_2 concentrations in both gas and aqueous microenvironments. The C_4 plant is capable of lowering its internal CO_2 gas concentration well below atmospheric thus creating a substantial CO_2 gradient between the atmosphere and mesophylic cells. However, at the same time the CO_2 concentration in the aqueous phase of the bundle sheath cells of the plant is increased well above the atmospheric CO_2 gas-water equilibrium value for subsequent assimilation of carbon by the C_3 pathway. Thus CO_2 is not rate-limiting for C_4 plant photosynthesis, as it is for C_3 plants, and C_4 species would be expected to respond less than C_3 plants to increasing atmospheric CO_2 concentrations. C_4 plants include some of the most productive crops, including maize, sugarcane, and sorghum, but also weeds such as purple nutsedge, Bermuda grass, and pigweed.

The CAM pathway has been hypothesized to be a metabolic adaptation to drought or otherwise dry conditions (e.g., Ting, 1994). The adaptation has resulted in these plants closing their stomata during the day and opening them during the night when evaporative transport is low. Another hypothesis is that CAM is an adaptation to low daytime CO_2 levels because many, if not most, CAM plants are tropical epiphytes

C₃ photosynthesis

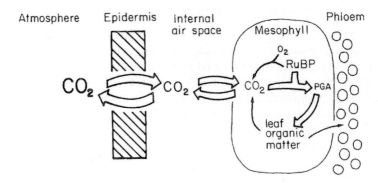

C₄ photosynthesis

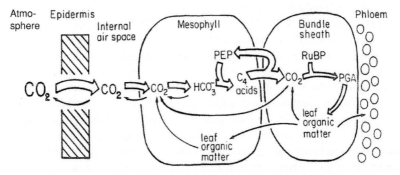

Figure 6.2   Photosynthetic pathways of C_3 and C_4 plants (from O'Leary, 1988; by permission of the author and American Institute of Biological Sciences). In the C_3 pathway, enzyme Rubisco (ribulose 1,5-biphosphate carboxylase) catalyzes a reaction between CO_2 and RuBP (ribulose 1,5-biphosphate) that produces two phosphoglycerate molecules (PGA), a 3-carbon compound giving the name to the C_3 pathway. In C_4 photosynthesis, CO_2 combines with PEP (phospho-enolpyruvate, a 3-carbon compound), catalyzed by PEP carboxylase, to form oxaloacetate and malate, both C_4 acids.

that live in forest canopies that have low atmospheric CO_2 concentrations during the day when photosynthesis is at a maximum. However, the stomata of the plants are open at night when atmospheric CO_2 concentrations are relatively elevated due to respiration. CAM plants fix carbon derived from HCO_3^- into organic acids such as malate that accumulates as malic acid during the night. During the day malic acid is decarboxylated to free CO_2 and a 3-carbon compound. Because the stomata are closed during the day, the CO_2 is trapped and water retained, and the CO_2 is assimilated by the C_3 pathway. CAM plants respond positively to elevated CO_2 levels but generally display lower photosynthetic rates than C_3 or C_4 plants (Allen and Amthor, 1995).

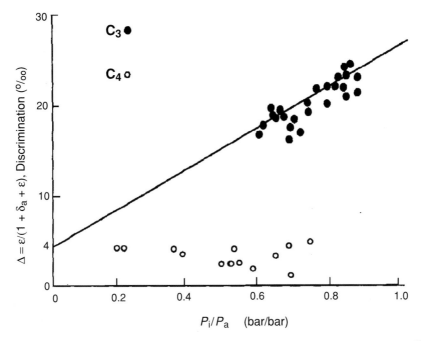

Figure 6.3   Discrimination of ^{13}C by C_3 and C_4 plants, equation (6.18). δ_a is the value of $\delta^{13}C$ (fraction) in the atmosphere. P_i/P_a is the internal to ambient CO_2 partial pressure ratio. From Farquhar *et al.* (1989), by permission of G. D. Farquhar.

Differences between the C_3 and C_4 plants in the mechanisms of carbon isotope fractionation are shown in Fig. 6.3. In C_4 plants, the main fractionation of atmospheric $^{13}C/^{12}C$, about -4 ‰, occurs in the smaller intake pores by diffusion (Table 6.4). In C_3 plants, the fractionation is a function of the intracellular to ambient CO_2 partial

Table 6.4   Isotopic fractionation of $^{13}C/^{12}C$ in steps leading to CO_2 fixation in plants, α defined as in equations (6.9) and (6.10); $\varepsilon = \alpha - 1 = \delta^{13}C_{product} - \delta^{13}C_{reactant}$ (from Farquhar *et al.*, 1989; O'Leary 1993; and Table 6.2)

	Process	Fractionation factor (ε, ‰)
1.	Diffusion of CO_2 in air through the stomatal pores	-4.4
2.	Diffusion of CO_2 in air through the boundary layer to the stomata	-2.9
3.	Diffusion of dissolved CO_2 through water	-0.7
4.	Net C_3 fixation with respect to internal P_{CO2}	-26.5
5.	Fixation of gaseous CO_2 by Rubisco from higher plants	-29
6.	Fixation of HCO_3^- by PEP carboxylase	-2
7.	Fixation of gaseous CO_2 by PEP carboxylase (at equilibrium with HCO_3^- at $25°C$)	5.7
8.	Equilibrium hydration of CO_2 at $25°C$	9.0
9.	Equilibrium dissolution of CO_2 into water	-1.1

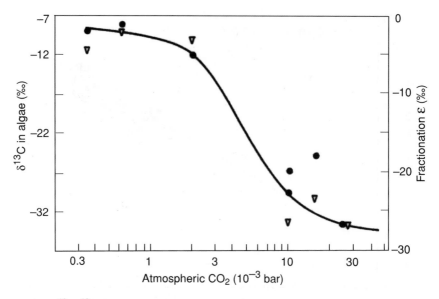

Figure 6.4 $^{13}C/^{12}C$ fractionation by species of algae grown at different CO_2 concentrations in experiments one year apart, in Pretoria, South Africa (circles and triangles; altitude 1380 m, atmospheric pressure 0.866 bar). Fractionation ε is the difference $\delta^{13}C_{product} - \delta^{13}C_{source}$. After Vogel (1993).

pressures. As mentioned previously, a characteristic feature of C_4 plants is retention of CO_2 in their cells that may be advantageous during periods of relatively low atmospheric CO_2 concentration (Cerling, 1992, 1999; Deines, 1980a). Discrimination of ^{13}C by plants drawing their carbon from the environment was defined by Farquhar *et al.* (1989) as:

$$\Delta = \frac{R_a}{R_p} - 1$$

$$= \frac{\varepsilon}{1 + \delta_a - \varepsilon} \tag{6.18}$$

where R_a is the ambient $^{13}C/^{12}C$ ratio and R_p is the same ratio in plants; δ_a for the ambient value of $^{13}C/^{12}C$ and fractionation factor ε are fractions, as defined in equation (6.17) and Table 6.4. Because the fractional values in the denominator of equation (6.18) are small (atmospheric δ_a is -0.006 to -0.008, and $\varepsilon = 0.01$ to 0.027), the fractionation factor ε is approximately equal to the discrimination factor Δ: $\varepsilon \approx \Delta$ (Fig. 6.3).

An experimental study of the isotopic fractionation in species of algae grown at different atmospheric CO_2 concentrations (Fig. 6.4) shows very little fractionation relative to atmospheric CO_2 in the range of 300 to 600 ppmv CO_2, but fractionation increases to -27 ‰ at the elevated atmospheric CO_2 levels. The kinetic fractionation in photosynthesis can also be represented as a sum of fractionation in the CO_2 gas diffusion

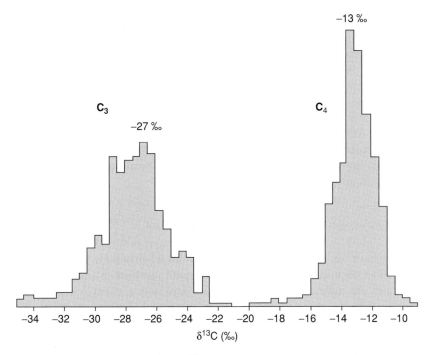

Figure 6.5   Frequency distributions of $\delta^{13}C$ values in C_3 and C_4 plants, based on about 1000 analyses done in five different laboratories (O'Leary, 1988; by permission of the author and American Institute of Biological Sciences). Earlier frequency distribution data published by Deines (1980a) and Vogel (1980).

through the pores (ε_d) and the fractionation in the enzyme-mediated production of organic compounds within the cell (ε_c) that can be written as (e.g., Vogel, 1993):

$$\varepsilon = \varepsilon_d + (\varepsilon_c - \varepsilon_d) \cdot C_i / C_a \quad (\text{‰}) \qquad (6.19)$$

where C_i is the intracellular CO_2 concentration and C_a is the ambient concentration, and the fractionation factors are negative numbers. For example, using the data from Table 6.4, $\varepsilon_d = -4.4$ and $\varepsilon_c = -29$, the net fractionation near $C_i / C_a \approx 0.8$ is about $\varepsilon = -24.6$ ‰, comparable to the values shown in Fig. 6.3. Variants of equation (6.19) are extensively used in studies of isotopic fractionation by plants and aquatic organisms, where the modeling of the photosynthesis allows for several different fractionation steps on the way from the CO_2 entry into the cell to the final reaction site. In such models, several fractionation terms are used to describe a photosynthesizing cell as a multicompartment system of successive reactions, more detailed and complex than the system shown in Fig. 6.2.

   The differences in the mechanisms and magnitudes of the carbon-isotope fractionation between the two major plant groups, C_3 and C_4, are reflected in the $\delta^{13}C$ values (Fig. 6.5). Despite the relatively wide ranges of variation within each group, the C_4

plants with the modal value of $\delta^{13}C \approx -13$ ‰ are clearly isotopically heavier by 14 ‰ than the C_3 plants that are about -27 ‰.

Extensive data from laboratory studies and natural occurrences show that different components of photosynthetically or heterotrophically produced organic matter differ in their $\delta^{13}C$ values from the whole product (Hayes, 2001). The four main groups of organic compounds that constitute organic matter—nucleic acids, proteins and amino acids, mono- and polysaccharides, and lipids—have different fractionation factors $\varepsilon^{13}C$ relative to the CO_2 source, as given in equation (6.17). The overall composition of the organic matter is a weighted mean of the individual organic component fractions. In the individual steps of biosynthesis, the biological isotope effects are characterized by values of ε that may vary from 8 to 35 ‰.

4.3   Physiological Responses to Atmospheric CO_2

Plant photosynthesis depends on the availability of carbon dioxide in the atmosphere, in addition to the availability of water and mineral nutrients in soil, under suitable temperatures and sufficient light. There exists a large body of experimental data on the effects of carbon dioxide on the metabolism of plants (e.g., Cullen *et al.*, 2001; Tolbert and Preiss, 1994; Woodwell and Mackenzie, 1995). Early studies dealt with the effects of elevated CO_2 concentration on individual plants; more recent studies have tried to determine CO_2 effects on communities of organisms in which portions of ecosystems have been subjected to changing atmospheric CO_2 levels. The growth of vascular plants increases with increasing atmospheric CO_2 to levels of 550 to 1000 ppmv (55 to 100 Pa partial pressure), but it stops in experimental studies when the ambient CO_2 concentration is about 100 to 150 ppmv or about 25 to 40% of the present-day atmospheric CO_2 (Drake *et al.*, 1997; Polley *et al.*, 1993). Lower atmospheric CO_2 concentrations of less than 10 ppmv can be tolerated by C_4 plants (Caldeira and Kasting, 1992). These CO_2 concentrations may be compared with the ambient 185 ppmv CO_2 at the Last Glacial Maximum, about 18,000 years ago, 280 ppmv at the end of pre-industrial time, about 300 years ago, and 380 ppmv in the year 2005.

Figure 6.6 illustrates the general response of plants to changing CO_2 concentrations. Photosynthesis, respiration, and transpiration are the terrestrial plant processes most directly affected by changing concentrations of CO_2. However, a number of interactive changes in plant growth resulting in negative and positive feedbacks accompany these primary changes. In general, if atmospheric levels of CO_2 were rising without concomitant changes in temperature or water regimes, then one might expect higher rates of photosynthesis. This is simply a result of the increased CO_2 gradient between the atmosphere and the air spaces inside the leaf that would promote the diffusive transfer and absorption of CO_2 into the chloroplasts and its conversion to carbohydrate. As intercellular CO_2 within the leaf increases, leaf photosynthetic CO_2 uptake rates increase. Note, however, that not all plants respond similarly. For cotton there appears to be no effect of increasing ambient CO_2 on uptake rate; for cabbage both the initial slope of the photosynthetic rate and the CO_2-saturated photosynthetic rate were decreased by elevated CO_2; and for soybeans both the initial slope and CO_2-saturated

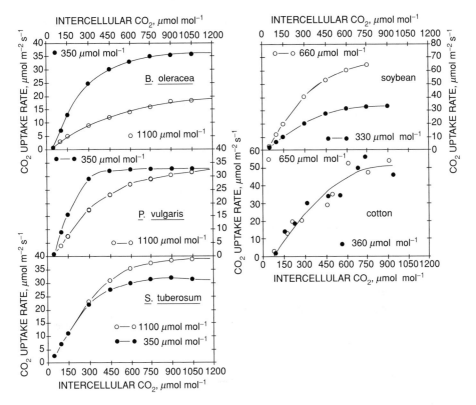

Figure 6.6   Leaf photosynthetic CO_2 uptake rates versus intercellular CO_2 concentration for five types of responses to elevated CO_2 concentrations in cabbage (*Brassica oleracea* L.), kidney bean (*Phaseolus vulgaris* L.), potato (*Solanum tuberosum* L.), soybean (*Glycine max* L. Merr.), and cotton (*Gossypium hirsutum* L.). Different symbols refer to the ambient CO_2 concentrations during plant growth (after Allen, 1994).

photosynthetic rate were increased. Ainsworth *et al.* (2002), from an analysis of the effects of elevated CO_2 on 25 variables from 111 studies for soybean physiology, growth, and yield, have shown that many factors can affect yield, even the pot size of the individual experiment. Very large pots and containers failed to predict the stimulation in yield of plants grown in the ground at elevated CO_2 concentrations. The authors concluded that their analysis of soybean supported several current paradigms of plant responses to increasing CO_2 concentrations. Stimulation of photosynthesis was greater for soybean plants that fix nitrogen than for those that do not fix nitrogen. Thus it appears that as a generality, photosynthetic capacity decreases when plant growth is nitrogen limited but not if the plants have adequate nitrogen. Ainsworth and co-workers also found that the root/shoot ratio of the soybean plants did not change with growth under elevated CO_2 conditions. The conclusion from the above studies is that there is no single type of acclimation of leaves or terrestrial plants in general to increasing CO_2 concentrations. However, as discussed in Chapter 11, rising CO_2 levels have probably been in part responsible for fertilization of terrestrial ecosystems, leading to a negative feedback on rising atmospheric CO_2 concentrations.

5   Isotopic Fractionation and ^{13}C Cycle

Biological fractionation of carbon isotopes is to a very large degree responsible for the isotopic composition of the major reservoirs of the carbon cycle on the Earth's surface that is shown in Fig. 6.7. The mean carbon isotopic composition of the Earth's surface reservoirs may be expected to be the same as that of the upper mantle. This reasoning leads to the estimates of $\delta^{13}C = -5$ to -7 ‰ for the mantle (Cartigny, 2005; Fuex and Baker, 1973; Garrels and Lerman, 1981; Hayes *et al.*, 1992; Schidlowski, 1988; Schidlowski *et al.*, 1983). A different approach is based on the carbon isotopic composition of diamonds, kimberlites, and carbonatites forming in the mantle. The $\delta^{13}C$ values of diamonds cluster in the range from -3 to -8 ‰, with the frequency distributions tailing off to less negative values and, for diamonds, to much more negative

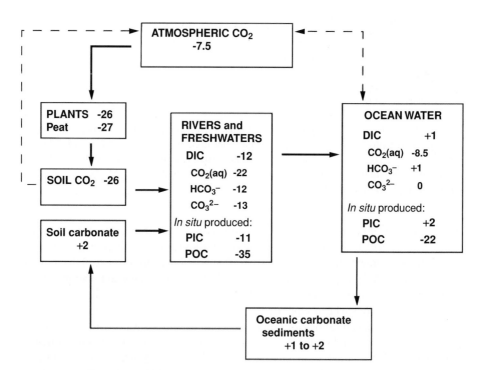

Figure 6.7 δ^{13}C values (‰) of Earth's surface reservoirs (VPDB scale, from Mook and Tan, 1991). Values of *in situ* produced particulate inorganic carbon (PIC) are for $CaCO_3$ forming at equilibrium with dissolved inorganic carbon (DIC) in water, and those of particulate organic carbon (POC) are for photosynthetically produced organic matter.

but much less frequent values, some as low as -30 ‰ (Fig. 6.1; Deines, 1980b; Hoefs, 1997). The bulk of the sampled materials, between -3 ‰ and -8 ‰, might represent the isotopic composition of the mantle carbon if there were no fractionation in the process of mineral formation from the melt or gas. Faure (1986) pointed out that evidence indicates that fractionation between CO_2, diamond, graphite, and calcium carbonate extends above $1000°C$ into the range where diamonds are believed to form. The calculated fractionation of $^{13}C/^{12}C$ for combinations of pairs of CO_2, diamond, graphite, and $CaCO_3$ (Bottinga, 1969a,b) indicates that the mineral phases are depleted relative to CO_2, and $\delta^{13}C$ of $CaCO_3$ is lower by about -2 ‰ (at $700°C$) and of diamond by -4.5 ‰ (at $1000°C$) than the CO_2 value. Thus for the isotopic composition of diamonds and carbonatites to be that of the mantle carbon, the case of no isotopic fractionation between them must be considered as a limiting case. Then, however, $\delta^{13}C = -5 \pm 3$ ‰ for the mantle carbon is comparable to the mean $\delta^{13}C$ of the surface reservoirs.

One-thousand years before present, atmospheric CO_2 had the value of $\delta^{13}C \approx -6.5$ ‰, as measured in the air bubbles in Antarctic ice cores (Indermühle et al., 1999; Fig. 6.8). Combustion of fossil fuels and, possibly, increased oxidation of soil organic

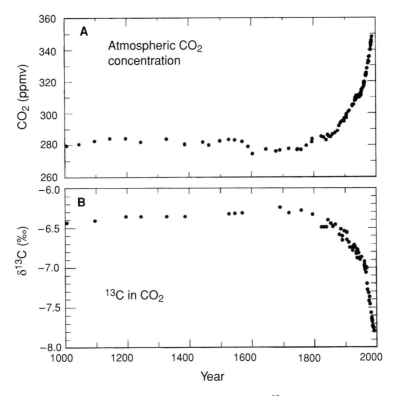

Figure 6.8   Atmospheric CO_2 concentration and its isotopic ^{13}C-composition since the year 1000, from ice cores and Mauna Loa, Hawaii, atmospheric record (after Trudinger et al., 1999).

matter due to human land use in the last 200 to 300 years have been adding isotopically lighter CO_2 to the atmosphere, of $\delta^{13}C$ between -18 ‰ and perhaps -29 ‰. By the mid-1950s, atmospheric CO_2 became diluted with isotopically light carbon to the level of $\delta^{13}C \approx -7$ ‰, and this dilution was due to the increase in the concentration of atmospheric CO_2, as shown in Fig. 6.8. Since 1980, an average decrease in $\delta^{13}C$ has been about 0.02 ‰ per year, leading to the projected value at the beginning of the 21st century of $\delta^{13}C \approx -8$ ‰ (Keeling et al., 1979, 1984; Mook et al., 1983). The dilution of atmospheric CO_2 with carbon of lower $^{13}C/^{12}C$ ratio is similar to the dilution of the ^{14}C concentration, known as the Suess effect and named after its discoverer. Hans E. Suess, working at the U. S. Geological Survey at the time, reported in the first half of the 1950s that carbon in parts of trees that were formed in mid- to late-19th century contained more radioactive ^{14}C than in the samples from the 1930s to 1950s (Suess, 1955). Because the half-life of ^{14}C is 5715 years, old fossil fuels and carbon dioxide formed from their combustion are essentially free of this nuclide.

The increase in atmospheric CO_2 concentration from the pre-industrial level of 280 ppmv to 350–380 ppmv in the late 1990s and early 2000s represents an increase in mass of up to 35%, making the pre-industrial carbon mass a fraction 0.75 and the added amount a fraction 0.25 of the present-day atmosphere. It should be recalled that the total mass of CO_2 emissions from fossil fuel burning and land-use change since the beginning of industrial time is greater than the increase in the atmosphere, and the balance is variably attributed to the ocean or both the ocean and the land organic carbon reservoir (Chapters 5, 10, and 11). An atmosphere containing 25% of CO_2 added from fossil fuels and organic matter on land, of $\delta^{13}C$ between -23 ‰ and -27 ‰, and 75% of pre-industrial CO_2, of $\delta^{13}C = -6.5$ ‰, would be a mixed atmosphere with $\delta^{13}C = -10.6$ to -11.6 ‰. However, because atmospheric CO_2 is isotopically heavier, $\delta^{13}C = -7$ to -8 ‰, there must have been additional exchange of CO_2 with a reservoir isotopically heavier than fossil fuels and land organic carbon, such as the ocean. The following simple estimate demonstrates the magnitude of such an exchange, as based on the work of Kroopnick (1985), Druffel and Benavides (1986), Quay et al. (1992), Hoefs (1997), and Böhm et al. (2002). In surface ocean water, DIC had a mean value of $\delta^{13}C \approx +2$ ‰ around the year 1970, although the variation between different oceanic sites is a few tenths of one permil. Since the years 1800–1820 to 1970, the $\delta^{13}C$ of the surface water layer declined by -0.5 to -0.6 ‰, as was determined from the composition of aragonite secreted by the sclerosponge Ceratoporella nicholsonii (Hickson 1911, 1912), indicating a pre-industrial value of $\delta^{13}C \approx +2.5$ ‰ for surface-ocean DIC. Since 1970 to 1990, Pacific Ocean surface water at different locations declined further by -0.3 to -0.6 ‰, and the aragonite record of Jamaican sponges shows a decline of about -0.4 ‰. Taking $\delta^{13}C = 2.5$ ‰ as representative of the most abundant carbonate species, HCO_3^-, the gaseous species CO_2 at equilibrium with it may be about 9 ‰ lighter and have $\delta^{13}C \approx -6.5$ ‰, as in the pre-industrial atmosphere (Table 6.2, reaction 4). An exchange of 20 to 25% of CO_2 in the mixed atmosphere with this oceanic source would result in an atmosphere of $\delta^{13}C \approx -7.5$ ‰. Although this calculation is simplistic because it does not describe the continuous two-way flows of CO_2 between the atmosphere and surface ocean layer nor the exchange of DIC between the surface and deep ocean, where DIC is isotopically lighter, at least in part due to remineralization of the settling organic matter, it does point to the importance of the carbon exchange between surface ocean water and the atmosphere in the global carbon balance.

Efforts to detect the past changes in the atmospheric $^{13}C/^{12}C$ ratio in the carbon of old trees have been only variably successful (Broecker and Peng, 1982; Deines, 1980a). On land, trees fractionate $^{13}C/^{12}C$ differently in their different morphological parts and in different classes of organic-biochemical compounds, and their fractionation has also been reported to depend on such environmental factors as temperature of the growing season and atmospheric precipitation. This variability makes it difficult to assign a sufficiently accurate representative value to land plants for tracing the industrial perturbation of atmospheric CO_2. The decrease in atmospheric $\delta^{13}C$ could conceivably be counteracted by a greater biological fixation of atmospheric CO_2 and its storage on land, effectively removing more light carbon from the atmosphere. Despite the reports of increased carbon storage in temperate and boreal climatic zone forests in the last decades of the 1990s, there is little evidence to point to greater sequestration of organic carbon on land during most of the past century. To the contrary, the evidence available suggests net loss of organic carbon from land to the atmosphere and ocean for much of this centurial period of time. Only during recent decades has the land biosphere been at times a net sink of atmospheric CO_2, albeit an erratic sink (Chapter 11).

The land part of the global carbon cycle (Figs. 1.4, 6.1, and 6.7) bears a strong imprint of carbon derived from the biomass that is isotopically light. An important feature of the present-day cycle is the low $\delta^{13}C$ values in the riverine input to the ocean. These reflect the production of isotopically light CO_2 in soils from organic matter that is evidently greater than the contribution of dissolved carbonate from the isotopically heavier marine limestones exposed on land. Reported values of $\delta^{13}C$ in land drainage and river waters vary with the terrain (Mook and Tan, 1991): West European rivers and the Mackenzie draining mixed-lithology terrains and soils range mostly from -6 to -12 ‰; the Amazon River had a range in the late 1970s from -13 to -23 ‰, probably due to a large contribution from isotopically light organic matter in a tropical climate. The value of -12 ‰ for DIC in continental waters indicates a contribution of about 70% from soil CO_2 and organic matter and about 30% from dissolution of limestones (see p. 251).

Most land plants and marine phytoplankton overlap in their $\delta^{13}C$ range of -22 to -28 ‰, which is also the range of sedimentary organic matter, including coal. This explains perhaps the tendency to use the value of -25 ‰ as a representative average for sedimentary organic matter (Fig. 6.1). In the ocean, the process called the "biological pump" (Chapters 1 and 10) takes light carbon from surface water by photosynthesis, transfers it to the deep ocean by settling organic matter of dead organisms and fecal pellets, and returns it in part by decomposition to ocean water. Aquatic photosynthesizing plants usually use CO_2 dissolved in water rather than HCO_3^- or CO_3^{2-} as their source of carbon, whereas the fractionation factor $\varepsilon^{13}C$ usually refers to $\delta^{13}C$ of total dissolved inorganic carbon (DIC) rather than to its CO_2 fraction. The spread of $\delta^{13}C$ values for different groups of planktonic and benthic primary producers is considerable (see also Fig. 6.1): -10 to -22 ‰ for algae and macrophytic plants, and -18 to -31 ‰ for plankton. However, a value of -20 to -21 ‰ is usually considered a representative average for marine plankton. Particulate organic carbon in the oceans (POC) is slightly lighter, -22 ‰, than dissolved organic carbon (DOC), -20 ‰, which may possibly indicate additional biological fractionation. $\delta^{13}C$ values in the range of -10 to -15 ‰, higher than the -20 to -25 ‰ of sedimentary organic matter, have been reported in marine near-shore sediments, such as in the Gulf of Mexico (Sackett, 1989). It may be that this occurrence and similarly elevated values of organic carbon $\delta^{13}C$ in other near-shore locations reflect inputs of organic matter from coastal marshes and wetlands where C_4 plants of higher $\delta^{13}C$ are more abundant.

There has been a considerable amount of work done on the relationships between the biological fractionation of $^{13}C/^{12}C$ by oceanic phytoplankton and such environmental factors as water temperature, dissolved CO_2 concentration, and physiology of the organisms. Such studies have focused on particulate organic matter in the oceans, sediments, and individual organic compounds (e.g., Arthur *et al.*, 1985; Madigan *et al.*, 1989; Popp *et al.*, 1989; Rau *et al.*, 1992; Thompson and Calvert, 1994). More recent laboratory studies have further demonstrated that the $^{13}C/^{12}C$ fractionation by marine phytoplankton depends on such physiological and morphological factors as the cell growth rate and cell shape, in addition to the dissolved CO_2 concentration (Laws *et al.*, 1985, 1997; Popp *et al.*, 1998). As a whole, these results indicate that carbon isotopic fractionation by marine phytoplankton is very variable because of the diverse physiological factors, and at this time few reliable inferences of dissolved or atmospheric CO_2 concentrations can be made from the $^{13}C/^{12}C$ values of the planktonic organic matter.

6   Long-Term Trends

6.1   Sedimentary Record

A long-term consequence of the production and sedimentary storage of organic matter is a trend toward isotopically heavier CO_2 in the Earth's surface reservoirs that may be balanced by the weathering of old organic matter and return of the isotopically light carbon to the environment. The $\delta^{13}C$ values of organic matter and carbonate sediments are shown in Fig. 6.9 since Late Proterozoic time, approximately 800 Ma ago.

The difference between the $^{13}C/^{13}C$ ratio of marine organic matter and marine limestones during the Phanerozoic fluctuates, in units of $\delta^{13}C$, around -30 ‰. Variations in this value have been interpreted as possibly due to changes in the mechanisms of biological fractionation through time. In the Late Proterozoic, large variations in the isotopic values of organic and carbonate carbon may correspond to the times of three glaciations that occurred between 800 and 600 Ma ago, when rates of storage of organic carbon might have been relatively low (Hayes *et al.*, 1999). A decline in $\delta^{13}C_{carb}$ since the beginning of the Cenozoic Era about 65 Ma ago is usually attributed to the rise in the elevation of land and emergence of the major mountain terrains, such as the Alps and the Himalayas, that exposed more of the old organic carbon in shales and thereby enabled its faster weathering. At the same time, however, the reasons behind

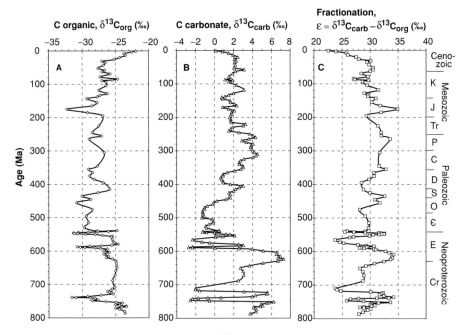

Figure 6.9   Carbon isotopic composition (δ^{13}C, in ‰) in Late Proterozoic (age >570 Ma) and Phanerozoic time of organic matter (A) and carbonates (B). Fractionation between carbonate and organic carbon, ε, in ‰, (C). Hayes *et al.* (1999). Data provided J. M. Hayes (Woods Hole Oceanographic Institution, Woods Hole, Mass., June 2003) and plotted by his permission.

a decrease in biological fractionation during the same period are not clear. Chapter 7 further discusses the carbon isotopic compositional changes during geologic time.

6.2   Sedimentary Isotopic Balance

The isotopic composition of carbon that is transported from land to the oceans places a constraint on the removal of $CaCO_3$ and organic carbon from ocean water into sediments. In an ocean that is in a quasi-steady state with respect to the input of carbon from the land and its removal into sediments (fluxes F), the balance can be written as (e.g., Hayes *et al.*, 1999):

$$F_{\text{input-C}} = F_{\text{carbonate-C}} + F_{\text{organic-C}} \qquad (6.20)$$

$$\delta^{13}C_{\text{input}} = f\delta^{13}C_{\text{carb}} + (1-f)\,\delta^{13}C_{\text{org}} \qquad (6.21)$$

$$\delta^{13}C_{\text{input}} = \delta^{13}C_{\text{carb}} - (1-f)\,\varepsilon^{13}C \qquad (6.22)$$

where f is the $CaCO_3$ mass fraction of total carbon removed by sedimentation and ε^{13}C is the fractionation factor in production of organic matter, as defined in Fig. 6.9C:

$$\varepsilon^{13}C = \delta^{13}C_{\text{carb}} - \delta^{13}C_{\text{org}} \qquad (6.23)$$

Input of carbon from land consists of three components: atmospheric CO_2 dissolved in water, dissolved carbon from the oxidation of organic matter in soils and old sediments, and carbon from the dissolution of limestones. A value of -5 ‰ for input to the ocean was used by Kump (1991), based on the weathering proportions of about 0.2 organic matter and 0.8 limestones, as representative of the geologic record. Values of $\delta^{13}C \approx -12$ ‰ in rivers are not at equilibrium with the atmosphere of -6 to -7 ‰, as follows from the equilibrium fractionation factors in Table 6.2: in land surface waters where the bicarbonate ion accounts for a large fraction of DIC, an equilibrium $\delta^{13}C$ value would be higher. Lower values indicate the presence of dissolved organic carbon (DOC) from soil organic matter and inputs from soil CO_2 that forms from respiration of organic matter and is therefore isotopically lighter than the atmosphere. For riverine input taken as $\delta^{13}C_{input} = -5$ to -7 ‰, the composition of carbonate sediments as $\delta^{13}C_{carb} = 1.5$ to 2 ‰ during the last 150×10^6 years (Fig. 6.9B), and a biological fractionation factor $\varepsilon^{13}C = 25$ to 30 ‰ (Hayes et $al.$, 1999), the $CaCO_3$ sediment fraction f that satisfies the steady-state condition is from equation (6.22):

$$f = 0.66 \text{ to } 0.78 \qquad\qquad (6.24)$$

The preceding result applies to a steady-state system where uplift and erosion produce average carbon input to the ocean that is a weighted sum of the carbonate and organic carbon reservoirs. The mass of carbonate sediments preserved in the geologic record, as given in Table 1.2, is 84% of the total sedimentary carbon, and this value of 0.84 is somewhat greater than the estimate of 0.66–0.78 in equation (6.24). Shales containing organic carbon, and limestones and dolomites containing carbonate-carbon are likely to be weathered at different rates (Chapter 8), and the generalized two-reservoir model of the sedimentary carbon cycle, as given in equation (6.20), does not take into account other processes within the cycle.

The present-day value of DIC input to the ocean from rivers, $\delta^{13}C_{input} \approx -12$ ‰ (Fig. 6.7), is not consistent with the preceding estimate of the carbonate (0.66–0.78) and organic-carbon (0.34–0.22) mass fractions in sediments at the same biological isotopic fractionation $\varepsilon^{13}C$ of 25 to 30 ‰. However, the present-day production of organic carbon by photosynthesis and of carbonate carbon by $CaCO_3$-secreting plankton is in a ratio very different from their proportions in the geologic record: the mass ratio C_{carb}/C_{org} in the geologic record is $5/1$ to $4/1$, whereas in the surface layer of modern oceans it is between $1/4$ and $1/4.8$ (Broecker and Peng, 1982, p. 476). These modern-ocean ratios correspond to the carbonate-carbon fraction $f = 0.2$ to 0.17. Calcite of benthic and planktonic organisms of ages <1 Ma has a mean $\delta^{13}C \approx 1.4$ ‰ (with a range from -1.0 to $+3.1$ ‰; Veizer et $al.$, 1999). Isotopic fractionation that is needed to obtain these mass fractions of carbonate-carbon and organic carbon from modern-river input is from equation (6.23):

$$\varepsilon^{13}C \approx \frac{1.4 - (-12)}{0.80 \text{ to } 0.83} = 16 \text{ to } 17 \text{ ‰} \qquad\qquad (6.25)$$

The latter value is lower than approximately 22 ‰ that is shown in Fig. 6.9C.

The main factors that are likely to control long-term variations in the $\delta^{13}C$ record are combinations of tectonic processes and biological production and storage of organic matter in sediments. One of the tectonic controls is, in general, the ocean floor spreading and possible addition of CO_2 from the mantle, even if the transfer fluxes of carbon between the Earth's surface and the upper mantle are approximately balanced (Chapter 2). Another control is the weathering of fossil sedimentary organic matter on the continents: faster erosion and oxidation of organic carbon in shales transfers isotopically light CO_2 to the atmosphere and waters. And yet another tectonic or physiographic control is the topography of the oceanic basins that allows a greater or smaller degree of storage of organic carbon that is produced in the ocean and transported from land. The global biological control is primary production that preferentially uses the light isotope in the formation of organic matter, resulting in the enrichment in ^{13}C of the carbon remaining in the source. It should be noted that some correlation exists in the geologic record between the periods of isotopically heavier carbonate deposition and greater mass storage of organic carbon in sediments (cf. Fig. 7.12B). This relationship is pronounced in particular in the Late Paleozoic that was a period of formation of major coal deposits, when preferential withdrawal of ^{12}C left the ocean enriched in ^{13}C, as evidenced by the higher values of $\delta^{13}C$ in the carbonates of that time, near 260 Ma (Fig. 6.9B), and near the Cretaceous-Tertiary boundary, near 65 Ma before present. Further discussion of the isotopic record of the sedimentary carbonate rocks is given in Chapter 7, and the material presented in this chapter prepares the reader for that discussion.

Chapter 7

Sedimentary Rock Record and Oceanic and Atmospheric Carbon

In Chapter 2 we discussed some aspects of the beginnings of the carbon cycle and Precambrian atmospheric CO_2 concentrations. The major conclusions, among others, were that (1) atmospheric CO_2 levels were relatively high at that time and they provided a necessary greenhouse warming in the period of the faint young Sun, in addition to the possible presence of higher concentrations of other potential greenhouse gases and aerosols (e.g., CH_4 and organic volatiles) in the Hadean and early Precambrian atmosphere; (2) because of the high atmospheric CO_2 levels, the early oceans were rich in dissolved inorganic carbon and total alkalinity, had a relatively low pH, and might have been depleted in dissolved Ca^{2+}; and (3) as Precambrian time progressed, carbon was removed from the atmosphere and stored mainly in the large geochemical reservoirs of inorganic and organic carbon. These reservoirs presently contain 99.9% of all the carbon in the exogenic system comprising the sediment, ocean, atmosphere, and biota (Fig. 1.4). In this chapter, we consider chemical, mineralogical, and isotopic attributes of sedimentary rocks and their relevance to the interpretation of the behavior of the ocean-atmosphere-sediment system during the geologic past, with most attention paid to the past 542 million years (the Phanerozoic Eon) of Earth's history and ocean evolution. In Chapter 10 we present an extended discussion and model of the Phanerozoic and earlier atmosphere-ocean carbon cycle evolution that is consistent with much of the proxy data presented in this chapter.

1   Geologic Time Scale and the Sedimentary Record

As this chapter deals mainly with the Phanerozoic, the last 542×10^6 years of the Earth's history, the geologic nomenclature and timing of intervals of time for this Eon are shown in Fig. 7.1. Only 12% of geologic time is represented by the Phanerozoic Eon, the rest by Precambrian time (see Fig. 1.1). However, in terms of preserved sedimentary rock mass, the Phanerozoic contains about 60–70% of the total rock mass and much of

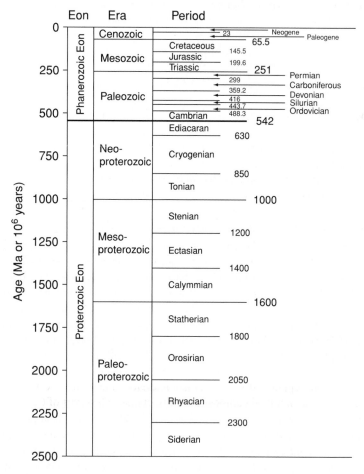

Figure 7.1   The post-Archean geologic time scale. See also Fig. 1.1. Data from Gradstein *et al.* (2004).

this mass, although altered by diagenesis and metamorphism in some areas or buried at currently inaccessible depths, contains retrievable information on the mineralogy, chemistry, and isotopic composition of the deposited sediments. On the other hand, the distribution with age of the sedimentary mass in the Precambrian is very uneven, being mainly concentrated at time intervals of important mountain-building events. In addition, Precambrian sediments are commonly altered to at least lower temperature and pressure grades of metamorphism, although there are some very well preserved carbonate sediments in the Precambrian. It should also be kept in mind that for the carbonate mass, it is biased in terms of preservation of various types of sedimentary rocks: pelagic, deep-sea carbonate deposits (the calcareous pteropod, foraminiferal, and coccolithophorid oozes) representative of post-Permian time are not found in the Paleozoic and Precambrian, and shallow-water, continental carbonates characterize these intervals of geologic time. These shallow-water carbonates were commonly deposited in vast seaways on the Paleozoic and Precambrian paleo-continents, so-called epeiric or epicontinental, platformal seas, for example, of the Ordovician and Cretaceous. These carbonate deposits can differ significantly in their mineralogical, chemical, and isotopic compositions from the coeval carbonates deposited in the much greater volume of the open ocean. For example, present-day, shallow-water carbonates are characterized by organo-detrital, aragonite, and magnesian calcite mineralogies with little pure calcite, while the deep-sea pelagic oozes are nearly pure calcite and aragonite. This has been the case for at least much of the Cenozoic. In addition and of interest to the discussion later in this chapter on the carbon isotopic composition of carbonates through geologic time (Section 7), the dissolved inorganic carbon (DIC) of seawater overlying shallow-water areas of carbonate deposition may differ from that of the open ocean. For example, the DIC of Florida Bay waters is significantly depleted in ^{13}C owing to respiration of organic matter as compared to the DIC of open ocean surface water. This implies that carbonates deposited in the bay will have lower $\delta^{13}C$ values than those deposited in the open ocean. Thus the $\delta^{13}C$ values of pre-Mesozoic carbonates that are largely shallow-water deposits may not reflect that of the ocean as whole. The above considerations must be kept in mind as the attributes of carbonate rocks and their relevance to ocean-atmosphere evolution are discussed in the following sections.

2   The Beginnings of Sedimentary Cycling

Sedimentary rocks are ultimately derived to a significant extent from the weathering and erosion of the upper continental crust. Therefore the extant global sedimentary mass should have a chemical composition comparable to this part of the crust. This is the case for most elements (Veizer and Mackenzie, 2004; Fig. 7.2), and the present-day composition of average global sediments for most elements does not generally deviate more than ±33% from that of the upper continental crust. Exceptions are the enrichments in B, Ca, U, Cr, Co, Ni, and Cu and the depletion in Na. In addition, the sediments are also strongly enriched in the excess volatiles of such elements as H, O, N, S, and C (Chapter 2), and have a higher oxidation state (Drever et al., 1988;

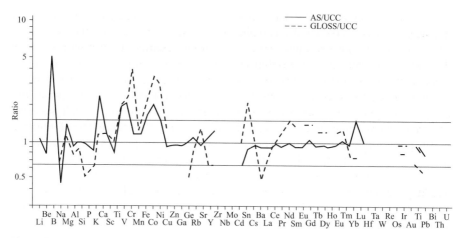

Figure 7.2   Elemental composition of the average sediment normalized to the composition of the upper continental crust. AS, average sediment composition; UCC, upper continental crust composition; and GLOSS, average composition of subducting sediments. The lines above and below that of 1 represent the upper and lower bounds of ±33% variation (data sources in Veizer and Mackenzie, 2004).

Goldschmidt, 1933). The anomalous enrichment in calcium and depletion in sodium are most likely a consequence of hydrothermal exchanges between the ocean floor and seawater. The general overall absence of large anomalies in the normalized average elemental composition of global sediments suggests that the exogenic and endogenic inputs and sinks for most major elements, except possibly calcium and sodium, were nearly in balance throughout most of geological history. This conclusion is supported in part by the fact that the average composition of the sediments that are subducted approaches that of the continental crust (Fig. 7.2). In actual fact, it is likely that we are dealing with a partially open system, because some sediments are being subducted while others are being formed at the expense of primary igneous and metamorphic rocks. In addition, large continental land masses supplying dissolved constituents and detritus to the ocean probably did not exist prior to about 3000 Ma ago. Thus at least the earlier part of the Archean was a time when the sedimentary mass was mostly growing by addition of first-cycle sediments from erosion of nearly contemporaneous, relatively young (≤250 Ma old) igneous precursors. Following the large-scale events that formed the ancient cratons and the establishment of a substantial global sedimentary mass by 2500 ± 500 Ma ago, cannibalistic sediment recycling became the dominant feature of sedimentary evolution. Estimates based on Sm/Nd systematics indicate that the sedimentary cycle is currently 90 ± 5% cannibalistic, attaining its near present-day steady state around the Archean-Proterozoic boundary of 2500 Ma ago (Veizer, 1988; Veizer and Mackenzie, 2004).

By the end of the Precambrian, the masses of carbon in the sedimentary organic and inorganic reservoirs were largely established (Des Marais *et al.*, 1992; Garrels and

Mackenzie, 1971b, 1972; Garrels and Perry, 1974). Since then only small amounts of carbon have been added to these reservoirs involving exogenic-endogenic system exchanges and the carbon fluxes into and out of the sedimentary reservoirs have not changed greatly over time. In other words, the exogenic system involving carbon has been relatively closed to new inputs from the endogenic (mantle and deeper Earth) system and hence cannibalistic. This statement is not explicitly true since there have been carbon exchanges between the deeper Earth and the exogenic system through processes involving basalt-seawater reactions at mid-ocean ridge and off-ridge sites, subduction of oceanic crust and sediments, and volcanism, but the overall mass transfer of new carbon into the exogenic system has been relatively small (pp. 28–30). In addition, the small, but significant, changes in the sedimentary fluxes of organic and inorganic carbon are to some extent responsible for atmospheric CO_2 and O_2 fluctuations and for the changing carbon and sulfur isotopic signatures preserved in sedimentary materials during Phanerozoic time (e.g., Garrels and Lerman, 1981, 1984).

3   Broad Patterns of Sediment Lithologies

The lithologic composition and the relative percentages of sedimentary and volcanic rocks preserved within the confines of present-day continents in crustal rocks of various ages are shown in Fig. 7.3. It should be remembered that with increasing geologic age, the total sedimentary rock mass diminishes and a given volume percentage of rock 3 billion years old represents a much smaller mass than an equal percentage of rocks 200 Ma old. Despite this limitation, some general trends in lithologic rock types agreed on by most investigators are evident.

The outcrops of very old Archean rocks are few and thus may not be representative of the original sediment compositions deposited. Nevertheless, it appears that carbonate rocks are relatively rare in the Archean. Veizer (1973) concluded that Archean carbonate rocks are predominantly limestones. During the early Proterozoic, the abundance of carbonates increases markedly and for most of this Era, the preserved carbonate rock mass is typified by the presence of early diagenetic, and perhaps primary, dolomites (Grotzinger and James, 2000; Veizer, 1973). In the Phanerozoic, carbonates constitute about 25% of the total sedimentary mass, with sandstones and shales accounting for most of the rest. Other highlights of the lithology-age distribution of Fig. 7.3 are: (1) a marked increase in the abundance of submarine volcanogenic rocks and immature sandstones (graywackes, sedimentary rocks with abundant fine-grained matrix between sand-sized grains) with increasing age; (2) a significant percentage of arkoses (potash feldspar-rich, sand-sized sedimentary rocks) in the early and middle Proterozoic, and an increase in the importance of mature sandstones (quartz-rich, sand-sized sedimentary rocks) with decreasing age; red beds are significant rock types of Proterozoic and younger age deposits; (3) a significant "bulge" in the relative abundance of banded iron-ore formations (jaspilites or BIF, discussed further in Chapter 10), silica-iron associations with coeval iron carbonate and iron sulfide lithologies, in the early Proterozoic; and (4) a lack of evaporitic sulfate (gypsum, anhydrite) and salt (halite, sylvite) deposits

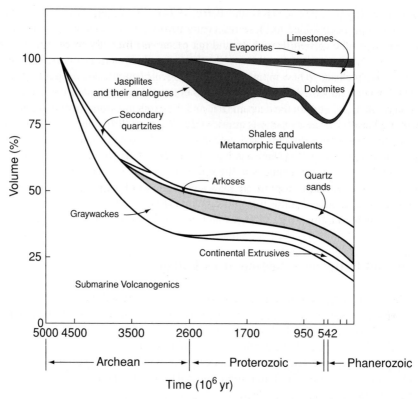

Figure 7.3   Volume percent of major lithologies of sedimentary and volcanic rocks as a function of geologic age. The extrapolation beyond 3 billion years must be considered hypothetical. Notice in particular the lack of carbonates in the Archean and the "bulge" during the latter part of the Proterozoic and the Paleozoic (after Ronov, 1964).

in sedimentary rocks older than about 800 million years, due in part to their more rapid recycling owing to dissolution by meteoric waters (Chapter 8).

An important feature of the mass-age distribution of carbonate rocks is the relative scarcity of dolomites in Cenozoic and Mesozoic sequences (Fig. 7.3), compared to their Paleozoic and particularly their Proterozoic counterparts. Carbonate sediments are mostly, but not exclusively, associated with low-latitude sedimentary environments presently and throughout Earth's history. However, for the Phanerozoic, the dolomite/total carbonate ratio within the tropical belt increases polewards towards arid climatic zones (Berry and Wilkinson, 1994). This suggests that the process of dolomitization is probably mainly a near-surface phenomenon. Therefore these dolomitic carbonates are either primary marine precipitates, or more likely, they are early diagenetic products of stabilization and dolomitization of carbonate mineral precursors (Chapter 4). Pore waters at this stage of dolomite formation would still be in contact with the overlying seawater and(or) would contain an appreciable seawater component. The high frequency of dolomites in ancient sequences may to some extent reflect the presence of shallow-water shelf settings preserved from Paleozoic and Proterozoic times. Changes in seawater chemistry, such as Mg/Ca ratio, sulfate content, and carbonate saturation state, may have been complementary factors leading to dolomite formation. Each of these changes in seawater chemistry individually or in concert with one another are found in such dolomite- producing environments as the Persian Gulf sabkha brines, the evaporated seawaters of the Coorong district of Australia and the Atlantic Coast of Brazil, Bonaire Island lagoon in the Caribbean Sea, Sugar Loaf Key carbonate sediments in Florida, supratidal carbonate sediments on Andros Island in the Bahamas, interstitial pore waters of carbonate reefs, anoxic marginal shelf and slope marine sediments, saline soda lakes, and in dilute ground waters where dolomite has been reported to be formed by methanogenic bacteria.

The trends in lithologic features of the sedimentary rock mass are a consequence of evolution of the surface environment of the planet as well as recycling and post-depositional processes, and both secular and cyclic processes have played a role in generating the lithology-age distribution we see today (Mackenzie, 1975; Veizer, 1973, 1988a). For the past 1.5 to 2.0 billion years, the Earth has been in a near present-day steady state and the temporal distribution of rock types since then has been controlled primarily by recycling in response to plate tectonic processes.

4   Differential Cycling of the Sedimentary Mass and Carbonates

Sedimentary rocks are formed by depositional processes involving principally the agents of water and wind and are destroyed when eroded or transformed chemically by metamorphism into other kinds of rocks, such as gneiss. The sedimentary rock mass today, as estimated from geochemical mass balance methods, is 2 to 3 $\times$ 10^{24} g (Table 2.2, with references to a range from 1.7 to 3.2 $\times$ 10^{24} g), with 85 to 90% of the mass lying within the continental and shelf region of the globe and 15 to 10% a part of the deep ocean floor. This rock mass estimate includes the classic lithologies of sandstone, shale, and carbonate, as well as their metamorphic equivalents of quartzite, slate, phyllite, low-grade schist, and marble, and volcanogenic sediments. The current mass is that preserved, not the total mass deposited throughout geologic time. Total sedimentary deposition over the last 3.5 billion years of Earth's history has been at least 13 $\times$ 10^{24} g.

Like human populations, sedimentary rock masses can be assigned birth rates and death rates, and they can be subdivided into age groups (Dacey and Lerman, 1983; Garrels and Mackenzie, 1971b; Veizer and Mackenzie, 2004). Gregor (1985) demonstrated that the sedimentary mass-age distribution for Carboniferous and younger sediments has a log-linear relationship:

$$\log S = 10.01 - 0.24t \tag{7.1}$$

where S is survival rate, defined as mass in metric tons of a stratigraphic System divided by duration in years of the corresponding Period, and t is the median age of the mass in units of 10^6 yr. At time t_0, S is equivalent to the modern rate of accumulation of sedimentary materials in the ocean in units of 10^{15} g/yr.

The survival rate of the total sedimentary mass and those of the carbonate and dolomite masses for different Phanerozoic Periods are plotted in Fig. 7.4. The difference between the survival rate of the total carbonate mass and that of dolomite is the mass of limestone surviving per interval of time. Equation (7.1) implies a 130 million

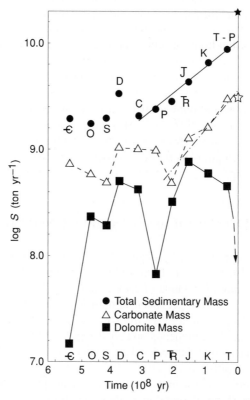

Figure 7.4   Total sedimentary and carbonate rock mass-age relationships during the Phanerozoic expressed as the logarithm of the survival rates in tons of total sediment, total carbonate, and dolomite per year (S) versus time in 10^8 years. The differences between the survival rates for the carbonate mass and those of the dolomite mass are the survival rates of the calcite mass. The straight lines are best fits to the Period-averaged data for the post-Devonian total mass (solid line; Gregor, 1985) and for the post-Permian carbonate mass (dash-dot line). The black star is the value of the modern total river flux of materials to the ocean of 15 to 20 × 10^{15} g/yr and the open star is the value of today's chemical and detrital total inorganic carbonate flux of 3.0 × 10^{15} g/yr. These values based on assessments of observational data are reasonably close to the time t_0 intercepts representing the modeled accumulation rates (after Mackenzie and Morse, 1992).

year half-life for the post-Devonian mass, and for a constant sediment mass with a constant probability of destruction, a mean sedimentation rate since Post-Devonian time of about 10×10^{15} g/yr. The modern global erosional flux is about 15 to 20×10^{15} g/yr and dissolved flux about 3 to 4×10^{15} g/yr; of this total, particulate and dissolved carbonate (as HCO_3^- in rivers) at about 3×10^{15} g/yr accounts for 15 to 20% of the erosional transport. Although the data are less reliable for the survival rate of Phanerozoic carbonate sediments than for the total sedimentary mass, a best log-linear fit to constitute the post-Permian preserved mass of carbonate rocks is:

$$\log S_{carb} = 9.55 - 0.36t \qquad (7.2)$$

where S_{carb} is carbonate survival rate, defined as mass of carbonate in metric tons of a stratigraphic System divided by the duration in years of the corresponding Period, and t is the median age of the carbonate mass in units of 10^6 yr. This corresponds to a half-life for the post-Permian carbonate mass of 86 million years, and a mean sedimentation rate of post-Permian carbonate sediments equivalent to about 3.5×10^{15} g carbonate per year. The present-day total carbonate flux from land to the ocean is estimated at about 3×10^{15} g/yr or less, 2×10^{15} g/yr, if only 53% of the bicarbonate ion in rivers is derived from the weathering of carbonate rocks (Chapter 8). The difference in half-lives between the total sedimentary mass, which is principally sandstone and shale, and the carbonate mass is a consequence of the more rapid recycling of the carbonate mass at a rate about 1.5 to 2 times the total mass.

The preceding statement is not an unlikely situation. The relative proportion of carbonate rocks within the continental realm generally increases with increasing age during the Phanerozoic, and the Mesozoic and Cenozoic deficiency of carbonate rock mass (Fig. 7.3) is due in part to a tectonic cause, the ubiquity of transient immature tectonic settings of carbonate deposition. Another probable reason is the general northward drift of the continents since the early Mesozoic, which resulted in a progressive decline in the shelf areas that fell within the confines of the tropical climatic belt (Bluth and Kump, 1991; Kiessling, 2002; Veizer and Mackenzie, 2004; Walker et al., 2002). Furthermore, in the course of the Mesozoic and Cenozoic, the locus of carbonate sedimentation migrated from the shelves to the pelagic realm, mirroring the evolution of calcareous plankton and the role that calcareous shells of foraminifera, pteropods, and coccolithophorids started to play in the carbonate budget of the ocean. With the advent of abundant carbonate-secreting, planktonic organisms in the Jurassic, the site of carbonate deposition shifted significantly from shallow-water areas to the deep sea. A graduate shift in carbonate deposition from shallow-water environments to the deep sea would increase further the rate of destruction, by eventual subduction, of the global carbonate mass relative to the total sedimentary mass from Jurassic time on. This environmental shift may have been accompanied by deepening of the carbonate saturation depth (Ross and Wilkinson, 1991) and may have led to enlargement of oceanic areas that were sufficiently shallow for preservation of pelagic carbonates.

Thus it appears that the carbonate component of the sedimentary rock mass has a cycling rate different than that of the total sedimentary mass. Although resistance to weathering may play some role in the selective destruction of sedimentary rocks, it is likely that differences in the recycling rates of different tectonic regimes in which sediments are deposited are more important.

5   Sedimentary Carbonate System

This section addresses the distributions of some attributes of carbonate rocks through Phanerozoic time and their meaning for the evolution of the near surface of the Earth. The interpretations of the behavior of the carbonate system through Phanerozoic time are based on data and discussions from a number of sources including Gregor, 1985; Hay, 1985; Mackenzie and Morse, 1990; Morse and Mackenzie, 1992; Veizer and Mackenzie, 2004; Wilkinson and Algeo, 1989; and Wilkinson and Walker, 1989.

5.1   Phanerozoic Calcite/Dolomite Ratios

For several decades it has been assumed that the Mg/Ca ratio of *cratonic* carbonate rocks increases with increasing Phanerozoic rock age. Early portrayals of this trend in cratonic carbonate rocks deposited principally in shallow marine waters and mainly from stratigraphic sequences of the Russian Platform and North America are shown in Fig. 7.5. One should keep in mind when examining these trends that Jurassic to

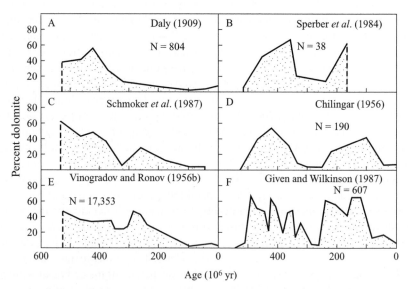

Figure 7.5   Some estimates of the percent dolomite in Phanerozoic, shallow-water, cratonic carbonate rocks as a function of age. The estimates on the left generally show an erratic increase in the percentage of dolomite within the carbonate rock mass with increasing age of the rock. The ones on the right suggest that the distribution of dolomite in the Phanerozoic carbonate rock record with age exhibits a cyclic pattern (after Wilkinson and Algeo, 1989).

present-day deep sea calcareous deposits (the pelagic oozes) are mainly composed of the calcitic skeletal remains of foraminifera and coccolithophorids (the calcite oozes) and the aragonitic skeletal remains of pteropods (the aragonite oozes).

The mass ratio Mg/Ca of carbonate rocks and the weight percent fraction of dolomite are often used interchangeably. However, it should be recognized that these two parameters are interrelated only qualitatively and, for example, an increase in the dolomite rock fraction by a factor of two does not necessarily indicate an increase in the Mg-content of the carbonate rock by the same factor. This difference is due to the fact that the lithologic units identified as dolomite and limestone are usually not pure minerals dolomite $CaMg(CO_3)_2$ and calcite $CaCO_3$, but they contain variable admixtures of non-carbonates. The Mg/Ca mass ratio of a carbonate rock that consists of dolomite and limestone strongly depends on the weight fractions of $CaMg(CO_3)_2$ in the dolomite-rock component and of $CaCO_3$ in the limestone-rock component. In pure $CaMg(CO_3)_2$, the mass ratio Mg/Ca is 0.6, and it is 0 in pure $CaCO_3$. In a mixture of equal mass proportions of the two minerals, the Mg/Ca ratio is 0.21, but it may be higher or lower depending on the mineral composition of the lithologic units of dolomite and limestone.

The trends on the left side of Fig. 7.5 (A, C, and E) represent a general, but erratic, decline in the calcium content and increase in the magnesium content of cratonic carbonate rocks with increasing age. The magnesium content is relatively constant in these carbonates for about 100 million years, then it increases gradually. In addition, the beginning of the gradual increase in magnesium content of cratonic carbonates appears to be at a geologic age that is very close to, if not the same as, the age of the beginning of the general increase in the Mg content of pelagic limestones (100 million years; Renard, 1986). The dolomite content of deep-sea sediments also increases erratically with increasing age back to about 125 million years before present (Lumsden, 1985). Thus the increase in magnesium content of carbonate rocks with increasing age into at least the Early Cretaceous appears to be a global phenomenon.

In the late 1980s, the accepted truism that dolomite abundance, as reflected in the Mg/Ca ratio of carbonate rocks, increases relative to limestone with increasing Phanerozoic age was challenged by Given and Wilkinson (1987). They reevaluated all the existing data on the composition of Phanerozoic carbonates and concluded that dolomite abundances do vary significantly throughout the Phanerozoic but do not necessarily increase systematically with increasing age. Figure 7.5F shows the results of Given and Wilkinson's assessment of the Phanerozoic carbonate mass-age relations to compare with the estimates shown on the left-hand side of the figure, along with those of Chilingar (1956) and Sperber et al. (1984) (Figs. 7.5B, D). These latter three assessments of the mass-age distribution of dolomite in the Phanerozoic Eon all suggest a cyclic pattern in the distribution of dolomite with age. The meaning of these abundance curves and their actual validity are still somewhat controversial because of sampling bias (Zenger, 1989). In addition, some investigators believe that the more than 17,000 samples analyzed by Vinogradov and Ronov (1956) from the Russian Platform (Fig. 7.5E) are of a sufficient sampling density, compared with the other estimates of dolomite abundance as a function of age that are shown in Fig. 7.5, to conclude that the trend of continuous but erratic increase of dolomite abundance with increasing

Phanerozoic rock age is the correct one and not the cyclic trend. Later in this chapter, we come back to the mass-age distribution of dolomite and its meaning in terms of ocean-atmosphere compositional evolution.

A voluminous literature exists on the "dolomite problem" (e.g., Hardie, 1987, and Holland and Zimmerman, 2000, for discussion, and Chapter 4) and shows that the reasons for the high magnesium content of carbonates are diverse and complex. Petrographic, chemical, and isotopic evidence suggests that: some dolomitic rocks are primary precipitates; others were deposited as $CaCO_3$ and then converted entirely or partially to dolomite before deposition of a succeeding layer; and still others were dolomitized by migrating underground waters tens or hundreds of millions of years after deposition. It is important to know the distribution of the calcite/dolomite ratios of carbonate rocks through geologic time because this information has a bearing on the origin of dolomite, as well as on changes in atmosphere-hydrosphere environmental properties through geologic time. If we accept the cyclical nature of the dolomite abundance, the mass ratio of the carbonate components as a function of Phanerozoic age can be calculated using Given and Wilkinson's (1987) percent dolomite mass-age curve.

Figure 7.6 illustrates the distribution of Phanerozoic carbonate rock masses and their calcite and dolomite contents on a Period-averaged basis. It can be seen that, as with the total sedimentary mass, the mass of carbonate rock preserved is pushed toward

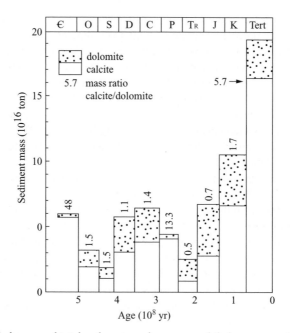

Figure 7.6   Period-averaged total carbonate rock masses and their average calcite to dolomite mass ratios during the Phanerozoic. Data from Given and Wilkinson (1987).

the front of geologic time within this mass-age trend. The Tertiary, Carboniferous, and Cambrian periods appear to be times of significant carbonate preservation, whereas the preservation of Silurian and Triassic carbonates is minimal. It also can be seen in Fig. 7.6 that the Period-averaged mass ratio of calcite to dolomite is relatively high for Cambrian, Permian, and Tertiary System rocks, whereas this ratio is low for Ordovician through Carboniferous age sediments and rises in value from the Triassic through the Recent.

The generalized sea level curve of Vail *et al.* (1977) and also that of Hallam (1984) correlate crudely with the calcite/dolomite ratio of carbonate rocks through Phanerozoic time. The Phanerozoic starts out with sea level rising, and Cambrian carbonate strata are enriched in calcite. For much of the Paleozoic, when sea level was high or declining from its Ordovician maximum, the calcite/dolomite ratio remains about 1 to 1.5, increasing sharply in the Permian to about 13. It then decreases into the Triassic, that has a ratio of 0.5. As global sea level rises toward the maximum Cretaceous transgression, the calcite/dolomite ratio remains low, but tends toward the higher ratios of the Tertiary and Quaternary as sea level falls, and dolomite becomes less and less abundant in the more modern sedimentary record. These cycles in calcite/dolomite ratios correspond crudely to Fischer's (1984) two Phanerozoic super-cycles and the Mackenzie and Pigott's (1981) oscillatory and submergent modes.

5.2   Ooids and Ironstones

Ooids are sand-sized, clastic grains of $CaCO_3$, usually aragonite and less commonly magnesian calcite. The aragonitic ooids have an internal, laminated, concentric structure in which fine-grained needles of aragonite are deposited tangentially to the roughly spherical surface of the ooid. The textures of ooids appear to vary during Phanerozoic time. Sorby (1879) first pointed out the petrographic differences between ancient and modern ooids: ancient ooids commonly exhibit relic textures of a calcite origin, whereas modern ooids are dominantly made of aragonite. Sandberg (1975) reinforced these observations by his study of the textures of some Phanerozoic ooids and a survey of the literature. His approach, and that of others who followed, was to employ the petrographic criteria of Sorby: if the microtexture of the ooid is preserved, then the ooid originally had a calcite mineralogy; if the ooid exhibits textural disruption, its original mineralogy was aragonite. The textures of originally aragonitic fossils are usually used as checks to deduce the original mineralogy of the ooids. Sandberg (1975) observed that ooids of inferred calcite composition are dominant in rocks older than Jurassic. Following this classical work, Sandberg (1983, 1985) and several other investigators attempted to quantify further this relationship. Figure 7.7 is a schematic diagram representing a synthesis of the inferred mineralogy of ooids during the Phanerozoic.

It appears that while originally aragonite ooids are found throughout the Phanerozoic, an oscillatory trend in the relative percentage of calcite versus aragonite ooids may be superimposed on a long-term evolutionary increase in ooids with an inferred original calcite mineralogy with increasing age of the rocks. Although the correlation

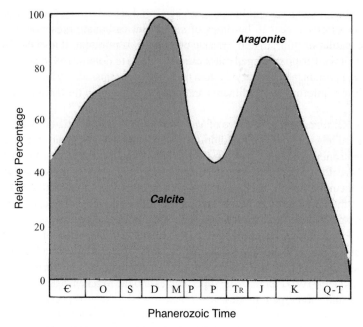

Figure 7.7   Trend in relative percentage of originally calcite and aragonite ooid mineralogies during the Phanerozoic. Note especially the cyclic pattern in ooid mineralogy with time (courtesy of B. H. Wilkinson, as redrawn in Mackenzie and Agegian, 1989, with data from Mackenzie and Pigott, 1981, and Sandberg, 1983, 1985).

is not strong, the two major maxima in the sea level curves of Hallam (1984) and Vail *et al.* (1977) appear to coincide with times when calcite ooids were important seawater precipitates. Wilkinson *et al.* (1985) found that the strongest correlation between the various data sets representing global eustasy and ooid mineralogy is that of inferred mineralogy with percentage of continental freeboard. The more flooded the continents, the more likely that calcite ooids were originally deposited. The less flooded the continents, the greater the likelihood of deposition of aragonite ooids. Sandberg (1983) also concluded that the cyclic trend in ooid mineralogy correlates with a cyclic trend observed for the inferred mineralogy of aragonite and calcite cements in carbonate rocks. Calcite cements characterize periods of inferred ooid calcite mineralogy and aragonite cements are found at times of inferred aragonite ooid mineralogy. Sandberg coined the term "calcite seas" for times of formation of calcite ooids and cements and "aragonite seas" for times of formation of aragonite ooids and cements. Van Houten and Bhattacharyya (1982) and Wilkinson *et al.* (1985) further showed that the distribution of Phanerozoic ironstones (hematite and chamosite oolitic deposits) exhibits a definite cyclicity that also appears to covary with the generalized sea level curve. Minima in ironstone abundance appear to coincide with times of sea level withdrawal from the continents.

5.3   Calcareous Shelly Fossils

In some similarity to the cyclic trends observed for marine ooids and cements, and the calcite/dolomite ratio of carbonate rocks and ironstones, the mineralogy of Phanerozoic carbonate skeletal components also shows a cyclic pattern (Fig. 7.8), with calcite being particularly abundant during high sea levels of the Early to Mid-Paleozoic and the Cretaceous. Notice also in Fig. 7.8 that times of calcite abundance coincide with times when Sr concentrations in biologically produced calcite were high. Overall there tends to be a general increase in the diversity of major groups of calcareous organisms during Phanerozoic time with the emergence of coccolithophorids, pteropods, reef-building hermatypic corals, and coralline algae in the later part of the Phanerozoic. It is noteworthy that the major groups of pelagic and benthic organisms contributing to carbonate sediments in today's ocean first appeared in the fossil record during the Middle Mesozoic and progressively became more abundant.

Because carbonate sediments of a skeletal origin are such an important part of the Phanerozoic carbonate rock record, this increase in mainly metastable mineralogies

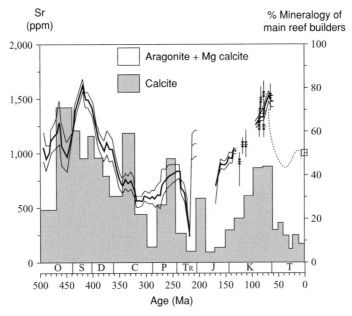

Figure 7.8   Inferred skeletal mineralogy of main reef-building organisms in percent and the Sr concentrations found in calcite of biological origin during the Phanerozoic. Calcite mineralogy is represented by the shaded area and aragonite and Mg calcite mineralogies by the white area. The solid dark line represents the mean trend for Sr in biological calcite with age of the rock units; the upper and lower bounds are the range about that mean trend. Mineralogy from Kiessling (2002) and Sr data from Lear et al. (2003) and Steuber and Veizer (2002) (after Veizer and Mackenzie, 2004).

(see Chapter 4) played an important role in the pathway of diagenesis of carbonate sediments. The ubiquity of low magnesian calcite skeletal organisms in the Paleozoic led to production of calcitic skeletal clasts in sediments whose original bulk chemical and mineralogical composition was closer to that of their altered and lithified counterparts of Mesozoic and, particularly, Early Cenozoic age.

6   Evaporites and Fluid Inclusions

Marine evaporites provide important clues to the history of seawater composition, including dissolved carbon species, because they can be used to place compositional bounds on the two major seawater components that are part of carbonate minerals, calcium and magnesium. Evaporites owe their existence to a unique combination of tectonic, paleogeographic, and sea-level conditions. Seawater bodies must be restricted to some degree, but also must exchange with the open ocean to permit large volumes of seawater to enter these restricted basins and evaporate, leaving precipitated salts behind. Environmental settings of evaporite deposition may occur on cratons or in rifted basins. Figure 7.9 illustrates that because evaporite deposition requires an unusual combination of circumstances, a "geological accident" (Holser, 1984), the intensity of deposition has varied significantly during geologic time.

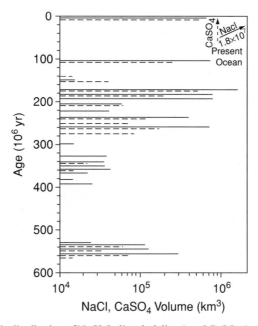

Figure 7.9   Sporadic distribution of NaCl (halite, dark lines) and $CaSO_4$ (gypsum and anhydrite, dashed lines) in marine evaporates during the Phanerozoic. There are approximately 1.8×10^7 km^3 of NaCl and 9×10^5 km^3 of $CaSO_4$ in the modern ocean water (after Holser, 1984).

There is no systematic decrease with time in the volume of preserved evaporites, which suggests that there was no *continuous* recycling during the 550 Ma of the Phanerozoic. The volume is about the same in the youngest 200 Ma, and it then decreases during the next 200 Ma; however, near the base of the Phanerozoic, it is significantly greater than in some parts of the later geologic record. This suggests that the volume preserved of NaCl and $CaSO_4$ per unit of time in the Phanerozoic reasonably reflects the volume deposited. As the oceans are the main reservoirs for these components, such large variations in the rate of NaCl and $CaSO_4$ output from the ocean to evaporites imply changes in the salinity and chemical composition of seawater. The near constancy of seawater composition during the Phanerozoic has been advocated by a number of investigators who at the same time recognized that there could be deviations in its composition (Garrels and Mackenzie, 1971; Holland, 1978; Holser, 1963; Mackenzie and Garrels, 1966). Calculations based on mineral sequences found in evaporite rocks and the development of a new method for extracting brines from fluid inclusions and determining their composition using ion chromatography provided some confirmation for Holser's (1963) earlier finding that the Mg/Cl and Br/Cl ratios in brines extracted from fluid inclusions in Permian halite deposits were similar to those of modern brines (Hardie, 1991; Harvie and Weare, 1980; Holland, 1978; Horita *et al.*, 1991; Lazar and Holland, 1988). These studies implied that Permian seawater composition was similar to that of today and lent some support to the assertion that seawater composition has been conservative during the Phanerozoic. Were we in for a surprise! The choice of a comparison between the Permian and modern seawater was unfortunate because both in atmospheric CO_2 concentration and in seawater chemistry, Permian time was indeed similar to the present with relatively low atmospheric CO_2 (Fig. 3.7) and a seawater chemistry similar to the modern situation (Fig. 7.10). The similarity between Permian and modern seawater (and, indeed, atmospheric CO_2) turned out to be the exception rather than the rule.

Despite potential problems arising from the alteration of the composition of fluid inclusion brines owing to diagenesis and the fact that marine halite deposits are sporadic and unevenly distributed throughout the Phanerozoic (Fig. 7.9), it appears at this stage that fluid inclusion data from carefully selected marine halite sequences provide a reasonably good guide to the composition of the Phanerozoic parent seawater from which the halite formed. However, one must take care in the interpretation of trends with time because brine chemistry is far removed from the parent seawater and reconstruction of the chemical evolution of these brines is a considerable challenge (Holland, 2004). Nevertheless, Fig. 7.10 illustrates Phanerozoic seawater concentrations of Mg^{2+}, Ca^{2+}, and SO_4^{2-} based on analyses of fluid inclusions in marine halite. One can see from the trends with age that seawater concentrations of Mg^{2+}, Ca^{2+}, and SO_4^{2-} have varied by a factor of two or slightly more during Phanerozoic time. The variation of Ca^{2+} and Mg^{2+} in seawater implies that the carbon chemistry of seawater has changed over time, including its saturation state with respect to carbonate minerals (see Chapters 5, 9, and 10). In addition, the variation in SO_4^{2-} concentrations most likely affected the formation of dolomite because the precipitation of this mineral may be promoted by lower sulfate concentrations in seawater-like fluids (McKenzie, 1991).

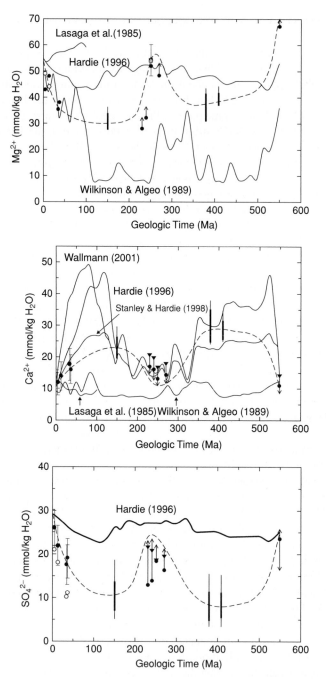

Figure 7.10   Concentrations of Mg^{2+}, Ca^{2+}, and SO_4^{2-} in seawater during Phanerozoic time based on analyses of fluid inclusions in marine halite evaporite deposits (solid bars, triangles and circles). Dashed lines are estimates of the concentrations versus age curves. The solid lines are modeling results from Hardie (1996), Lasaga *et al.* (1985), Stanley and Hardie (1998), Wallmann (2001), and Wilkinson and Algeo (1989) (after Holland, 2004, and Horita *et al.*, 2002).

7 Isotopic Trends

7.1 Strontium

Strontium is an important element found in carbonate minerals and rocks, particularly in the phase aragonite (Chapter 4). Both the Sr/Ca ratio and strontium isotopic composition of marine precipitates have been determined through geologic time. It has been observed that the Sr/Ca ratios of planktonic foraminifera and the $^{87}Sr/^{86}Sr$ ratios of sedimentary precipitates increase irregularly though the Cenozoic with decreasing age of the sediment indicating, if these signals are primary and not altered by diagenetic processes, a change in seawater Sr content and its isotopic composition during this time period. These changes are most likely related to the magnitude of the river plus groundwater input of Sr to the ocean, and their mean $^{87}Sr/^{86}Sr$ ratio, relative to the exchange fluxes involving the reaction of seawater with oceanic basalts. The Sr/Ca ratio may also be related to the observed increase in the accumulation of Sr-poor, calcite-rich carbonate sediments in the pelagic realm of the ocean over this time period, resulting in a slight enrichment of ocean water in Sr.

In the modern oceans the concentration of Sr is ∼8 ppm and its residence time is 4 to 8 Ma (Holland, 1984). The present isotopic ratio $^{87}Sr/^{86}Sr$ is 0.7092 (McArthur, 1994) and is controlled essentially by two fluxes, from the mantle and from the continents by rivers. The former represents Sr exchanged between seawater and oceanic crust ($^{87}Sr/^{86}Sr = 0.703$) in hydrothermal systems on the ocean floor. The latter, reflecting the more fractionated composition of the continental crust, feeds into the oceans more radiogenic ^{87}Sr, with an average isotopic ratio for rivers of about 0.711. The latter flux may vary from 0.703 to 0.730 or more, depending on whether the river is draining young volcanic terrain or an old granitic shield. The third input is the flux of Sr from diagenesis of carbonates, which results in expulsion of some Sr from principally aragonite as it recrystallizes to calcite during diagenesis. This flux is not large enough to influence the isotopic composition of seawater. A simple balance calculation based on isotopes shows that Sr in modern seawater is a mixture of approximately 75% "river" flux and 25% "mantle" flux, generating the modern value of 0.7092. The above considerations illustrate that the Sr isotopic composition of seawater is controlled essentially by tectonic evolution, that is, by relative contributions from weathering processes on continents and by the intensity of submarine hydrothermal systems. However, during geologic time, the isotopic compositions of these two fluxes have evolved, because ^{87}Sr is a decay product of ^{87}Rb:

$$\left(\frac{^{87}Sr}{^{86}Sr}\right)_p = \left(\frac{^{87}Sr}{^{86}Sr}\right)_0 + \left(\frac{^{87}Rb}{^{86}Sr}\right)(e^{\lambda t} - 1) \qquad (7.3)$$

where p is present ratio; 0 is initial ratio at the formation of the Earth 4.5 Ga ago (0.699); λ = decay constant of ^{87}Rb (1.42×10^{-11} yr^{-1}); and t is time since the beginning, such as the formation of the Earth 4.5 Ga ago.

From equation (7.3) it is evident that the term ($^{87}Sr/^{86}Sr)_p$ for coeval rocks originating from the same source ($^{87}Sr/^{86}Sr)_0$ depends only on their Rb/Sr ratios. Since this ratio is about 6 times larger for the more fractionated continental rocks than for the basalts (0.15 to 0.027; Faure, 1986), the $^{87}Sr/^{86}Sr$ of the continental crust at any given time considerably exceeds that of the mantle and oceanic crust, increasing to the present day value of about 0.730 for the average continental crust, as opposed to 0.703 for the oceanic crust. The rivers draining the continents are less radiogenic (about 0.711) than the crust itself due to the fact that much of riverine Sr originates from the weathering of carbonate rocks rather than from their silicate counterparts. The carbonates as marine sediments inherited their Sr from seawater which contains also the less radiogenic Sr from hydrothermal sources.

The Sr isotopic trend with age of the rock mass during the Precambrian is similar to mantle values until about the Archean-Proterozoic transition (Veizer and Mackenzie 2004). From then on it deviates toward higher radiogenic values, reflecting the increasing dominance of input from the continental crust. Thus the primary control on the Sr isotopic composition of seawater on the billion-year scale is the pattern of growth of the continental crust. Model calculations by Goddéris and Veizer (2000) of seawater $^{87}Sr/^{86}Sr$ evolution show that the measured observational data fit well a pattern of continental growth in which the continents were generated episodically during geologic time, with major phases of continent formation in the late Archean and early Proterozoic, and attainment of a near modern extent by approximately 1.8 Ga ago. This scenario is also compatible with the evolution of sediments and their chemistry. The Archean oceans were most likely buffered to an important extent by reactions involving vigorous circulation of seawater through basalts via submarine hydrothermal systems. With the exponential decline of Earth's internal heat dissipation, the intensity of the hydrothermal system also declined and at the same time the flux of radiogenic Sr from growing continents brought in by rivers started to assert itself. This tectonically controlled transition from mantle- to river-buffered oceans across the Archean-Proterozoic transition is a first order feature of Earth's evolution, with consequences for other isotope systematics, for the redox state of the ocean-atmosphere system, and for other related phenomena. The transition certainly involved a change in the carbon cycle because dissolved inorganic carbon in the ocean during hydrothermal circulation of seawater through basaltic rocks of the mid-ocean ridges is converted to CO_2 in a reaction involving the formation of chlorite or deposition as calcite. With a smaller continental area available for weathering in the early Precambrian, the former reaction to chlorite might have been more predominant for addition of CO_2 to the ocean-atmosphere system than later in geologic time.

The resolution of the $^{87}Sr/^{86}Sr$ database is reasonably good for the Phanerozoic. The latest version of the Phanerozoic trend shows a decline in $^{87}Sr/^{86}Sr$ values from the Cambrian to the Jurassic, followed by a steep rise to modem values, with superimposed oscillations at 10^7 year frequency (Fig. 7.11). In general, it is probably tectonics that is the cause of the observed Phanerozoic trend, with the mantle input of greater relative importance at times of the troughs and the river flux dominating in the Tertiary and early Paleozoic. Nevertheless, it is difficult to correlate the overall trend or the superimposed oscillations with specific tectonic events. The problem simply arises from the fact that model solutions do not produce unique answers. Nevertheless, the river flux is likely

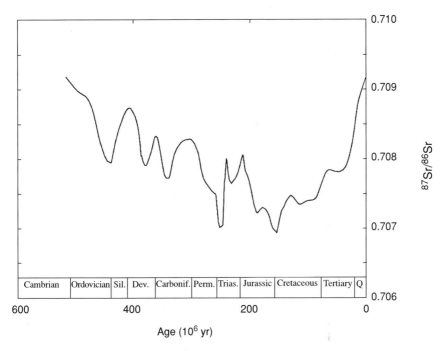

Figure 7.11   Generalized strontium isotopic curve based on 4,055 samples of brachiopods, belemnites, and conodonts during the Phanerozoic (after Veizer *et al.*, 1999).

the major reason for the observed $^{87}Sr/^{86}Sr$ oscillations because the changes in sea floor spreading rates are relatively sluggish and the Sr isotope ratio of the mantle flux is relatively constant at about 0.703. However, the river flux may vary widely in both Sr elemental flux and its isotope ratio. For example, the rapid Tertiary rise in $^{87}Sr/^{86}Sr$ (Fig. 7.11) is commonly interpreted as reflecting the uplift of the Himalayas. Such uplift and a consequential increase in weathering may be in part one reason for the Cenozoic drawdown in atmospheric CO_2 (Fig. 3.7).

7.2   Carbon

In Chapter 6 we discussed in detail the isotope systematics of carbon, and the isotopic fractionation of $^{13}C/^{12}C$ in biological production of organic matter by photosynthesis and in the carbon transfer between the atmosphere, continental and ocean waters, and carbonate sediments. Here we briefly review these concepts in terms of the changes in the isotopic composition of the rock record with geologic age. The two dominant exogenic reservoirs of carbon are carbonate rocks and organic matter in sediments (Fig. 1.4). They are linked in the carbon cycle via atmospheric CO_2 and the carbon species dissolved in the hydrosphere. Isotope ^{12}C in carbon dioxide is preferentially utilized by photosynthesizing organisms for production of organic carbon, causing enrichment in ^{13}C that accounts for the near-surface ocean waters being usually heavier that the deep waters (Kroopnick, 1980; Tan, 1988). To reiterate, organic carbon is strongly depleted in ^{13}C relative to its carbon source. The labile or reactive fraction of organic matter is easily remineralized to CO_2 inheriting a ^{13}C-depleted signal.

The δ^{13}C of mantle carbon is approximately -5 to -7 ‰ (Chapter 6) and in the absence of life and its photosynthetic capabilities, this would also be the isotopic composition of seawater. Yet, as far back as 3.5 to 4 billion years ago, carbonate rocks, and hence seawater, had a δ^{13}C of about 0 ‰ (Shields and Veizer, 2002). This suggests that a reservoir of reduced organic carbon accounted for approximately 20% of the entire exogenic carbon existing some 4 Ga ago, raising the δ^{13}C of the residual 4/5 of carbon, present in the oxidized form in the ocean-atmosphere system, from -5 to 0 ‰. This is an oxidized/reduced partitioning of carbon that is very similar to that of the Phanerozoic sediments.

It appears that life with its photosynthetic capabilities can be traced almost as far back as we have a rock record (Chapters 2 and 3). This photosynthesis or biosynthesis may or may not have been generating oxygen as its byproduct, but it was essential to give rise to seawater δ^{13}C values similar to modern values and higher than those of the mantle. In order to sustain seawater δ^{13}C at this level during much of geological history, the inputs and outputs in the carbon cycle had to have the same δ^{13}C. Since the input from the mantle via volcanism and hydrothermal systems has a δ^{13}C of -5 ‰ and the subducted carbonates are near 0 ‰, the subduction process must involve also a complementary ^{13}C-depleted component, that is organic matter that occurs in oceanic sediments. Today, organic matter is preserved in part even in the fully oxygenated system of the ocean and it is conceivable that oxygen-generating photosynthesis may have been present as far back as we have a geologic record without necessarily inducing total remineralization of organic matter in the early ocean-atmosphere system (Lasaga and Ohmoto, 2002). If the organic matter were completely oxidized, then the subduction loss of ^{13}C-enriched limestone carbon, coupled with the addition of mantle carbon, would slowly force the δ^{13}C of seawater back toward mantle values. In such a scenario, in order to sustain the near 0 ‰ of seawater during much of geologic history, the input of mantle carbon into the ocean-atmosphere system would have to be smaller and progressively diminishing the impact of hydrothermal and volcanic activity over geologic time.

In the Proterozoic, occurrences of heavy δ^{13}C values in carbonates at $\sim$2.2 Ga ago have been interpreted as the beginning of oxygen-generating photosynthesis that resulted in the sequestration of large quantities of organic matter into coeval sediments (Karhu and Holland, 1996). Later, in the Neoproterozoic, negative δ^{13}C values following a glaciation at the time of the proposed Snowball Earth (Hoffman *et al.*, 1998) suggest a lack of, or much reduced, organic production when the Earth was possibly frozen and covered by ice, including much of the ocean area of the time.

The sampling density and temporal resolution of δ^{13}C in the Phanerozoic enable the delineation of a well-constrained secular curve that exhibits a maximum in the Permo-Carboniferous (Fig. 7.12A). However, even in this case, we are dealing with a band of data, reflecting the fact that the δ^{13}C of DIC of seawater is not uniform in time and space,

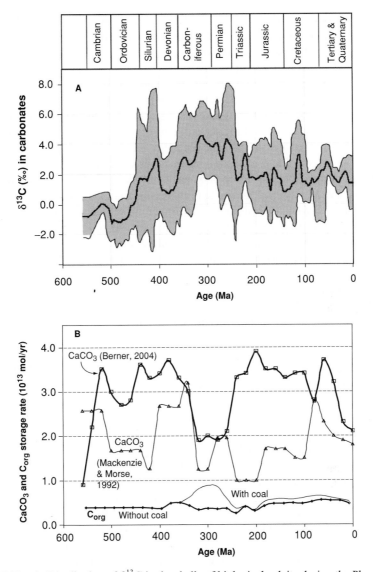

Figure 7.12   A. Distribution of δ^{13}C in the shells of biological calcite during the Phanerozoic based on approximately 4,500 samples. The black line is a running mean and the shaded area includes the 95% confidence limit about the mean (after Veizer *et al.*, 1999). B. Calculated CaCO$_3$ and organic carbon storage rates (accumulation rates) during the Phanerozoic (after Locklair and Lerman, 2005).

that organisms can incorporate metabolic carbon into their shells exhibiting a "vital effect", and that some samples may also contain a diagenetic overprint. As mentioned in Chapter 6, the main factors that control the long-term variations in the δ^{13}C record in carbonates are biological production of organic matter and subsequent storage in sediments of some portion of the ^{13}C-depleted organic carbon produced. In addition, isotopically lighter CO_2 can be added from the Earth's interior to the ocean-atmosphere system, and from the processes of weathering and erosion of sedimentary organic matter in continental rocks, isotopically lighter CO_2 can be added to the atmosphere and to the ocean via river runoff of ^{13}C-depleted organic carbon. Also, storage of ^{13}C-depleted organic carbon in the ocean depends to some extent on the variations in the topography of the ocean basins and in sea level and on the degree of oxidation of the oceans as a whole. On the time scale of tens of millions of years, these variations are controlled by the intensity of plate tectonics and the rate of sea floor spreading: as mid-ocean ridge volume increases during times of more rapid rates of spreading and volcanism on the seafloor, ocean water is displaced onto the continents; during times of reduced seafloor volcanism, seawater drains from the continents and sea level falls. On the glacial-integlacial time scale of continental glaciations of 10 to 100 ka within the Pleistocene, Permo-Carboniferous, and the Neoprotreozoic, such variations are controlled by the buildup and advancement of continental glaciers and sea level fall, and by the melting and retreat of the glaciers and sea level rise. Frakes *et al.* (1992) proposed that the δ^{13}C of carbonates (hence seawater) becomes particularly heavy at times of glaciations, and that such times are also characterized by low CO_2 levels. The coincidences of the δ^{13}C peaks with the Late Ordovician and Permo-Carboniferous glacial episodes appear to support this proposition, but the Mesozoic-Cenozoic record is divergent (Fig. 7.12A).

Figure 7.12B shows the $CaCO_3$ and organic carbon storage or accumulation rates in sedimentary rocks through Phanerozoic time. Note the differences in the trends for the storage rates of $CaCO_3$ between the Berner (2004) and the Mackenzie and Morse (1992) curves. The former is based mainly on calculations in which the δ^{13}C of inorganic carbonates through Phanerozoic time is used to obtain the storage rates; the latter is based on the preserved record of the Phanerozoic carbonate rock mass-age distribution and a cycling model. Nevertheless, note that the Permo-Carboniferous, a time of continental glaciation and ^{13}C-enriched carbonates, is a time of relatively high organic carbon to inorganic carbon burial rates, as shown by both curves. The important burial at this time of ^{13}C-depleted organic matter in sediments would drive the ocean toward higher δ^{13}C values as the ^{12}C was preferentially removed from the ocean water DIC pool and incorporated in organic matter. The Permo-Carboniferous is the time of the emergence and spread of lowland plants living in coastal marine settings and these events probably account for much of the enrichment of ^{13}C-enriched inorganic carbon in carbonates. These great swamps of the Permo-Carboniferous would eventually become the vast reserves of coal that we mine today through processes involving the maturation of the organic matter buried in them. In addition, the emergence and spread of land plants in the late Paleozoic also played an important role in the drawdown of atmospheric CO_2 at this time (see Fig. 3.7).

7.3   Oxygen

The geologically long-term oxygen isotope record of limestones and calcareous fossil samples shows a clear trend of ^{18}O depletion with age of the rocks (Fig. 7.13).

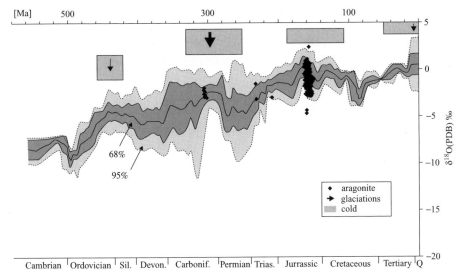

Figure 7.13   Phanerozoic variation in the oxygen isotope composition of marine carbonates of calcitic and aragonite shells, based on 4,500 samples from about 100 localities worldwide. The central dark line shows the mean trend of the data whereas the light-grey and dark-grey areas represent the 95 and 68% confidence levels of the data, respectively. Periods of inferred cold climate and glaciations are also shown (after Veizer *et al.*, 1999).

This isotope record in ancient marine carbonates and also cherts and phosphate deposits is one of the more controversial topics of sedimentary isotope geochemistry. The controversy centers on the issue of whether or not the secular trend is primary or post-depositional in origin (Land, 1995, as opposed to Veizer, 1995). An extensive review of $\delta^{18}O$ and $\delta^{13}C$ data for numerous taxonomic groups secreting $CaCO_3$ shells and skeletons (Land, 1989) led to the conclusion that the spread of values is much greater than can be accounted for by temperature-dependent fractionation or variation of the isotopic ratios in ocean water, and that diagenetic alteration of biogenic carbonates involves dissolution and *in situ* reprecipitation of calcite rather than a reaction in the solid state.

Undoubtedly, diagenesis and other post-depositional phenomena reset the $\delta^{18}O$ signature, usually to more negative values, during stabilization of original metastable carbonate phases (aragonite and high-Mg calcite) into the stable phase, low-Mg calcite (Bathurst, 1974; Morse and Mackenzie, 1990). When carbonate rock is subjected to this stabilization process, much of its original chemistry and mineralogy can be altered. An exception to this can be the original calcite shells low in magnesium of some organisms, such as brachiopods, belemnites, and foraminifera. In addition, the rocks once diagenetically altered become relatively unreactive and inert to further resetting of the $\delta^{18}O$ signature, and the overall bulk rock depletions relative to the stable phases are generally only 2 to 3 ‰. For Late Precambrian limestones this is not the case and the overall depletions in ^{18}O can be as high as -7 ‰ or more. The observed Precambrian

δ^{18}O secular trend is likely real, albeit shifted by 2 to 3 ‰ to lighter values, and reflects the changing δ^{18}O of seawater and perhaps higher Earth's surface temperatures of the Precambrian. The relationship between the ^{18}O/^{16}O ratio in CaCO$_3$ and in water where carbonate precipitates, is a function of temperature (Friedman and O'Neil, 1977, Fig. 13; O'Neil et al., 1969; Savin, 1980):

$$\delta^{18}O_{calcite} - \delta^{18}O_{H_2O} = 2780/T^2 - 0.00289$$

where T is temperature in kelvin. Calcite is enriched in ^{18}O relative to water (on the PDB or VPDB scale, explained in Box 6.1) by about 2 ‰ at 5°C and depleted by −4.4 ‰ at 35°C. This corresponds to ±1 ‰ of δ^{18}O change in the enrichment of calcite for every ∓4.7°C change in temperature, when calcite is at an equilibrium with water of a constant ^{18}O/^{16}O ratio. Among the limitations on the universality of the paleotemperature method are the variations in the ^{18}O/^{16}O ratio of ocean water through geologic time, alteration of the isotopic composition by diagenetic changes and reactions with other waters, and the different responses to temperature by different groups of calcite and aragonite secreting organisms.

The reasons for believing that the Phanerozoic trend shown in Fig. 7.13 is essentially a primary trend are discussed in detail by Veizer et al. (1999). If we accept that the δ^{18}O of past seawater was evolving toward ^{18}O-enriched values (Wallmann, 2001) and if we account for this evolutionary trend in the data of Fig. 7.13 accordingly, the superimposed second-order structure of the curve appears to correlate reasonably with the Phanerozoic paleoclimatic record (Veizer and Mackenzie, 2004). Therefore the observed structure in the δ^{18}O secular trend to some degree likely reflects paleotemperature. However, the structure does not correlate well with the paleoatmospheric CO$_2$ curve (Fig. 3.7): the δ^{18}O curve suggests four periods of cooling in the Phanerozoic whereas if CO$_2$ alone is the major driving force of climatic change, one would conclude there are only two major coolings during which there were continental glaciations (see further discussion in Chapter 10).

7.4   Sulfur

Sulfur is found in carbonate minerals as a minor constituent. Its importance to the history of the carbon cycle stems mainly from the fact that the isotopes of sulfur in sedimentary precipitates provide boundary conditions on the evolution of some organismal groups over geologic time and the relative accumulation rates of sulfate-sulfur to sulfide-sulfur on the sea floor. The latter is tied to the carbon cycle through the fact that organic matter contains important amounts of sulfur primarily strengthening protein bonds and sulfur as dissolved sulfate in sediment pore waters is an important oxidizing agent of organic matter (Chapter 9).

As with carbon, the isotopes of sulfur are strongly fractionated by biological processes, particularly during dissimilatory bacterial reduction of sulfate to sulfide. The laboratory results for this step are anywhere from δ^{34}S $= +4$ to -46 ‰ (relative to the standard Canyon Diablo Troilite meteorite, denoted CDT), but even larger fractionations have been observed in natural systems. The geologic record is characterized by a lack of Precambrian evaporitic sequences, including sulfate evaporites. Layered barites ($BaSO_4$) do exist, but they may be, at least in part, of hydrothermal origin. Most Archean sulfides, such as pyrites, contain $\delta^{34}S$ close to 0 ‰, a value typical of the mantle, rather than the expected highly negative values characteristic of bacterial dissimilatory sulfate reduction. These observations were interpreted (e.g., Hayes et al., 1992; Schidlowski et al., 1983) as being due to biological evolution, where it was proposed that the appearance of oxygen-generating photosynthesis and of bacterial sulfate reduction were relatively late developments in evolutionary history. Following this line of argument, most of the sulfur in Archean host phases would have originated from mantle sources and carried its isotopic signature. Only with the onset of these two biological processes in the Late Archean or Early Proterozoic was there sufficient oxygen generated to stabilize sulfate in seawater and to initiate its bacterial reduction to H_2S, the latter eventually forming sulfide minerals, such as pyrite. This development resulted in the burial of large quantities of sulfides depleted in ^{34}S in the sediments, causing the residual sulfate in the ocean to shift toward heavier values.

The above interpretation of the $\delta^{34}S$ secular trend is reasonable and compelling. However, Goddéris and Veizer (2000), using the same scenario of continental growth that generated the Sr isotope trend (Section 7.1) have also generated the observed $\delta^{34}S$ pattern and the growth of sulfate in the oceans. Briefly, a possible explanation is that the early mantle-buffered oceans had a large consumption of oxygen resulting from chemical reactions in submarine hydrothermal systems (Holland, 2004; Wolery and Sleep, 1988) because they operated at considerably higher rates than today. The capacity of this sink has probably declined somewhat in the course of geologic history, reflecting the decay in the dissipation of the heat from the core and mantle as mentioned earlier. As a result, the buffering of the ocean was taken over by the continental river flux. Thus it is possible that bacterial dissimilatory sulfate reduction may have existed at the time of the shift in the $\delta^{34}S$ record, but the isotope data do not provide a definitive answer as to the timing of this evolutionary process.

The $\delta^{34}S$ of sulfate variations in Phanerozoic evaporites and sulfur structurally bound in calcitic shells (and hence oceans) form an overall trough-like trend similar to Sr isotopes (Fig. 7.14). However, there are large age uncertainties and temporal gaps between the evaporitic sequences investigated. This is due to their episodic occurrence and uncertain chronology and is part of the reason for the large spread in the coeval $\delta^{34}S$ values despite the fact that the $\delta^{34}S$ sulfate in seawater is spatially essentially homogeneous. Another reason for this large spread in the $\delta^{34}S$ values is the evolution of sulfur isotopes in the course of the evaporative process, from Ca-sulfate to Na-chloride to later rocks bearing complex sulfate and chloride salts. A recent development of the technique that enabled measurement of $\delta^{34}S$ in structurally bound sulfate in carbonates (Kampschulte, 2001; Kampschulte and Strauss, 1998) has yielded the Phanerozoic secular curve of Fig. 7.14 which has a much greater temporal resolution.

The $\delta^{34}S$ of sulfate and $\delta^{13}C$ of carbonate secular curves correlate negatively, suggesting that it is the redox balance (exchange of oxygen between the carbon and sulfur

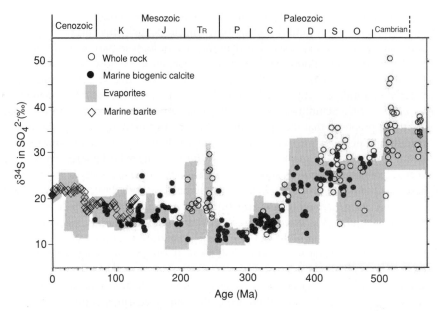

Figure 7.14   Sulfur isotopic composition of sulfate in evaporites and in sulfur structurally bound in calcite shells during the Phanerozoic. The open circles represent data from whole rock analyses of $\delta^{34}S$ whereas the filled black circles are data from biologically produced calcite deposited under marine conditions. The shaded grey area represents data from evaporite deposits. This trend reflects the sulfur isotopic composition of Phanerozoic seawater (after Kampschulte, 2001, as shown in Veizer and Mackenzie, 2004). Marine barite from Paytan *et al.* (1998, 2004).

cycles) that controls the $\delta^{34}S$ variations in Phanerozoic rocks and hence oceans. If the redox balance is indeed a major control mechanism, it suggests that the withdrawal of ^{32}S due to pyrite burial in sediments was twice as large as today in the early Paleozoic and about one half as large as today in the late Paleozoic (Kump, 1989).

8   Summary of the Phanerozoic Rock Record in Terms of Ocean Composition

One conclusion to be drawn from the mass-age relationships discussed in this chapter is that these observations are the result of changing atmosphere-hydrosphere environmental conditions through the Phanerozoic. Another is that the ooid observations are not statistically significant, as argued by Bates and Brand (1990), or that the Phanerozoic dolomite mass-age relationship is not cyclic (Holland and Zimmermann, 2000). However, the latter authors do accept that there is generally lower dolomite abundance in carbonate sediments deposited during the last 200 Ma. These qualifications notwithstanding, a number of previously described parameters (Sr/Ca, Mg/Ca, SO_4^{2-}, aragonite/calcite, possibly dolomite/calcite, skeletal mineralogy of fossil reefs, and frequency of ooids and sedimentary iron ores) appear to be related in some degree to sea level stands, the latter at least in part a reflection of plate tectonic activity during the Phanerozoic.

Earlier studies assumed that the chemical composition of seawater during the Phanerozoic was comparable to present-day composition. Subsequently, experimental data on fluid inclusions in halite and on carbonate cements suggested that at least Mg, Ca, and Sr, and their ratios in Phanerozoic seawater have been variable (Cicero and Lohmann, 2001; Horita *et al.*, 1991, 2002; Lowenstein *et al.*, 2001). Steuber and Veizer (2002) have assembled a continuous record of Sr variations in biological calcite for the Phanerozoic and hence its oceans (Fig. 7.8) that covaries positively with the rate of accretion of the oceanic crust and negatively with the Mg/Ca ratio. Such a covariance suggests that we are dealing with coupled phenomena and that all these variables are ultimately driven by plate tectonics, specifically ridge accretion and seafloor spreading rates that, in turn, control the associated hydrothermal and low-temperature alteration processes, as has been suggested earlier (Morse and Mackenzie, 1990; Veizer and Mackenzie, 2004). Since the hydrothermal alteration of young oceanic crust efficiently exchanges Mg in seawater for Ca in basalt, the accretion of the oceanic crust would modulate the Mg/Ca and SO_4/Ca ratios of seawater. At high accretion rates, the low Mg/Ca and SO_4/Ca ratios favor precipitation of calcite and as a result, higher retention of Sr in seawater. At slow accretion rates, the higher Mg/Ca and SO_4/Ca ratios favor aragonite precipitation and a higher rate of Sr removal from seawater.

It appears that the first-order changes in sea level are driven by the accretion of ridges: high accretion rate, high sea level; low accretion rate, low sea level. Regardless of the tie between oceanic plate production and sea level, extended times of global high sea level may have been times of enhanced atmospheric CO_2 levels, higher temperatures (not necessarily related solely to atmospheric CO_2 concentrations), probably lower seawater Mg/Ca and SO_4/Ca ratios, different saturation states of seawater, and different seawater sulfate and strontium concentrations than at present. The converse is true for first-order global sea level low stands. It appears that the conditions for early dolomitization, calcite ooid and cement formation, reef skeletal mineralogy enriched in calcite, and perhaps ironstone deposition are best met during extended times of global high sea level when the continents are flooded (the calcite seas of Sandberg, 1983). The flooding provides ample area for shallow-water carbonate deposition and accumulation. The chemical state of seawater would favor precipitation of less soluble carbonate minerals, low magnesian calcite ooids and cements, rather than high magnesian calcite and aragonite phases, in regions of carbonate deposition. The deposition of predominantly calcite would leave seawater enriched in strontium. Dolomitization of precursor calcite and aragonite phases either in marine waters or in mixed continental-marine waters may be enhanced under these conditions if the SO_4^{2-} concentrations in seawater, that may be an inhibitor of dolomite formation, were low during extended periods of high sea level. Although contrary to expectation, the lower magnesium content of seawater might actually aid dolomite formation because of lessened competition with the soluble magnesian calcite and aragonite phases. Furthermore, the potentially lowered pH of marine waters during times of enhanced atmospheric carbon dioxide would favor syndepositional or later dolomitization in mixed marine-meteoric waters because the range of seawater-meteoric compositional mixtures over which calcite could be dissolved and dolomite precipitated is expanded (Plummer, 1975). Perhaps superimposed on the hypothesized Phanerozoic cyclic dolomite/calcite ratio is a longer term trend in which dolomite abundance increases with increasing age, particularly in rocks older than 200 million years, due to favorable environmental conditions as well as to advancing late diagenetic and burial dolomitization. During the last 150 Ma, this magnesium has been transferred out of the dolomite reservoir ("bank" of Holland and Zimmermann, 2000) into the magnesium silicate reservoir by dissolution of dolomite and precipitation of silicates, and to a lesser extent into the ocean reservoir, thus accounting in part for the increasing Mg/Ca ratio of seawater during the last 150 million years.

Rowley (2002) argues that the rate of oceanic plate production may not have varied significantly for the latest 180 Ma. If so, this may have major implications for our understanding of the linkages of plate tectonics to sea level change, atmospheric CO_2 variations, seawater chemistry, and related phenomena. Nevertheless, while the 30 to 40% variations in seafloor spreading rates during the last 100 Ma, as inferred by Delaney and Boyle (1986), may be too high, we cannot at present dismiss entirely the proposition that bydrothermal exchange between seafloor and ocean, and presumably the rate of plate generation, varied during this interval of time and for the Phanerozoic. Indeed in terms of atmospheric CO_2 and seawater compositional variations in the Phanerozoic, model calculations show that the accretion rate is a relatively insensitive parameter and its near constancy does not prohibit atmosphere and ocean compositional changes during Phanerozoic time (Arvidson et al., 2006; Berner, 2004).

It is probable that the first-order changes in sea level can still be driven in part by the accretion rates of the ridges that in turn are linked to submarine volcanism and ridge and off-ridge volume changes and thus to sea level change. Relative sea level change may also be linked to the development of the great super-continent of Pangea in the later part of the Carboniferous and its break-up in the Late Triassic, a prolonged period of relatively low global sea level like that of today. The development of the super-continent would insulate the underlying mantle, heat it, and lead to some slight expansion under the super-continent. This in turn would lead to a relative fall in sea level about the super-continent. As the continents broke up and accretion rates increased into the Mesozoic, the cooling continental masses and the enhanced accretion rates would lead to a rise in sea level and the flooding of the continents during the Cretaceous worldwide transgression of the sea. The apparent trends in Phanerozoic carbonate chemical, mineralogical and isotopic composition are linked to such sea level changes and are indicators also of changes in ocean-atmosphere conditions that are driven in part by sea level changes tied to plate tectonic mechanisms.

In order to discuss more fully the carbonate sedimentary record as an indicator of the history of the Earth's surface and the behavior of the global carbon cycle through geologic time, as given in Chapter 10, we must first address the weathering, erosion, and transport of the continental sediments (Chapter 8) and the processes that operate in the shallow oceanic coastal ocean, the major recipient of weathered and eroded materials from the land (Chapter 9).

Chapter 8

Weathering and Consumption of CO$_2$

Carbon dioxide dissolved in rain and soil and ground waters is the main acid that reacts with minerals in the sedimentary and crystalline crust, releasing the ionic constituents of river water and producing the bicarbonate ion, HCO$_3^-$, in the process. The consumption of CO$_2$ in mineral weathering reactions is a major transfer path of CO$_2$ from the atmosphere to the land in the carbon cycle. On land, CO$_2$ is bound in organic matter by plant photosynthesis, and released into the soil pore space by the remineralization of organic matter. The direct uptake of CO$_2$ by waters reacting with the carbonate and silicate minerals represents a drain on the atmosphere. Dissolved inorganic carbon (DIC), consisting mainly of the bicarbonate ion in river waters that is a product of the CO$_2$-mineral reactions, is transported to the surface ocean. In the ocean, CO$_2$ is generated by the precipitation of CaCO$_3$ and its storage in sediments (Chapter 5), by the remineralization of organic matter produced *in situ* and brought in from the land, and by reverse weathering reactions (Mackenzie and Garrels, 1966b) where degraded aluminosilicates, cations, and HCO$_3^-$ react to make clay minerals and CO$_2$. The production of CO$_2$ in seawater enables the CO$_2$ flux from the surface ocean to the atmosphere that compensates in part for the CO$_2$ consumed in weathering reactions. This very brief description of the CO$_2$ consumption and input to the atmosphere highlights the importance of the mineral weathering flux in the global carbon cycle.

In this chapter we discuss various aspects of the weathering of rock minerals that consume mainly CO$_2$ and produce HCO$_3^-$ and dissolved cations and use some simple models to derive a quantitative picture of the mineral and rock sources of materials in river waters and the amount of CO$_2$ consumed during weathering. We conclude the chapter with a brief look at the issue of increased chemical weathering due to anthropogenic emissions of sulfur gases to the atmosphere and increased deposition of sulfuric acid on the Earth's surface.

1   Weathering Source: Sedimentary and Crystalline Lithosphere

1.1   Sediments and Continental Crust

The term weathering describes in general the processes occurring in the sedimentary and crystalline rocks in the outer shell of the Earth that come in contact with the atmosphere and waters. Physical weathering or physical denudation of the Earth's surface is the disintegration of the rocks, minerals, and organic matter into smaller particles and their removal by wind and running water, also known as erosion. Chemical weathering includes the chemical reactions taking place between minerals and waters and, to a lesser extent, between minerals and gases that result in the dissolution of minerals and formation of new minerals. Because the dissolved products are transported by running water from the land surface to the ocean, the result of the chemical weathering is chemical denudation of the Earth's surface. Bacterial activity plays an important role in chemical weathering where different bacteria convert organic matter to CO_2 and water, and oxidize and reduce the inorganic species of nitrogen, sulfur, iron, and trace elements that occur in various oxidation states in soil and ground waters. As a whole, chemical weathering is a process that drives much of the transport of gaseous and dissolved materials from the sedimentary and crystalline lithosphere to the atmosphere and oceans. Chemical reactions between natural waters and many minerals of soils, sediments, and continental crust are effected by such acidic species in solution as CO_2, sulfate, and nitrate from atmospheric and soil sources, and organic acids from decomposition of organic matter in soils. The source of CO_2 for weathering reactions that is physically close to the minerals is the pore space of soils where the concentration of CO_2 is usually higher than that in the overlying air because of the continuous respiration or oxidation of organic matter. Below the root zone, the CO_2 concentrations in soil air have been reported as high as 10,000 to 15,000 ppmv. The higher CO_2 partial pressure in the soil gas phase accounts for the CO_2 flux from the soil to the atmosphere, and the ^{13}C-depleted carbon isotopic composition of soil CO_2 reflects its source in the respiration of ^{13}C-depleted organic matter (Chapter 6). Concentration profiles of CO_2 in soil vary with the surface vegetation type, season of the year, and the state of the soil surface, such as its wetness or snow cover, as shown in Fig. 8.1. In the example shown of a grass field, concentrations in the upper 1 m of the soil pore space may be about 20 times higher than the atmospheric concentration. The concentration gradient of CO_2 in soils is responsible for its diffusion to the atmosphere and the higher concentrations of dissolved CO_2 in rain water percolating through the soil and reacting with minerals.

The sediments occurring on the continents account for 90% of the total sediment mass that had to a large extent been deposited or formed in the oceans of the geologic

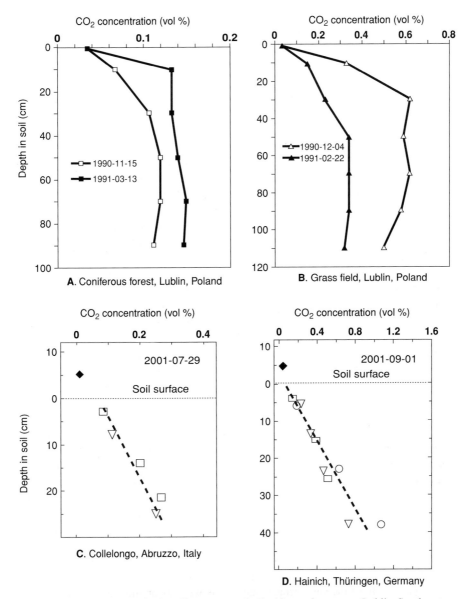

Figure 8.1 CO$_2$ concentration in soil pore space. A. Coniferous forest, near Lublin, Southeastern Poland; mean δ^{13}C of soil air is in the range -17 to -20 ‰. B. Grass field, same locality; mean δ^{13}C of soil air is in the range -21.5 to -24 ‰ (modified from Dudziak and Halas, 1996). C and D. CO$_2$ concentrations in soil pore space and in the air, 5 cm above soil surface (from data of V. Hahn and N. Buchmann, *personal communication*, 2005, by the authors' permission).

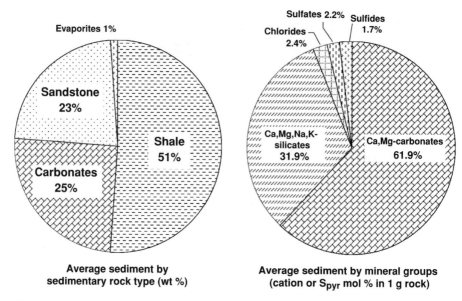

Average sediment by
sedimentary rock type (wt %)

Average sediment by mineral groups
(cation or S_{pyr} mol % in 1 g rock)

Figure 8.2   Composition of an average sediment by sedimentary rock types (weight %) and by mineral groups based on chemical composition: carbonates (Ca- and Ca,Mg-carbonates) are mol % of Ca and Mg of the total of cations and S in pyrite that make 100 mol %; silicates of Ca, Mg, Na, and K, showing mol % of cations; sulfates, mostly $CaSO_4$, as mol % Ca; chlorides, mainly NaCl with minor additions of $MgCl_2$ and KCl (mol % of cations); sulfides taken here as pyrite, FeS_2, showing mol % S. From data in Li (2000).

past. Sediments are usually classified by *sedimentary rock types* that are based on mixed criteria of their mineral abundance, sediment texture, and origin, such as the lithologic groups of carbonates, shales, sandstones, and evaporites. Each of these sedimentary rock types contains varying proportions of minerals that according to their chemical composition can be classified primarily into carbonates, silicates, sulfates, sulfides, chlorides, oxides, and hydroxides. Thus the composition of an average sediment that is based on the rock types differs from its composition based on the mineral abundance, as shown in Fig. 8.2.

The most abundant sedimentary rocks are the mainly detrital rock types of shales and sandstones, which are mainly silicates in their mineral composition. However, because carbonate rocks (limestones and dolomites) contain silicate minerals in the form of clays and sands, and shales and sandstones contain carbonate minerals calcite and dolomite, the picture of the mineral abundances is different from that of the lithologic types: the most abundant mineral group is the carbonates, $(Ca,Mg)CO_3$, followed by the silicates containing Ca, Mg, Na, and K. Ca in sulfates, Na in chlorides, and reduced sulfur occurring mainly in pyrite (FeS_2) each accounts for about 2 mol % in 1 gram of average sediment.

The composition of the sedimentary rocks and upper continental crust in terms of their main mineral components is given in Table 8.1. The carbonate mineral

Table 8.1 Major cations (Ca^{2+}, Mg^{2+}, Na$^+$, and K$^+$) and pyrite-sulfur (S$_{pyr}$) in four main sediment classes and upper continental crust. Note that carbonate, silicate (Ca-, Mg-, Na-, and K-silicates), and evaporite minerals (CaSO$_4$ and NaCl) occur in all sediment classes. Ratio ψ described in Section 2

	Mass (10^{24} g) and sediment mass fraction (%)[a]	Carbonates 0.48 25.3%	Shales 0.96 50.5%	Sandstones 0.44 23.2%	Evaporites[b] 0.02 10%	Average sediment 1.90 100%	Upper crust[c] 1.12	Ave. sediment 63 wt% and crust 37 wt% 3.02
	Mineral components				10^{-3} mol cation/g rock			
CaCO$_3$	Calcite and dolomite component	6.510	0.6782	0.6782	0.2215	2.147	—	1.352
MgCO$_3$	Dolomite component	1.102	0.1148	0.1148	0.0375	0.3635	—	0.2290
CaAl$_2$Si$_2$O$_8$	In albite-anorthite series and other Ca silicates	0.5091	0.0153	0.1387	—	0.1685	0.7989	0.4017
Mg$_5$Al$_2$-Si$_3$O$_{10}$(OH)$_8$	Chlorite in sediments and Mg-silicates	0.1059	0.5178	0.3516	—	0.3698	0.6575	0.4763
NaAlSi$_3$O$_8$	Albite	0.1272	0.3181	0.4639	0.0445	0.3007	1.099	0.5962
KAlSi$_3$O$_8$	K-feldspar (orthoclase)	0.1274	0.6412	0.4162	0.0004	0.4525	0.6646	0.5310
S$_{pyr}$	Pyrite (FeS$_2$)	0.0374	0.0873	0.0624	—	0.06802	—	0.0429
CaSO$_4$	Gypsum or anhydrite	0.1162	0.0412	0.0212	3.140	0.08814	—	0.0555
NaCl	Halite	0.0116	0.0240	0.0169	7.116	0.09386	0.0042	0.0607
KCl	Sylvite				0.0679	0.0007107	—	0.0004
MgCl$_2$					0.1114	0.001172	—	0.0007
MgSO$_4$					0.0799	0.0008407	—	0.0005
ψ = (CO$_2$ consumable)/(HCO$_3^-$ producible)		0.53	0.75	0.74	0.54	0.61	1.00	0.73

[a] Concentrations of cationic components from weight percent of their oxides (except for S$_{pyr}$) and masses of the sediment classes as given by Li (2000).

[b] Mineral composition of evaporites was balanced by addition of 3.0% Cl (2.23×10^{-4} mol/g rock) and 2.6% S (8.99×10^{-5} mol/g) to make KCl, MgCl$_2$, and CaSO$_4$.

[c] Upper crust of composition 2/3 granite and 1/3 basalt (Li, 2000). Mass of 1.12×10^{24} g is based on density 2.8 g/cm^3, thickness 2.3 km, occurring over the area of continents and shelves, 177×10^6 km^2. Other estimates of thickness and chemical composition variants given by Wedepohl (1995) and Rudnick and Gao (2003).

components, $CaCO_3$ and $MgCO_3$, monovalent- and divalent-cation silicates, pyrite, and calcium sulfate occur in all the sediment types. NaCl occurs in all the sediments as well as in the igneous crust. The overlap in mineral composition between the igneous crust and sediments is the silicates that share the same cations (Ca^{2+}, Mg^{2+}, Na^+, and K^+) in structurally different mineral groups, such as, for example, feldspars, plagioclases, olivines, pyroxenes, amphiboles, micas, and clays.

1.2   Organic and Inorganic Carbon

Carbon occurs in the weathering source rocks mainly in the form of Ca- and Mg-bearing carbonate minerals (Chapter 4) and as organic matter. The erosion of the land surface and weathering reactions produce four types of carbon-containing materials (Chapter 1), designated as particulate inorganic carbon (PIC), particulate organic carbon (POC), dissolved inorganic carbon (DIC), and dissolved organic carbon (DOC). Particulate inorganic carbon (PIC) is the product of erosion of limestones and dolomites that have not dissolved in river water that is undersaturated with respect to calcite (Chapter 5). POC is derived from soil humus and older organic matter and it consists of a reactive fraction that undergoes a relatively fast remineralization to CO_2 in transport and in the coastal ocean, and a refractory fraction that is stored in oceanic sediments and is more resilient to remineralization. DOC is a product of decomposition of soil humus and older organic matter (Fig. 1.4) that is also destined to be remineralized at shorter or longer time scales in the ocean. DIC is the sum of the CO_2, bicarbonate, and carbonate ions in solution that form in oxidation of organic matter and dissolution of carbonate minerals. While the weathering reactions between minerals and CO_2 and transport of solid organic matter to the ocean by rivers and water runoff represent a drain on atmospheric CO_2, remineralization of organic matter and precipitation of $CaCO_3$ return some of the CO_2 that was taken from the atmosphere by plant photosynthesis and weathering. The total transport of dissolved mineral solids by rivers is 2.8 to 4.4 Gt/yr, and a much larger mass is carried as particulate load in suspension and on the river bed, 13 to 20 Gt/yr (Berner and Berner, 1996; Garrels and Mackenzie, 1971; Holland, 1978; Meybeck, 1979, 1984; Meybeck and Ragu, 1995; Milliman and Syvitski, 1992). This transport, as shown in Fig. 8.3, varies with latitude because of the size of the land drainage areas and river discharge that are larger in the Northern Hemisphere, in the latitudinal band between the equator and latitude N60°.

Particulate inorganic carbon (PIC) as a product of erosion of limestones and particulate organic matter (POC) were carried by rivers in pre-industrial time at the rate of about 2 Gt/yr, as explained below, which represents 10 to 15% of the riverine total particulate load, as shown by the data in Fig. 8.4.

The particulate inorganic carbon flux, mostly as $CaCO_3$, is about 0.18 Gt C/yr or 1.5 Gt $CaCO_3$/yr. Refractory and reactive particulate organic carbon fluxes are each 0.1 Gt C/yr, which translates into a flux of 0.5 Gt CH_2O/yr for organic matter of composition CH_2O. Most of the carbon flux, at the pre-industrial value of 0.7 Gt C/yr, is dissolved inorganic carbon (DIC), 0.38 Gt C/yr, and dissolved organic carbon (DOC), 0.22 Gt C/yr. DOC accounts for a relatively large fraction of about $1/3$ of total dissolved

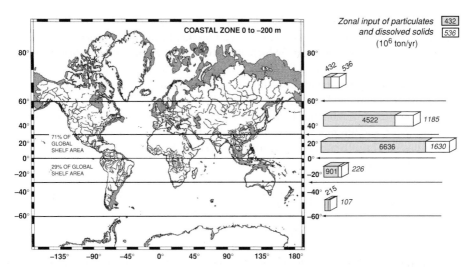

Figure 8.3   World land, rivers, coastal zone, and open ocean in a Mercator projection. Riverine transport of dissolved and suspended solids, latitudes N85° to S75°, in latitudinal zones North 90°−60°, 60°−30°, 30°−0°, and South 0°−30° and 30°−60° (river data from Meybeck and Ragu, 1995, adjusted for the global river flow by multiplying their data by 3.74/2.65, see p. 234).

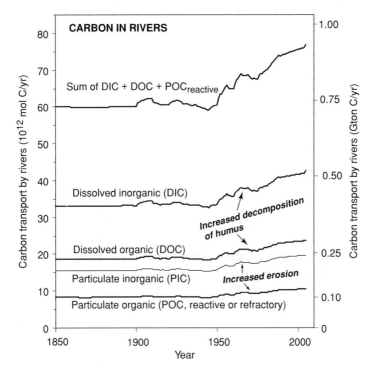

Figure 8.4   Riverine transport of different forms of carbon from land to the coastal ocean since pre-industrial time to year 2005 (after Mackenzie *et al.*, 2002).

carbon flux and it is remineralized to CO_2 in the ocean over a longer period of time. It should be noted in Fig. 8.4 that the reactive carbon flux from land changed little during the decades prior to 1950, after which it increased in the subsequent half century by nearly 30%, from 0.7 to 0.9 Gt C/yr. One-half of this increase, or 0.1 Gt C/yr, is accounted for by the rising fluxes of PIC and POC that are driven by land erosion. The enhancement of land erosion by human industrial and agricultural activities is due to a great extent to the reduction of the vegetation cover and exposure of land to wind and running water, and ultimately to greater dissolution of $CaCO_3$ and remineralization of humus in soil. The effects of the Industrial Age on land denudation are further discussed in Chapter 11.

2 Dissolution at the Earth's Surface

2.1 Chemical Denudation Rate

Chemical weathering acts on sedimentary rocks (shales, sandstones, limestones, dolomites, evaporites) and igneous and metamorphic minerals (mostly silicates), such that a world average river represents a pooled average of water-mineral reactions within a layer of sediments and crustal rocks of some thickness, extending over a part of the continental surface that is drained by rivers and groundwater. The chemical denudation rate is a measure of the loss of either the mass (W_m) or thickness (W_h) of the continental surface:

$$W_m = \frac{C_{flow} V_{flow}}{S_{land}} \quad (\text{g m}^{-2} \text{yr}^{-1}) \tag{8.1}$$

and

$$W_h = \frac{10^{-4} W_m}{\rho} \quad (\text{cm yr}^{-1}) \tag{8.2}$$

where C_{flow} is the concentration of dissolved solids in rivers (g/kg), V_{flow} is the annual water discharge to the oceans (kg/yr or liter/yr), S_{land} is the geographic surface area (m^2) of land that is drained to the oceans, and ρ is the mean bulk density (g/cm^3) of the source rocks undergoing weathering.

The concentrations of the major constituents of an average river water are shown in Table 8.2. The bicarbonate ion, HCO_3^-, is the most abundant anion in river waters containing seven major ionic species. The composition of an average river as given by Meybeck (1979), Mackenzie (1992), and Berner and Berner (1996) is variably corrected for the recycled oceanic salts carried through the atmosphere and the pollutant inputs of the Industrial Age. We refer to these rivers as representing pre-industrial time and their relationships to the mineral composition of the weathering source sediments and crystalline rocks are discussed in more detail in Sections 4 and 5 of this chapter. The land surface areas that drain to the oceans, also called the external, exorheic or peripheral areas, have been estimated at 76.1×10^6 km^2 (Meybeck, 2003), 88.6×10^6 km^2 (Berner and Berner, 1996; Milliman and Meade, 1983, with references to earlier estimates), and 99.9×10^6 km^2 (Meybeck, 1984, p. V-2), excluding Greenland and Antarctica. Annual water flow from the external drainage

Table 8.2   Dissolved constituents of five world average rivers. Concentrations in mg/kg or mg/liter reported by Holland (1978), Meybeck (1979), Drever (1988), and Berner and Berner (1996). Concentrations in mol/kg in *italics* adjusted to reduce the ionic charge imbalance to <0.1%

Chemical species	Global mean Livingstone (1963) Holland (1978) mg/kg	Holland (1978), corrected for atmospheric input mg/kg	Natural (unpolluted) Meybeck (1979), Drever (1988) mg/kg	10^{-3} mol/kg	Natural, corrected for pollution and recycling salts (Mackenzie, 1992) mg/kg	10^{-3} mol/kg	Natural, corrected for pollution and recycling salts (Berner and Berner; 1996, Tables 5.6, 5.10, 5.11) mg/kg	10^{-3} mol/kg
Na$^+$	6.3	4.8	5.15	0.2240	3.66	0.1590	4.47	0.1944
K$^+$	2.3	2.4	1.3	0.03325	1.25	0.03200	1.29	0.03299
Ca^{2+}	15	14.8	13.4	0.3343	13.39	0.3340	13.39	0.3341
Mg^{2+}	4.1	4.0	3.35	0.1378	3.22	0.1325	3.33	0.1370
HCO$_3^-$	58.4	54.3 (18.9)[a]	52 (16.7–18.1)[a]	0.8522	52.90 (18.3)[a]	0.8670	52.0 (16.2–16.7)[a]	0.8522
Cl$^-$	8.8	5.7	5.75	0.1622	3.19	0.09000	4.76	0.1343
SO$_4^{2-}$	11.2	6.7	8.25	0.08588	8.02	0.08350	5.19	*0.09185*
SiO$_2$	13.1	12.6	10.4		10.4		10.4	
TOTAL	118.2	105.3	99.6		96.0		94.8	
Total from source rock[b]	69.9	65.0	65.0		61.4		59.3	

[a] Lower HCO$_3^-$ concentration in parentheses is the fraction of total HCO$_3^-$ that is derived from dissolution of carbonates, (Ca,Mg)CO$_3$. See also Table 8.5.

[b] Carbonate-derived HCO$_3^-$, all other ionic constituents, and SiO$_2$.

area is 3.74×10^{16} kg/yr; and with the contribution of melting ice from Greenland and Antarctica it is 3.97×10^{16} kg/yr (Baumgartner and Reichel, 1975; Meybeck, 1979, 1984). The discharge of the rivers studied by Meybeck and Ragu (1995) adds to 2.65×10^{16} kg/yr, or 71% of the total river discharge. A considerably higher estimate of global discharge, 4.7×10^{16} kg/yr, is given by Shiklomanov (1993). From the dissolved solids concentrations in average rivers that are attributed to rock dissolution (Table 8.2), the mean chemical denudation rate is:

$$W_m = \frac{(59 \text{ to } 70) \times 10^{-3} \text{ g/kg} \times 3.74 \times 10^{16} \text{ kg/yr}}{(76 \text{ to } 100) \times 10^{12} \text{ m}^2} \approx 29 \text{ to } 26 \text{ g m}^{-2}\text{yr}^{-1}$$

A lower value of 22 g m^{-2} yr^{-1} was given by Holland (1978). The chemical denudation rate of basaltic rocks in Iceland is higher than the mean, 55 g m^{-2} yr^{-1} (Gíslason et al., 1996). The differences between the denudation rate per unit of geographic land surface area and per unit of mineral reactive surface area are addressed in Section 2.2.

The porosity of up to 50 vol % of soils and partly weathered regolith reduces the mineral density of 2.6 to 2.7 g/cm^3 by $\frac{1}{2}$, to a bulk density of about 1.3 g/cm^3 for dry material. On a global scale, the bulk densities of several thousand samples of different soils fall mostly in a range from 1.2 to 1.7 g/cm^3 (Batjes, 1996). The rate W_h, using mean bulk density $\rho \approx 1.3$ to 2.5 g/cm^3, is:

$$W_h = \frac{10^{-4} \times 27 \text{ g m}^{-2}\text{yr}^{-1}}{(1.3 \text{ to } 2.5) \text{ g cm}^{-3}} \approx 2.1 \text{ to } 1.1 \times 10^{-3} \text{ cm/yr}$$

or about 1 to 2 cm/1000 yr.

The fraction of the calcium in rivers that is derived from the weathering of carbonate rocks is 60 mol % (Table 8.7) and the fraction of carbonates in the surface rock outcrop area is 0.159, as estimated by Meybeck (1987). From these values, the rate of limestone weathering by chemical denudation may be estimated as follows:

$$W_{\text{CaCO}_3} = \frac{13.4 \times 10^{-3} \text{g Ca}^{2+}/\text{kg} \times (100/40) \times 0.60 \times 3.74 \times 10^{16} \text{ kg/yr}}{0.159 \times (76 \text{ to } 100) \times 10^{12} \text{ m}^2}$$
$$\approx 55 \pm 7 \text{ g m}^{-2}\text{yr}^{-1}$$

This mass denudation rate of carbonate rocks is equivalent to about 2 to 4 cm/1000 years or twice the global mean denudation rate. With the inclusion of Mg from carbonates in rivers, the mean denudation rate increase to 61 g m^{-2} yr^{-1}.

Propagation of the weathering front at the boundary between soil (saprolite) and parent rock of granite and schist has been estimated between 0.3 and 3.7 cm/1000 years (White, 1995, with references), the range of these rates bracketing the preceding estimate of chemical denudation rate. Other values of global and regional denudation rates, and relationships between the particulate and dissolved loads transported by individual rivers can be found in many publications (e.g., Berner and Berner, 1996; Drever, 1988; Garrels and Mackenzie, 1971; Lerman, 1988; and references cited therein).

2.2 Weathering Layer Thickness

Estimates of the chemical denudation rate, such as those given in the preceding section, often refer to the mineral mass dissolved per unit of area of the geographic land surface. The unit of the geographic surface area is used because the reactive surface area and mass of the minerals involved in weathering reactions are usually not well known. In this section we estimate the weathering layer thickness that corresponds to an average river water composition and certain parameters that reasonably characterize the layer texture. The mass of dissolved solids in water runoff from the land per unit area of land surface corresponds to the mass dissolved from the surface area of minerals occurring within a layer of some thickness. The thickness of such a weathering layer depends on the texture and distribution within it of interconnected joints or pore spaces that are water conduits, the mineral surface area that is a function of the grain size, and rates of mineral dissolution under the weathering conditions of the environment.

The dissolution rate per unit area of land surface, W_m in equation (8.1), is a product of mineral dissolution rate, R_i (grams per 1 m^2 of mineral surface per unit of time), and the mineral surface area that is, for model mineral particles conventionally taken as spheres:

$$S_i = 4\pi r^2 N \tag{8.3}$$

where r is the particle radius (cm) and N is the number of particles in a layer volume of 1 cm^2 base. The dissolution mass balance equation is therefore:

$$4\pi r^2 N R_i = W_m \tag{8.4}$$

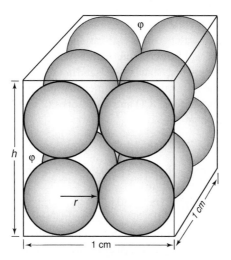

Figure 8.5   Spherical particles of radius r and interstitial pore space (porosity volume fraction φ) in a sediment column h cm high and 1 cm^2 cross-section.

N particles of radius r occupy a volume h cm high and 1 cm^2 base, of porosity φ (Fig. 8.5):

$$\frac{4}{3}\pi r^3 N = h \times 1 \times (1 - \varphi) \qquad (8.5)$$

The layer thickness h is from (8.4) and (8.5):

$$h = \frac{W_m}{3(1 - \varphi)R_i} \cdot r \qquad (8.6)$$

The geometric surface area of the particles may be increased by the roughness factor $\lambda > 1$ (White and Brantley, 2003), or it may be decreased due to the occurrence of unreactive sections of the surface or due to the packing texture, by some factor $\varepsilon < 1$. With these two possible corrections to the mineral reactive surface area, the weathering layer thickness becomes:

$$h = \frac{W_m}{3(1 - \varphi)R_i \lambda \varepsilon} \cdot r \qquad (8.7)$$

If the product $\lambda \varepsilon \approx 1$, then equations (8.6) and (8.7) are identical and they describe dissolution of a mineral geometric reactive surface area. It should be noted that the layer thickness is directly related to the particle size, its radius r, and it increases with increasing porosity φ. Cubic or hexagonal close packing of spheres of the same size has a volume porosity fraction of 0.2595 or 26%. At the maximum possible packing, the porosity fraction is 0.2216 or 22%. Random close packing of spheres gives greater porosities of the aggregates, from 35% and higher (Weisstein, 2005).

Mineral dissolution rates (R_i) that are reported in the literature from laboratory or field measurements may vary widely even when the measurements were done under similar conditions of temperature, solution composition, and pH. In particular, the experimentally determined dissolution rates of the carbonate minerals (Chapter 4) are much higher than can be compatible with the observational evidence of limestone weathering rates. In river waters, SiO_2 and variable fractions of the major ions Na^+, K^+, Ca^{2+}, Mg^{2+} are derived from dissolution of silicate minerals in sedimentary and crystalline rocks. In sediments, shales and sandstones are the main sources of the four cations from silicate minerals. The sedimentary silicates, contained in shales (51 weight % of sediment mass, Table 8.1) and sandstones (23 wt %), are characterized by dissolution rates in the range

$$R_i = 0.003 \text{ to } 0.014 \text{ g m}^{-2} \text{ yr}^{-1}$$

with the lower estimate probably more representative of the sedimentary weathering environment (Lerman and Wu, 2006).

The fractions of Ca^{2+}, Mg^{2+}, Na^+, and K^+ in an average river water that are derived from mineral silicate reactions with CO_2 are given in Table 8.7. These silicate-derived ion concentrations in river water may be stoichiometrically equated to masses of such common rock-forming minerals as Ca-plagioclase, Mg-olivine and(or) Mg-chlorite, Na-plagioclase, and K-feldspar, giving a combined mass weathering rate of 16 to 20

$g\ m^{-2}\ yr^{-1}$, as compared to the total rate of 26 to 29 $g\ m^{-2}\ yr^{-1}$. For an order of magnitude estimate of a thickness of the weathering layer of the silicate minerals in shales and sandstones, the mineral dissolution rate $R_i = 0.003\ g\ m^{-2}\ yr^{-1}$ and particle sizes from silt to fine sand, 60 microns to 1 mm in diameter ($r = 3 \times 10^{-3}$ to 0.05 cm), give from equation (8.7):

$$h \geq \frac{16\ to\ 20\ g\ m^{-2}yr^{-1}}{3 \times (1 - 0.26) \times 0.003\ g\ m^{-2}\ yr^{-1}} \times (0.003\ to\ 0.05\ cm) = 7.2\ to\ 150\ cm$$

The latter result for the weathering layer thickness is a lower bound because it is based on the silicate dissolution rates only and a reactive surface area that is identical to the geometric surface area of the mineral particles. As mentioned earlier, cementation and aggregation of particles may make the effective area smaller and(or) the particle size larger. Hypothetically, but perhaps not inconceivably, if only 10 to 2% of the geometric particle surface area is reactive ($\varepsilon = 0.1$ to 0.02), then the weathering layer thickness with an effective particle size of $r = 0.05$ cm would be much larger, 15 to 75 m.

One important conclusion that can be drawn from the preceding results is that the thickness of the weathering rock layer strongly depends on the surface area of the mineral particles and the texture of the source rock being weathered, as reflected in the roughness factor, possible reduction of the reactive area, and increase in the particle size owing to compaction and cementation or aggregation.

3 Mineral-CO₂ Reactions in Weathering

3.1 CO₂ Reactions with Carbonates and Silicates

As was mentioned at the beginning of this chapter, consumption of CO_2 in mineral dissolution and, more generally, weathering reactions is one of the major fluxes in the global carbon cycle that drives the weathering process and transport of dissolved solids and particulate materials from the land by surface runoff to the ocean. The role of CO_2 as a natural acid that reacts with carbonate and silicate minerals in sediments and continental crystalline crust is supplemented by other acids, such as the sulfuric acid forming in oxidation of reduced sulfur minerals (mainly pyrite, FeS_2), inorganic acids occurring in the atmosphere from volcanic eruptions or biogeochemical reactions on the land or in ocean waters (e.g., sulfuric, hydrochloric, and nitric acids), and organic acids forming by biological processes in soils. Reactions between dissolved CO_2 and minerals containing the common alkali and alkaline-earth metals Na, K, Mg, and Ca produce negatively charged bicarbonate and carbonate ions that neutralize the positive cation charges in solution and create alkalinity (Chapter 5). The most frequently used examples of this process are dissolution reactions of $CaCO_3$, as a generic proxy for carbonates, and $CaSiO_3$, as a generic notation for Ca and Mg in silicates. The dissolution of a divalent-metal carbonate consumes CO_2 and produces HCO_3^- in the ratio of $[CO_2]/HCO_3^-] = 1{:}2$ and dissolution of a divalent-metal silicate consumes CO_2 and produces HCO_3^- in the ratio of $[CO_2]/HCO_3^-] = 1{:}1$. If these two minerals occurred in equal proportions in a rock, their weathering would consume $3CO_2$ and produce $4HCO_3^-$, making the CO_2 consumption 75% of the bicarbonate or of the cation mol-equivalents produced (2 mol Ca^{2+} are 4 mol-equivalents):

$$\frac{CaCO_3 + CO_2 + H_2O \;= Ca^{2+} + 2HCO_3^-}{\underset{3CO_2}{CaSiO_3 + 2CO_2 + H_2O = Ca^{2+} + 2HCO_3^- + SiO_2}}$$

$$\underset{3CO_2}{} \qquad \underset{4HCO_3^-}{}$$

In dissolution of common silicate minerals, the mass of CO_2 consumed depends both on the stoichiometric composition of the mineral and its abundance, and many of these reactions are incongruent in which a new mineral plus dissolved constituents are formed, as has been demonstrated by such reactions as given below that are often cited in many textbooks (e.g., Berner and Berner, 1996; Drever, 1988, 1997; Faure, 1998; Garrels and Mackenzie, 1971):

$$NaAlSi_3O_8 + CO_2 + 2H_2O = Na^+ + HCO_3^- + 3SiO_2 + Al(OH)_3$$

where Na-silicate albite reacts with one CO_2 producing one of each Na^+ and HCO_3^-, and SiO_2 and solid $Al(OH)_3$ that balance the reaction as dissolved and(or) solid species. For monovalent-cation silicates, the ratio of CO_2 consumed to HCO_3^- formed is 1:1. Similarly, dissolution of Ca-feldspar, anorthite, and Mg-silicates, olivine and chlorite, shows the 1:1 relationship between the HCO_3^- formed and CO_2 consumed:

$$CaAl_2Si_2O_8 + 2CO_2 + 4H_2O = Ca^{2+} + 2HCO_3^- + 2SiO_2 + 2Al(OH)_3$$

$$Mg_2SiO_4 + 4CO_2 + 2H_2O = 2Mg^{2+} + 4HCO_3^- + SiO_2$$

$$Mg_5Al_2Si_3O_{10}(OH)_8 + 10CO_2 + 4H_2O = 5Mg^{2+} + 10HCO_3^- + 3SiO_2$$
$$+ 2Al(OH)_3$$

The preceding reactions are congruent dissolution reactions with respect to the cations: for example, if a K- or Mg-containing clay mineral forms as a weathering product of a K-feldspar or Mg-silicate and the clay also dissolves, then the balance of the K^+ or Mg^{2+}-ion released congruently would be the same and there would be no net effect on CO_2 consumption. However, if clay minerals were forming by removal of such main cations as Mg^{2+}, K^+, and Na^+ from water (Faure, 1998) and without further dissolution, then the cation masses released would be smaller and the potential uptake of CO_2 in silicate weathering would also be smaller. In sediments, it is generally poorly known how much and what kind of clays form from parent silicate minerals and it is therefore difficult to estimate any effects of the clay-water reactions on the CO_2 uptake in weathering.

Sediments contain reduced sulfur associated mostly with pyrite (FeS_2) that produces sulfuric acid when pyrite is oxidized (Stumm and Morgan, 1981):

$$FeS_2 + 3.75\, O_2 + 3.5\, H_2O = 2H_2SO_4 + Fe(OH)_3$$

Oxidation of pyrite involves transfer of 15 electrons in oxidation of Fe^{2+} to Fe^{3+} and S^{1-} to S^{6+} that are taken up by the 7.5 oxygen atoms. Sulfuric acid may react with carbonate and silicate minerals, releasing cations to solution and producing HCO_3^- or

CO$_2$ from carbonates in different stoichiometric proportions:

$$2CaCO_3 + H_2SO_4 = 2Ca^{2+} + 2HCO_3^- + SO_4^{2-} \tag{8.8a}$$

$$CaCO_3 + H_2SO_4 = Ca^{2+} + SO_4^{2-} + H_2O + CO_2 \tag{8.8b}$$

If all the pyrite-bound sulfur in sediments (S_{pyr}, Table 8.1) is oxidized to H$_2$SO$_4$ that reacts with carbonate and silicate minerals in the proportions of their abundance in an average sediment, reactions (8.8a) and (8.8b) would add SO$_4^{2-}$ and cations to the water with a concomitant formation of HCO$_3^-$ or CO$_2$ from the carbonate minerals. In this case, bicarbonate is produced by dissolution of carbonate minerals without CO$_2$ consumption.

Reactions of H$_2$SO$_4$ with divalent and monovalent-cation silicates produce cation/acid ratios of 1:1 and 2:1, without CO$_2$ consumption:

$$CaAl_2Si_2O_8 + H_2SO_4 = Ca^{2+} + SO_4^{2-} + 2SiO_2 + 2AlOOH$$

$$2NaAlSi_3O_8 + H_2SO_4 = 2Na^+ + SO_4^{2-} + 6SiO_2 + 2AlOOH$$

where SiO$_2$ and AlOOH are a shorthand notation for conservation of silica and alumina in mineral oxides and hydroxides. More generally, other acids, such as hydrochloric acid, HCl, from hydrothermal circulation or monoprotic organic acids, RCOOH, from decomposition of organic matter may also react with carbonates and generate HCO$_3^-$ without direct consumption of CO$_2$ or produce CO$_2$, similarly to reactions (8.8a, b):

$$2CaCO_3 \left|
\begin{array}{l}
+2HCl = 2Ca^{2+} + 2Cl^- + 2HCO_3^- \\
+2RCOOH = 2Ca^{2+} + 2RCOO^- + 2HCO_3^-
\end{array}
\right.$$

$$CaCO_3 \left|
\begin{array}{l}
+2HCl = Ca^{2+} + 2Cl^- + H_2O + CO_2 \\
+2RCOOH = Ca^{2+} + 2RCOO^- + H_2O + CO_2
\end{array}
\right.$$

3.2 CO$_2$ Consumption and HCO$_3^-$ Production

From the discussion in the preceding section, it is clear that the mass of CO$_2$ consumed, and the equivalent mass of HCO$_3^-$ produced, depends on the relative abundances of the different carbonate and silicate minerals in the weathering source rock. The ratio of the CO$_2$ consumed to the HCO$_3^-$ produced can be written in terms of the individual cation concentrations or p_i, for those in the silicates (where $i = 4$, for Ca, Mg, Na, and K, in mol cation/g rock), and p_j, for those in the carbonates ($j = 2$, for Ca and Mg). The cation mol fraction of carbonates, y, in a rock containing both carbonates and silicates is:

$$y = \frac{\sum\limits_{j=1}^{2} p_j}{\sum\limits_{i=1}^{4} p_i + \sum\limits_{j=1}^{2} p_j} \tag{8.9}$$

and the cation mol faction of silicates, with $i = 4$ for the four major cations, is:

$$1 - y = \frac{\sum\limits_{i=1}^{4} p_i}{\sum\limits_{i=1}^{4} p_i + \sum\limits_{j=1}^{2} p_j} \tag{8.10}$$

For Ca- and Mg-carbonates reacting with CO_2 and sulfuric acid, the CO_2 consumption and HCO_3^- production depend on the composition of a reacting mixture that contains CO_2 (mol fraction x) and sulfuric acid $(1 - x)$:

$$\psi = \frac{CO_2 \text{ consumed}}{HCO_3^- \text{ produced}} = \frac{x}{2x + (1 - x)} = \frac{x}{1 + x} \tag{8.11}$$

In the absence of any sulfuric acid in weathering, the CO_2 fraction is $x = 1$ and the ratio for carbonates is $\psi = 0.5$.

When CO_2 (x) and H_2SO_4 $(1 - x)$ react with a mixture of carbonates (fraction y) and silicates $(1 - y)$, the consumption/production ratio depends on whether HCO_3^- or CO_2 is produced in a reaction between carbonates and sulfuric acid, as discussed in the preceding section. If the product is HCO_3^- then its fraction $(1 - x)y$ is produced from H_2SO_4-$CaCO_3$ reaction, $2xy$ is produced from CO_2 and $CaCO_3$, and $x(1 - y)$ from CO_2 and silicates. Thus the following ratio ψ_1 describes the CO_2 uptake in mineral weathering as a function of the CO_2 (x) and acid $(1 - x)$ fractions reacting with the carbonates (y) and CO_2 reacting with the silicates (acid reactions with silicates consume no CO_2):

$$\psi_1 = \frac{CO_2 \text{ consumed}}{HCO_3^- \text{ produced}} = \frac{xy + x(1 - y)}{2xy + x(1 - y) + (1 - x)y}$$

$$= \frac{x}{x + y} \tag{8.12}$$

The effect of acid as part of a reactive mixture with CO_2 is shown in Fig. 8.6 on the CO_2 consumption/HCO_3^- production ratio for rocks varying in their mineral composition from pure carbonates to pure silicates. Equation (8.12) describes the case of $CaCO_3$ reaction with H_2SO_4 producing the bicarbonate ion HCO_3^-. In the absence of acid $(x = 1)$, pure carbonates $(y = 1)$ have a CO_2 consumption/HCO_3^- production ratio of 0.5, and pure silicates $(y = 0)$ have the same ratio of 1.0.

If the carbonate reaction with sulfuric acid produces CO_2, its fraction produced is $(1-x)y$, and a possible return of this CO_2 to the atmosphere reduces the CO_2 consumed from x to $x - (1 - x)y$. The fraction of HCO_3^- produced is $2xy$ from the CO_2-carbonate reactions and $x(1 - y)$ from CO_2-silicate reactions. In this case, the ratio of net CO_2

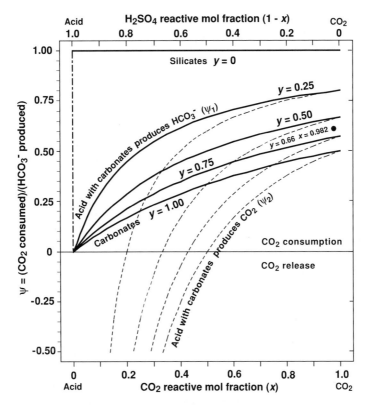

Figure 8.6   Consumption of CO$_2$ and production of HCO$_3^-$ in silicates and calcium carbonate reactions with water containing dissolved CO$_2$ and sulfuric acid, equations (8.5) and (8.6). y is the cation mol fraction of carbonates in the source rock, x is the CO$_2$ mol fraction in CO$_2$-H$_2$SO$_4$ mix. The point at $x = 0.982$, $y = 0.66$ represents an average sediment.

consumed to HCO$_3^-$ produced is:

$$\psi_2 = \frac{CO_2 \text{ consumed}}{HCO_3^- \text{ produced}} = \frac{x - (1-x)y}{2xy + x(1-y)}$$

$$= \frac{x - (1-x)y}{x(1+y)} \tag{8.13}$$

In the reaction path of equation (8.13), if the acid fraction is relatively large, then the CO$_2$ returned to the atmosphere significantly reduces or even exceeds the amount consumed ($\psi_2 < 0$). However, in a normal weathering environment where CO$_2$ is the main reactant ($x > 0.90$), the contribution of H$_2$SO$_4$ is small and the values of ψ_1 and ψ_2 shown in Fig. 8.6 are similar. The CO$_2$ consumption and HCO$_3^-$ production potential, ψ_1 in equation (8.12), is smaller for carbonate sediments and evaporites that contain more carbonate minerals than for shales and sandstones (Table 8.1). For an average continental sediment, consisting of carbonate rocks, shales, sandstones, and evaporites in the proportions as given in Table 8.1 and Fig. 8.2, the ratio of potentially consumable CO_2 to producible HCO_3^- is 0.61: the weathering of an average sediment requires 61% of HCO_3^- to be taken from atmospheric or soil CO_2 and the remaining 39% supplied by carbonate minerals. Of all the HCO_3^- produced, 73% is derived from carbonate minerals and 27% from silicates.

4 CO_2 Consumption from Mineral-Precipitation Model

The masses of minerals weathering in the global sedimentary cycle that produce the composition of a world average river have been studied along two main lines: the approach of a "backward precipitation" that withdraws minerals sequentially from the river water, thereby accounting for the mineral sources (Berner and Berner, 1996; Garrels and Mackenzie, 1967, 1971; Mackenzie, 1992), and "forward dissolution" that estimates the proportions of the different mineral components of the sedimentary and crystalline crust that account for the solution composition (Holland, 1978; Meybeck, 1987). More recent studies of river systems on large subcontinental scales have estimated the crustal and sedimentary contributions on the basis of the ionic and $^{87}Sr/^{86}Sr$ ratios in river waters (e.g., Dalai et al., 2002; Gaillardet et al., 1999; Huh et al., 1998; Jacobson et al., 2002; Millot et al., 2003; Mortatti and Probst, 2003; Oliver et al., 2003).

The "backward precipitation" method of mineral removal is applied here uniformly to each of the three world average rivers (Table 8.2) that are, in chronological order, those of Meybeck (1979), Mackenzie (1992), and Berner and Berner (1996). The sequence of mineral withdrawal is explained below and shown for one average river in Table 8.3, where the computational steps are conceptually similar, but not identical, to those of the authors cited above.

Precipitation of individual minerals from an average river water is constrained by the following abundance ratios in the weathering source, taken as an average sediment (Tables 8.1, 8.2, 8.4):

1. Sulfide (S_{pyr}) mol fraction of total sulfur (sulfate and sulfide): 0.244
2. Ca mol fraction of Ca-Mg-carbonates: 0.855
3. Ca-carbonate mol fraction of Ca-carbonates and Ca-silicates: 0.927

First, in step 1, all Cl^- is removed from river water as NaCl. Next, in step 2, part of SO_4^{2-} is removed as S^- into pyrite (S_{pyr}) in the proportion of the occurrence of S_{pyr} in an average sediment, 0.244 (a smaller fraction of 0.11 for pyrite-derived sulfate in rivers was given by Berner and Berner, 1996). This is equivalent to 2.035×10^{-5} mol SO_4^{2-}/kg (Table 8.4) that were produced by oxidation of pyrite and reacted with the carbonate and silicate minerals, releasing the metal cations to solution and forming some HCO_3^- from the carbonates. The amounts of cations dissolved from these minerals are in the stoichiometric proportions of the reactions given in Section 3.1. For example,

Table 8.3   Removal of minerals from dissolved ionic components of a world average river of Mackenzie (1992) (Table 8.2)

Mineral removal and solution balance	Ca^{2+}	Mg^{2+}	Na^+	K^+	Cl^-	SO_4^{2-}	HCO_3^-	Precipitated	Minerals	Fraction of total precipitated (%)	Fraction of HCO_3^- produced (%)
					mol/kg river water						
Start concentration	3.34E-04	1.33E-04	1.59E-04	3.20E-05	9.00E-05	8.35E-05	8.67E-04				
1. Removal of NaCl			-9.00E-05		-9.00E-05			9.00E-05	NaCl	13.28	
Remaining	3.34E-04	1.33E-04	6.90E-05	3.20E-05	0	8.35E-05	8.67E-04				
2. Removal of S$_{pyr}$ and carbonates and silicates dissolved by H$_2$SO$_4$											
a. Removal of S$_{pyr}$ as SO$_4^{2-}$ formed from pyrite						-2.03E-05		2.03E-05	S$_{pyr}$	3.00	
b. Removal of Ca & Mg carbonates dissolved by H$_2$SO$_4$	-2.30E-05	-3.89E-06				-1.34E-05	-2.59E-05	2.69E-05	(Ca, Mg)CO$_3$	3.96	3.10
c. Removal of cation silicates dissolved by H$_2$SO$_4$	-9.02E-07	-1.98E-06	-3.22E-06	-4.84E-06		-6.91E-06		1.09E-05	Ca, Mg, Na, K sil.	1.61	
Remaining	3.10E-04	1.27E-04	6.58E-05	2.72E-05		6.32E-05	8.40E-04				
3. Removal of Ca sulfates	-6.32E-05					-6.32E-05		6.32E-05	CaSO$_4$	9.32	
Remaining	2.47E-04	1.27E-04	6.58E-05	2.72E-05		0	8.40E-04				
4. Removal of Ca, Mg carbonates	-2.29E-04	-3.88E-05					-5.36E-04	2.68E-04	(Ca, Mg)CO$_3$	39.50	61.77
Remaining	1.80E-05	8.79E-05					3.04E-04				
5. Removal of Ca, Mg silicates	-1.80E-05	-8.79E-05					-2.12E-04	1.06E-04	Ca, Mg sil.	15.71	24.41
6. Removal of Na, K silicates			-6.58E-05	-2.72E-05			-9.29E-05	9.29E-05	Na, K sil.	13.71	10.72
Total removed	3.34E-04	1.33E-04	1.59E-04	3.20E-05	9.00E-05	8.35E-05	8.67E-04	6.778E-04		100	100
Remaining	0	0	0	0	0	0	0	0			

Table 8.4   Parameters and stoichiometric relationships for computation of S_{pyr} removed from river water and carbonate and silicate minerals dissolving by reactions with H_2SO_4. Reaction stoichiometry in Section 1. Data on sediment and river water composition in Tables 8.1 and 8.2. C_s is sulfide concentration

Sulfate and sulfide S

S_{pyr}/SO_3 in sediment (mol/mol):
$= (0.12/32.066)/(0.93/80.0642)$
$= 0.322$

S_{pyr} fraction of total S = 0.244
S_{pyr} from average river of Mackenzie (1992):
$C_s = 0.244 \times [SO_4^{2-}]$
$= 0.244 \times 8.350 \times 10^{-5}$
$= 2.035 \times 10^{-5}$ mol/kg water

Sequential no.	Mineral component		Cation mol fraction in sediment (n_i)	Fraction of C_s in river water reacting with cation		Cation removed with equivalent S_{pyr} (C_s)	
	Mineral	mol cation/g rock		$n_i C_s$	mol/kg water	$2n_i C_s$ / expression	mol/kg water
1	$CaCO_3$	2.147E-03	0.5646	$n_1 C_s$	1.149×10^{-5}	$2n_1 C_s$	2.298×10^{-5}
2	$MgCO_3$	3.635E-04	0.09562	$n_2 C_s$	1.946×10^{-6}	$2n_2 C_s$	3.891×10^{-6}
3	Ca-silicate	1.685E-04	0.04432	$n_3 C_s$	9.017×10^{-7}	$n_3 C_s$	9.017×10^{-7}
4	Mg-silicate	3.698E-04	0.09728	$n_4 C_s$	1.979×10^{-6}	$n_4 C_s$	1.979×10^{-6}
5	Na-silicate	3.007E-04	0.07911	$n_5 C_s$	1.610×10^{-6}	$2n_5 C_s$	3.219×10^{-6}
6	K-silicate	4.525E-04	0.1190	$n_6 C_s$	2.422×10^{-6}	$2n_6 C_s$	4.844×10^{-6}
Sum		3.802E-03	1.00		2.035×10^{-5}		

the mass of $CaCO_3$ that is dissolved by a reaction with H_2SO_4 is:

$$2CaCO_3 + H_2SO_4 = 2Ca^{2+} + 2HCO_3^- + SO_4^{2-}$$

where $2Ca^{2+}$ are produced for $1SO_4^{2-}$. Accordingly, removal of $1SO_4^{2-}$ from solution into pyrite-sulfur S_{pyr} is accompanied by precipitation of $2CaCO_3$. Using the sediment abundances and river water sulfate concentration (Tables 8.1, 8.2), the mass of Ca^{2+} precipitated is

$$2 \times 0.5646 \times 2.035 \times 10^{-5} \text{ mol } SO_4^{2-}/\text{kg} = 2.298 \times 10^{-5} \text{ mol } Ca^{2+} \text{ as}$$
$$CaCO_3/\text{kg}$$

This amount is shown as 2.30×10^{-5} mol Ca^{2+}/kg in Table 8.3, step 2b, and Table 8.4, line 1. The amount of HCO_3^- removed in this step is equal to the amount of Ca^{2+} and Mg^{2+} removed as carbonates from river water. Using a similar procedure, the amounts of silicates dissolved by reactions with H_2SO_4 are returned as divalent and monovalent-cation silicates to the weathering source.

Step 3 follows the removal of S_{pyr} and the carbonates and silicates dissolved by sulfuric acid: it is removal of the remaining SO_4^{2-} as $CaSO_4$. Because sulfates and chlorides dissolve without production of HCO_3^-, the preceding removal steps 1-3 determine the subsequent removal of the bicarbonate ion by the remaining divalent (Ca, Mg) and monovalent (Na, K) cations. In other words, the fractional abundances in the source rock of the cations not producing HCO_3^-, such as in the minerals NaCl and $CaSO_4$, affect the consumption of atmospheric CO_2 and production of HCO_3^- by other minerals. After the first three steps (1-3) of cation removal into chlorides and sulfates, as well as reconstitution of the silicates that reacted with H_2SO_4 producing no HCO_3^- and of carbonates that produce only little HCO_3^-, the remaining cations in river water determine the CO_2 consumption with little dependence on how these cations are partitioned between the carbonates and silicates. This is so in the precipitation model where SiO_2 and Al_2O_3 are not limiting and stoichiometrically available for making cation-aluminosilicates by reactions of cations with dissolved or amorphous SiO_2, cation-free aluminosilicate minerals such as kaolinite, and(or) Al-oxyhydroxides.

In the next step 4, Ca and Mg are removed as carbonates, in proportions of their occurrence in an average sediment. In Table 8.3, Ca^{2+} is removed from river water into $CaCO_3$ as its fraction of the Ca-carbonate and Ca-silicate in sediments: $0.927 \times 2.47 \times 10^{-4}$ mol/kg water $= 2.29 \times 10^{-4}$ mol Ca^{2+}/kg. Then Mg^{2+} is removed in the proportion of its occurrence with Ca^{2+} in sedimentary carbonates: $(2.29 \times 10^{-4}$ mol Ca^{2+}/kg$) \times 0.145/0.855 = 3.88 \times 10^{-5}$ mol Mg^{2+}/kg. In step 5, the remaining Ca and Mg are removed as silicates into the weathering source, and the final step 6 is the removal of Na and K silicates.

The relative proportions of the precipitated minerals, grouped into carbonates (Ca, Mg), silicates (Ca, Mg, Na, K), chlorides (NaCl), sulfates ($CaSO_4$), and sulfides (S_{pyr}), are shown in Fig. 8.7A. Carbonates are the most abundant minerals in an average sediment source and in the calculated sources of the three average rivers. However, inclusion of the continental crystalline crust in the weathering source rock, as 37

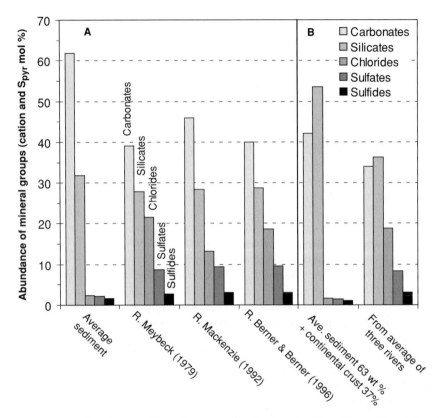

Figure 8.7   A. Mineral composition of an average sediment and of the sediment "precipitated" from three average rivers. Removal of elements from river water into minerals in the same proportions as in an average sediment (Tables 8.1, 8.2, and 8.3). B. mineral composition of a weathering source consisting of 63 wt % average sediment and 37% upper continental crust, and minerals "precipitated" from an average river. Removal of elements from river water into minerals in the same proportions as their occurrence in the weathering source.

wt % crust and 63 wt % average sediment, understandably increases the abundance of silicates, as shown in Fig. 8.7B. For this source, the mol fraction of Ca-carbonate in the combined Ca-carbonate and Ca-silicate minerals is smaller than in an average sediment source as given on p. 242, 0.927:

Ca-carbonate mol fraction of Ca-carbonates and Ca-silicates: 0.771

It should be noted that the mineral reactions with H_2SO_4 from pyrite oxidation account for about 5% of the silicate-mineral dissolution and about 10% of the carbonate dissolution. The results of CO_2 consumption and mineral sources given in Table 8.5 may be compared with those in Table 8.1. The differences between the average sediment

Table 8.5   Cation, S_{pyr}, and HCO_3^- sources in three world average rivers (mol %). Percentages shown are for concentrations in mol/g rock contributing to river water

	Meybeck (1987)	This study	Mackenzie (1992)	This study	Berner & Berner (1996)	This study
Cations (Ca^{2+}, Mg^{2+}, Na^+, K^+)						
Carbonates	46.7	39.1	45.2	43.5	39.7	39.9
from reaction with CO_2		35.4	41.5	39.5		35.9
from sulfuric acid dissolution		3.7	3.7	4.0		4.1
Silicates	38.4	28.9	29.6	30.9	30.7	28.7
from CO_2 and Ca-Mg-silicate reactions	11.1	14.8	13.7	15.6		15.3
from CO_2 and Na-K-silicate reactions	26.3	11.6	13.7	13.7		11.7
from sulfuric acid dissolution		1.5	2.1	1.6		1.7
Evaporites	15.9	30.3	22.2	22.6	29.6	28.3
NaCl	9.1	21.6	13.3	13.3		18.6
$CaSO_4$	6.8	8.7	8.9	9.3		9.6
Pyrite S^- (oxidized to SO_4^{2-})		2.8	3.0	3.0		3.1
HCO_3^- from carbonates	69.5	64.4	69.1	64.9	62.2	64.2
from Ca and Mg carbonates	65.6	61.2	66.3	61.8		60.7
from sulfuric acid dissolution	3.9	3.2	2.9	3.1		3.5
HCO_3^- from silicates	30.5	36.3	30.9	35.1	38.8	35.8
from Ca and Mg silicates		26.3	20.1	24.4		25.9
from Na and K silicates		10.0	10.7	10.7		9.9
HCO_3^- from atmosphere or soil CO_2	**65.3**	**66.2**	**65.4**	**66.0**	**68.9**	**66.2**
HCO_3^- from carbonates	**34.8**	**33.8**	**34.6**	**34.0**	**31.1**	**33.8**

and the sediment source calculated from the river-water composition may be attributed to the differences between the dissolution rates of individual minerals, as is discussed in the next section.

5 CO_2 Consumption from Mineral-Dissolution Model

It is shown in the preceding section that the proportions of the dissolving minerals that can account for an average river water are in general different from those in the weathering source rock. The mass of the continental sediments and upper part of the continental crystalline crust that amounts to 3×10^{24} g or about 6.5 km in thickness spread over the area of the continents and continental shelves is schematically shown in Fig. 8.8. Only a very small fraction of this mass is dissolved by waters discharging to the oceans, showing that on a global scale, chemical weathering or denudation of

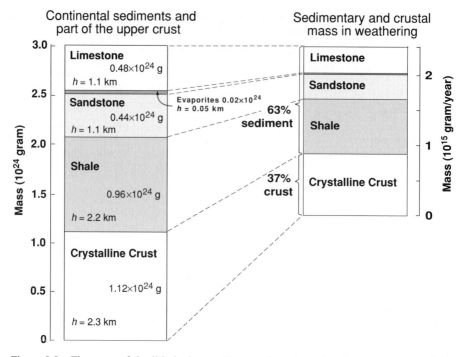

Figure 8.8   The mass of the lithologic constituents of continental sediments and part of the continental upper crust that are the weathering source in a dissolution model (Table 8.1, and Fig. 8.2; from data in Li, 2000). On the right is shown the mass transported as dissolved solids by world rivers. Note that the chemical denudation rate is much lower than the rates of physical denudation and recycling of the sediments (Chapter 7).

the continents is a much slower process than mechanical denudation and sediment recycling, as is shown in Fig. 8.3 and further discussed in Chapter 9. Figure 8.9A further shows that river waters are enriched relative to the source in Ca^{2+}, Mg^{2+}, Na^+, Cl^-, and SO_4^{2-} due to the differences in the dissolution rates of different minerals. No single mix of an average sediment and continental crust satisfies all the ion/HCO_3^- ratios of the average rivers. This suggests that river-water constituents are derived from minerals dissolving at different rates. A mass balance relationship between the supply of Ca to river water and its removal by river flow demonstrates the essential points of the dissolution model. In the weathering rock source, Ca occurs in Ca-carbonates, Ca-silicates, and Ca-sulfates, and dissolution of these minerals results in a Ca^{2+}-ion concentration in river water, C_{Ca}. A balance can be written as:

$$(X_{Ca\text{-}carb}k_{Ca\text{-}carb} + X_{Ca\text{-}sil}\,k_{Ca\text{-}sil} + X_{Ca\text{-}sulf}k_{Ca\text{-}sulf})M_{rock} = C_{Ca}F_{river} \qquad (8.14)$$

where $X_{Ca\text{-}mineral}$ is Ca concentration in the source rock (mol Ca/g rock), $k_{mineral}$ is

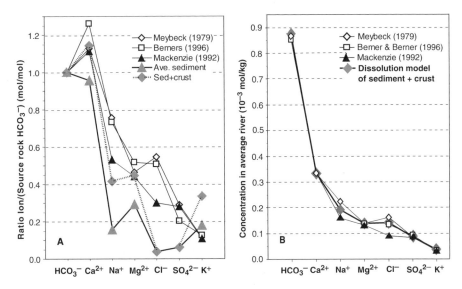

Figure 8.9   A. Ratios of CO$_2$ to cationic and anionic mineral constituents in an average sediment, in a composite weathering source consisting of 63 weight % average sediment and 37% continental upper crust, and ratios of HCO$_3^-$ derived from the CO$_2$ of sediment carbonates to other ions in three world average rivers (Table 8.2). B. Ion/HCO$_3^-$ molar ratios in the same average rivers and calculated (black line) from dissolution of a weathering source rock consisting of 63 mass % average sediment and 37% continental crust (after Lerman and Wu, 2006). Calculated as explained in the text.

a dissolution rate parameter of a mineral (yr^{-1}), M_{rock} is the rock mass undergoing weathering (g/yr), and F_{river} is water discharge (kg/yr or m^3/yr). From the known Ca-mineral concentrations that occur in different rock types and the Ca concentration in river water, the factors $k_{mineral} \times M_{rock}/F_{river}$ (g rock/kg river water) are determined empirically and the uptake of CO$_2$ in mineral dissolution and production of HCO$_3^-$ are calculated simultaneously. The consumption of CO$_2$ in weathering reactions and production of the bicarbonate ion are the results of the mineral dissolution computation, and their values provide an additional check of the procedure. By this procedure, the concentrations of the individual ions in river water, including HCO$_3^-$ that is only in part derived from the weathered rock, shown in Fig. 8.9B, are within the range of values of the different average river compositions. The computed concentrations correspond to mineral dissolution of the following proportions of the sedimentary and crystalline crustal rocks (Table 8.1): 63 wt % sediments and 37 wt % crust or, in rounded-off numbers, 16% carbonate rocks, 32% shale, 15% sandstone, 1% evaporites, and 37% upper continental crust. Furthermore, the agreement between the model results and the reported average river concentrations is good in light of the differences between the individual average rivers and the uncertainties involved in the composition of an

Table 8.6   Order of relative dissolution rates in the weathering of an average sediment and upper continental crust that produces the ionic concentrations in an average river shown in Fig. 8.9C. The dissolution rate of $(Ca,Mg)CO_3$ taken as 1. ψ_1 is the ratio of $(CO_2$ consumed)/(HCO_3^- produced), equation (8.12)

	Na-, Mg-, K-chlorides	Ca-sulfates	S_{pyr} to SO_4^{2-}	Mg-silicates	$CaCO_3$	$MgCO_3$	Ca-silicates	Na-silicates	K-silicates	ψ_1
Dissolution of average sediment + crust[a]	×15.0	×7.8	×3.3	×1.4	×1	×1	×1	×0.6	×0.4	0.72
Precipitation of average sediment + crust[b]	×13.7	×7.7	×3.2	×1.4	×1	×1	×1.0	×0.7	×0.4	0.72

[a] From average sediment 63 wt % and crystalline crust 37 wt % (16% carbonate rocks, 32% shale, 15% sandstone, 1% evaporites, 37% upper crust; Table 8.1).
[b] Removal or precipitation from an average river water of dissolved solids as minerals in the proportions of their occurrence in a weathering source consisting of 63 wt % average sediment and 37 wt % crystalline crust (Fig. 8.7).

average sediment. A relative dissolution rate of a mineral can be represented by a factor, greater or smaller than 1 (Table 8.6), in the following general sequence from the less resistant to weathering to the more resistant, as shown in Fig. 8.10:

Chlorides → Sulfates → Pyrite → Mg-silicates → Carbonates →
Ca-silicate → Na-silicate → K-silicate

The relative dissolution rate factors are analogous to the Goldich stability series of silicate minerals in the weathering cycle that was introduced by Goldich (1938; Lasaga, 1998; Lerman, 1979; Pettijohn, 1957) and further developed by Stallard (1988) from studies of weathering in the tropical Amazon River basin. The results show that the evaporite mineral chlorides and sulfates dissolve faster than carbonates, and pyrite is also oxidized faster than the rate of carbonate dissolution. For the silicate minerals, Mg-silicates weather somewhat faster than carbonates, Ca-silicates about the same as carbonates, and Na- and K-silicates weather slower than carbonates. Among the latter three, the order of stability agrees with the observations of Ca-rich plagioclases dissolving faster than Na-plagioclases.

The proportions of the main constituents in an average river water, derived from the dissolution model, are given in Table 8.7. The ratio of CO_2 consumed to HCO_3^- produced $\psi = 0.72$ is higher than that of 0.65 to 0.69 that was derived from the mineral precipitation model (Table 8.5). In the latter, an average sediment was taken as a weathering source rock, but for a source rock that is richer in metal silicates, more CO_2 is consumed, and this is reflected in the value of $\psi = 0.72$ both in the case of precipitation of a mixed source (Fig. 8.5B) and dissolution of a mixed source (Fig. 8.9B).

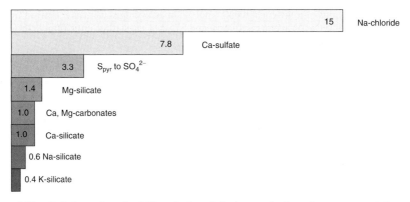

Figure 8.10   Relative order of stability of minerals in the weathering of a source consisting of 63 wt % average sediment and 37% upper continental crust. Rate of carbonate mineral dissolution, (Ca,Mg)CO$_3$, taken as 1. Table 8.6 and Fig. 8.9B.

On the basis of carbon isotopic data, Mook and Tan (1991) concluded that the main source of dissolved inorganic carbon in rivers (DIC) is "CO$_2$ derived from the decay of organic matter in continental soils and from the dissolution of carbonate, while in general the contribution of atmospheric CO$_2$ is negligibly small." From their data on the present-day δ^{13}C of soil CO$_2$ (-26 ‰) and HCO$_3^-$ in rivers (-12 ‰), the Phanerozoic sedimentary carbonates ($+1.5$ ‰), and fractionation factors between HCO$_3^-$ $_{(aq)}$ and CO$_{2(g)}$, it is possible to estimate the fraction of soil organic matter that contributes to the dissolved HCO$_3^-$ in rivers and shallow soil and ground waters.

The fractionation of ^{13}C/^{12}C in a gas-solution system makes HCO$_3^-$ $_{(aq)}$ heavier than CO$_{2(g)}$ by 9.5 ‰ at 10°C and by 7.9 ‰ at 25°C (Friedman and O'Neil, 1977; Mook *et al.*, 1974; Table 6.2). The following balance equation gives the fraction x of soil CO$_2$ in rivers:

$$[-26\text{ ‰} + (9.5\text{ to }7.9)\text{ ‰}]x + 1.5\text{ ‰}(1 - x) = -12\text{ ‰};$$
$$x = 0.69 \text{ to } 0.75 \tag{8.15}$$

Thus 69 to 75% of the HCO$_3^-$ in river water is derived from soil CO$_2$ that reacts with the silicate and carbonate minerals and the remaining 31 to 25% from the weathering and dissolution of carbonates.

To summarize, estimates of the ratio ψ range from 0.65 to 0.75, including the values computed from the precipitation and dissolution models (Tables 8.5, 8.7). They correspond to a global CO$_2$ uptake in mineral weathering reactions of 20 to 24 $\times$ 10^{12} mol C/yr, at the rate of the HCO$_3^-$ transport by rivers of 32 $\times$ 10^{12} mol C/yr. These estimates are close to those already in the literature and lead one to conclude that global weathering of rocks and minerals by CO$_2$-charged soil and ground waters and production of DIC are reasonably well understood processes.

Table 8.7   Proportions of ionic constituents (mol %) in an average river water derived from a source consisting of 63 wt % average sediment and 37 wt % crystalline crust (Fig. 8.9B and Table 8.1)

Balance	Ca^{2+}	Mg^{2+}	Na^+	K^+	Cl^-	SO_4^{2-}	HCO_3^-	
From atmosphere/ soil CO_2								72
From silicate minerals	18	71.3	27.6	85.9			46	
From carbonate minerals	59.8	24.4					53	28
From evaporite minerals	19.4	1.6	70.7	2.2	100	75.6		
From H_2SO_4 reactions	2.9	2.7	1.7	11.9		24.4	1	
Total	100	100	100	100	100	100	100	100

6   Environmental Acid Forcing

It was mentioned earlier that in addition to carbon dioxide, other acids in the environment may play variable roles in weathering, such as the organic acids in soils and nitric and sulfuric acids. The specific effects of such acids on a global scale are not well known, but their occurrence and behavior are often reported on local scales where pollution from industrial sources or sulfur emissions from volcanic eruptions is significant. During the Industrial Age of the last 200 to 300 years, fossil fuel burning has been the main source of nitrogen and sulfur oxide (NO_x and SO_x) emissions to the atmosphere. Indeed in some areas, such as the Eastern United States, most of the nitrogen and sulfur in precipitation are from the combustion of fossil fuels. Some of the N- and S-oxides released to the atmosphere become nitric and sulfuric acids, although not all the emitted nitrogen and sulfur return to the land surface and oceanic coastal zone. The parts that do return may not all be in the form of HNO_3 and H_2SO_4, and their involvement in the biological production and decomposition of organic matter leads to the utilization of nitrate and sulfate, and their subsequent reduction or modification by bacteria in soils and water. In the year 2000, the emissions of N and S to the atmosphere were of a comparable magnitude: about 70×10^6 ton S/yr (Mt S/yr) or 2.2×10^{12} mol S/yr, and about 30×10^6 ton N/yr or 2.1×10^{12} mol N/yr. One of the estimates of sulfur emissions in the year 2015 projects a rise to as high as 100 Mt S/yr or 3.1×10^{12} mol S/yr (Fig. 11.11C; Browne et al., 1997; Dignon, 1992; Dignon and Hameed, 1989; Hameed and Dignon, 1992; Ver et al., 1999). For the purpose of demonstrating the potential effect of H_2SO_4 on mineral weathering, the latter value may be considered an upper bound of anthropogenic addition of sulfuric acid to the environment. It should nevertheless be noted that even if only a fraction of sulfur emissions is converted to sulfuric acid, it is likely that at least some of this H_2SO_4 on the land surface is neutralized rapidly by reactions with minerals: addition of non-neutralized H_2SO_4 at a current rate of about 2×10^{12} mol S/yr to the volume of freshwater lakes of 125×10^3 km^3 and(or) the volume of soil water to depth of 10 m, 121×10^3 km^3 (Berner and Berner, 1996; Lerman, 1994; Ver, 1998), would add 1.6 to 3.3×10^{-5} mol H$^+$/kg to the continental waters and, if not neutralized, make them much more acidic than the range of 3×10^{-7} to 3×10^{-8} mol H$^+$/kg (pH $\approx$ 6.5 to 7.5). In local environments and where freshwater lakes occur in crystalline-rock drainage basins, the observed water acidification is likely to be a combined effect of acid deposition and slow reaction rates of silicate minerals. Lakes in Scandinavia, the Swiss Alps, parts of Canada, and the Eastern United States include many acidic lakes among several thousand lakes of surface area <20 km^2 (Berner and Berner, 1996; Lerman, 1979; Overton et al., 1986).

In comparison to the CO$_2$ consumed in weathering, the consumption of H$_2$SO$_4$ that forms from the pyrite oxidation is much smaller, $0.64 \pm 0.15 \times 10^{12}$ mol S/yr, based on the pyrite-derived fraction of 24.4% of total dissolved SO$_4^{2-}$ in the average rivers (Table 8.2). The projected flux of 3.1×10^{12} mol S/yr is five times greater than the natural H$_2$SO$_4$ flux from the oxidation of pyrite in sediments. An increasing input of anthropogenic H$_2$SO$_4$ to the weathering regime would result in additional dissolution of metal carbonates and silicates in the sediments and crust, and increase the concentration of dissolved solids and the sulfate-ion in river water (Fig. 8.11), as has been shown regionally for the Eastern United States (Bischoff et al., 1984). An estimated upper bound of the input of anthropogenic sulfuric acid to the weathering cycle would increase the H$_2$SO$_4$ fraction in the reactive CO$_2$-H$_2$SO$_4$ mix in an average sediment from the present value of 1.8 mol %, that is shown in Fig. 8.6, to 8.6%. However, this addition would have only a very small effect on the CO$_2$ consumption to HCO$_3^-$ production ratio. Thus the main effects of the environmental perturbation by acid deposition that are due to the possible addition of sulfuric acid to the continental surface are in the effects on local and regional water systems and the ecological communities of land plants,

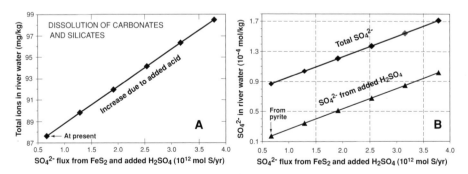

Figure 8.11   Computed results for an upper bound of addition of anthropogenic sulfuric acid from SO$_4$ emissions to the weathering cycle. A. Increase in concentration of total dissolved ions in an average river due to dissolution of the carbonates and silicates by the anthropogenic addition of H$_2$SO$_4$. B. Increase in total and acid-derived SO$_4^{2-}$ in an average river.

and on primary producers and animals on higher trophic levels in smaller bodies of water.

The inorganic and organic products of the chemical and physical denudation of the land surface are transported to the coastal ocean where they play an important role in the biogeochemical cycles and dynamics of carbon and nutrient nitrogen and phosphorus in this shallow-ocean domain, as is discussed in Chapter 9. The global role of the CO_2 consumption in continental weathering and the importance of this flux in the global carbon cycle are addressed in Chapter 10.

Chapter 9

Carbon in the Oceanic Coastal Margin

In this chapter we address the behavior of inorganic and organic carbon in the shallow coastal ocean where a large part of the biological production and sediment accumulation occur. At present, the coastal zone is more or less synonymous with the continental shelf that is covered by ocean water as a result of ice melting and sea level rise of 120 meters since the end of the Last Glacial Maximum about 18,000 years ago. In some intervals of the geologic past, shallow epicontinental seas were much more widespread during the periods of marine transgressions when the land was covered by seawater due to a rising sea level, caused by such factors as change in the relative elevation of land and(or) displacement of the ocean-water volume by the growth of spreading ridges on the ocean floor. In fact, a large part of marine sediments preserved on the continents was formed in shallow seas of the past that covered parts of the cratons of what are now different continents. The importance of the coastal ocean in the regulation of the global carbon cycle is primarily related to its position at the junction of the land, atmosphere, and open ocean, with all of which it interacts differently and modifies the transport fluxes of carbon. We emphasize in this chapter the inorganic and organic carbon cycles in the coastal ocean at the time scale of the Industrial Era, the last approximately 300 years, and up to three centuries into the future that are the time of increasing perturbations of the global carbon cycle by human activities.

1   The Global Coastal Zone

The coastal zone is the environment of continental and insular shelves to 180 to 200 m depth, including bays, lagoons, estuaries, and near-shore banks that occupy, in various estimates, 7 to 8% of the surface area of the ocean (24×10^6 to 29×10^6 km^2; Table 9.1). The mean depth of coastal zone water is approximately 130 m. The occurrence and extent of continental shelves are shown on the world map in Fig. 9.1 and on the hypsometric curve in Fig. 9.2. Estuaries are the main points of input from land to the coastal zone and their total area was estimated as about 1×10^6 km^2 (Borges, 2005) or less than 5% of the shelf area. The zone of 0 to -200 m accounts for about 5% of the global surface area. The continental shelves average 75 km in width, with a bottom slope of 1.7 m/km, and they are generally viewed as divisible into the interior or proximal shelf, and the exterior or distal shelf (Drake and Burk, 1974). The depth of the outer edge of the global continental shelf is usually taken as the depth of the break between the continental shelf and slope at approximately 200 m, although this depth varies throughout the world oceans. In the Atlantic, Emery and Uchupi (1984) gave the median depth of the shelf-slope break at 120 m, with the range from 80 to 180 m. The depths of the continental shelf are near 200 m in the European section of the Atlantic, but they are close to 100 m on the African and North American coasts.

The sea level stand at the Last Glacial Maximum 18 ka ago was about 120 m below the present level (Fig. 9.3), making the shelf ice-free area about 30% of the present area. Notice in Fig. 9.3 the relatively rapid rise in sea level of approximately 95 cm per century between 18,000 and 7000 years ago and the slowing down of sea level rise to 12 to 20 cm per century in younger time. The sea level curve of Fig. 9.3 from Barbados has been confirmed to a variable extent from other areas of the world, including other oceanic islands such as Bermuda. The flooding of continental shelves during the late Pleistocene and Holocene has been responsible for major changes in land-sea exchange of materials, including organic carbon and nutrients, and has modified the behavior of the global carbon cycle and increased the ocean area of air-sea exchange of atmospheric CO_2, as

Table 9.1   Estimates of the global continental shelf area

Area (10^6 km^2)	Reference
28.8	Rabouille et al. (2001)
28.3	Milliman (1993)
28.1	Lagrula (1966)
28.0	From ETOPO5 (Fig. 9.5)
27.1	Drake and Burk (1974)
27.0	Tsunogai et al. (1999)
26.0	Borges (2005)
25.2	Gattuso et al. (1998)
24.6	From P. W. Sloss (Fig. 9.2)
24.3	World Resources Institute (2000)

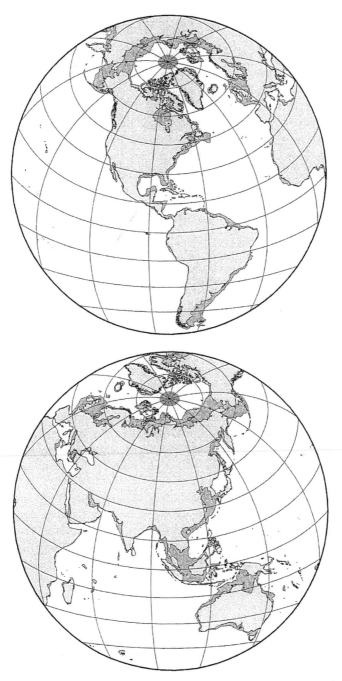

Figure 9.1   Global coastal zone of 0 to −200 m depth. Lambert azimuthal projection, from ETOPO5 data. The parallels are 15° apart and the meridians 30° apart. Coastal zone is dark shaded area.

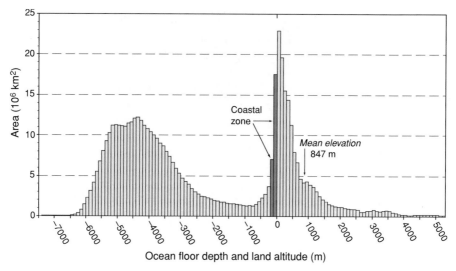

Figure 9.2   Global hypsometric curve showing the surface areas of the ocean floor and land surface in 100-meter intervals. Areas deeper than −7000 m and higher than 5500 m are too small to be seen in the graph. Total global surface area is 510×10^6 km^2. Coastal zone area between 0 and −200 m (darker shaded 2 bars) is 24.6×10^6 km^2. From P. W. Sloss (*personal communication* of unpublished data, 2003, based on ETOPO2 Gridded Elevation Data, NOAA National Geophysical Data Center, Boulder, Colorado http://www.ngdc.noaa.gov/mgg/fliers/01mgg04.html).

discussed later in this chapter. This concept of the coastal zone includes essentially all of the continental shelves and the points of input from land, yet it differs from some other definitions of the coastal zone that extend from some elevation on land above mean sea level to some depth on the continental shelf: for example, the international program on Land-Ocean Interactions in the Coastal Zone (LOICZ) defines the coastal zone as extending from +200 m on land to −200 m on the continental shelves (Pernetta and Milliman, 1995). Global continental shelves have a very uneven distribution around the continents, as the maps in Figs. 8.3 and 9.1 show. Bearing in mind that the Mercator projection of the map in Fig. 8.3 exaggerates the extent of the continental shelves in the Arctic Ocean, the prominence of the continental shelves is readily apparent in Northern Europe, East Asia, Southeast Asia, North Australia, and the Eastern coasts of the Americas.

The main drainage basins of the rivers are major areas of input from the continents to the coastal zone (Fig. 9.4), where it can also be noted that some of the larger river outflows occur on relatively narrow or poorly developed continental shelves (Northeast coast of South America, East and West coasts of Africa). This physiographic feature shows that the bigger and wider continental shelves are not merely physical links on the path between river mouths or deltas and the open ocean, but they play a broader role in the global transport and transformation of materials in the land-atmosphere-ocean system.

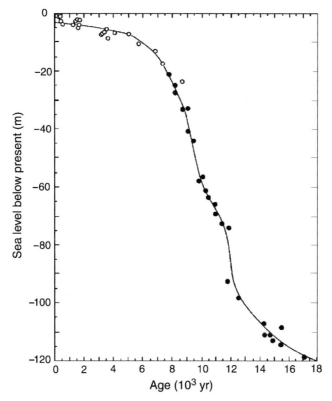

Figure 9.3   Rise in sea level since the Last Glacial Maximum, based on corrected radiocarbon ages (years before present, counted from year 1950 as zero) of corals on Barbados (filled circles) and other Caribbean locations (blank circles), and corrected for mean land uplift of Barbados of 34 cm/1000 yr (from Fairbanks, 1989; by permission of the author and *Nature Publishing Group*).

Estimated amounts of particulate matter transported from land by rivers at present are 13 to 20×10^9 ton/yr (Chapter 8). Riverine input of particulate organic carbon to the oceans increased from about 0.19×10^9 ton C/yr in pre-industrial time to about 0.24×10^9 ton C/yr in the year 2000, and this carbon is likely to be derived from soil humus of an average C:N:P atomic ratio of 268:17:1 (Lerman *et al.*, 2004; Mackenzie *et al.*, 2002). Thus the particulate organic carbon flux accounts for a fraction of 1 to 2% of the total particulate matter input by rivers to the oceans. The particulate inorganic carbon (PIC) flux has been estimated as 0.18 Gt C/yr.

The water volume of the coastal zone to 100 m mean depth is 2.2 to 3.3% of the volume of the surface ocean layer (300 m deep, 108×10^6 km³) and its surface area is approximately 8% of the ocean surface. The present-day coastal zone comprises important sites of deposition and regeneration of organic carbon and of calcium carbonate produced *in situ*: 45% of total carbonate and approximately 85% of total organic

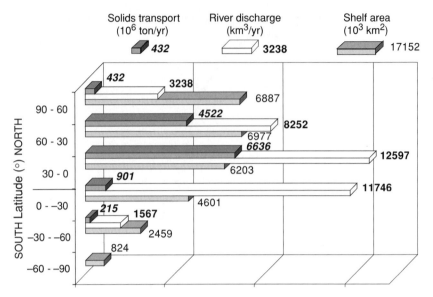

Figure 9.4   Continental shelf area (0 to −200 m depth), river water discharge, and delivery of suspended solids to the coastal ocean in 30°-latitudinal zones. No data for Antarctica. Water discharge from Baumgartner and Reichel (1975), suspended solids data from Meybeck and Ragu (1995, adjusted for the global river flow by multiplying their data by 3.74/2.65, see p. 234).

carbon accumulation in the ocean occur in the coastal margin, and 10 to 30% of total oceanic biological production takes place in this region (Mackenzie *et al.*, 2004; Milliman, 1993; Turner and Adger, 1996; Wollast, 1994, 1998). Eighty percent of the mass of riverine terrigenous materials reaching the ocean is deposited in the coastal zone (Milliman and Syvitski, 1994). Active depositional areas in the coastal ocean may have sedimentation rates as high as 30 to 60 cm/1000 yr, as compared to average rates for hemipelagic and pelagic sediments of 20 cm/1000 yr and 0.1 to 1 cm/1000 yr, respectively. It is thus understandable why the coastal zone is regarded as both a filter and a trap for natural as well as anthropogenic materials transported from the continents to the open ocean (Mantoura *et al.*, 1991). Coastal environments are also regions of higher biological productivity relative to that of average oceanic surface waters, making them an important reservoir in the global carbon cycle. The higher primary productivity is variably attributable to the nutrient inflows from land, as well as most importantly from upwelling of deeper ocean waters along certain sections of the global coastal margin. The global role of the coastal zone as a whole must be viewed through the geographic distribution of its latitudinal sections that is shown in Fig. 9.5.

In the Northern Hemisphere, above the latitude N60°, the shelves make a relatively large fraction of the total surface area, about 20%. This is also a region of major inputs by some of the rivers in North America and Siberia (Fig. 9.4). The shelves in the Northern Hemisphere occupy a much larger area than in the Southern Hemisphere. The climatic differences between the latitudinal zones affect primary production and

COASTAL ZONE BY LATITUDES: SHELF 0 to –200 m

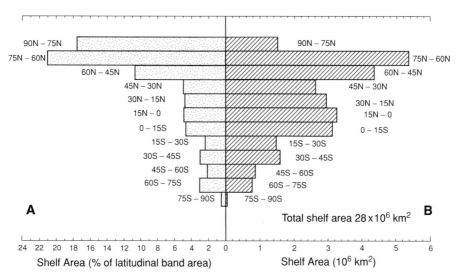

Figure 9.5   Distribution of the coastal zone in 15°-latitudinal bands. A. Shelf area as a fraction of the total area of each latitudinal band. Note that the coastal zone in the northern latitudes occupies relatively large parts of the total surface area. B. Shelf areas of the latitudinal bands showing the greater prominence of the shelves in the Northern Hemisphere.

biological calcification. Carbonate production on the shelves by corals, calcareous algae, and foraminifera is greater in the warmer latitudes, but it should be kept in mind that there are significant deposits of biogenic calcareous sediments at higher latitudes (cool-water carbonates). In addition, significant carbonate mineral production occurs in areas dominated by terrigenous inputs, such as in the river deltaic areas of the coastal ocean, but the carbonate accumulating there is diluted by the detrital input and difficult to quantify in terms of accumulation flux.

Despite its relatively small size, the coastal ocean is an important interface between the land and open ocean, and it is also in direct exchange with the atmosphere. Large river drainage basins connect the vast interiors of continents with the coastal zone through river and groundwater discharges. The ocean surface links the coastal ocean to the atmosphere via gas exchange at the air-sea interface, production of sea aerosols, and atmospheric deposition on the sea surface; substances released at the air-sea interface of the coastal zone may be subsequently transported through the atmosphere and deposited on land as wet and dry depositions; conversely, emissions from land to the atmosphere are in part deposited in the coastal zone. Additionally, physical exchange processes at coastal margins, involving for example coastal upwelling (water rising from the deeper ocean) and onwelling (water that moves on and across the shelf), and net advective transport of water, dissolved solids, and particles from the coastal zone offshore connect the coastal ocean with the surface and intermediate depths of the open ocean. The processes of settling, deposition, resuspension, remineralization of organic matter, dissolution and precipitation of mineral phases, and accumulation of materials connect the water column and the sediments of the coastal zone.

In general, interfaces between the larger material reservoirs (that is, the land, atmosphere, and ocean) are important in the control of the biogeochemical cycling of three of the major bioessential elements found in organic matter: carbon (C), nitrogen (N), and phosphorus (P), because they act as relatively fast modifiers of transport and perturbation processes at geologically short time scales. Over the past several centuries, activities of humankind have significantly modified the exchange of materials between the land, atmosphere, and ocean on a global scale. Humans have become, along with natural processes, agents of environmental change. For example, rapid population growth, increasing population density in the areas of the major river drainage basins and close to oceanic coastlines (about 40% of the world population lives within 100 km of the shoreline; Cohen *et al.*, 1997), and changes in land-use practices in past centuries have increased discharges of industrial, agricultural, and municipal wastes into oceanic coastal waters. Land-use activities include the conversion of land for food production (grazing land, agricultural land), for urbanization (building human settlements, roads, and other structures), for energy development and supply (building dams, hydroelectric plants, and mining of fossil fuels), and for resource exploitation (mining of metals, harvest of forest hardwood) (e.g., Mackenzie, 2003). These activities on land have contributed to increased soil degradation and erosion, eutrophication of river and coastal ocean waters through addition of chemical fertilizers to agricultural land and sewage discharge, degradation of water quality, and alteration of the coastal marine food web and community structure. It is estimated that only about 20% of the world's drainage basins have pristine water quality at present (Meybeck and Ragu, 1995). Estuarine and coastal regions showing much human-induced change are located, for example, along the coasts of the North Sea, the Baltic Sea, the Adriatic Sea, the East China Sea, and the East and South coasts of North America (Fig. 9.4; De Jonge *et al.*, 1994; Richardson and Heilmann, 1995).

2   Carbon Cycle in the Coastal Ocean

2.1   Cycle Structure and Main Processes

For the global coastal ocean as described in the preceding section, a conceptual model of the broad features of the carbon cycle is shown in Fig. 9.6. The inorganic carbon reservoir exchanges CO_2 with the atmosphere ($\pm F_{CO_2}$) and receives inputs of dissolved inorganic carbon from land via rivers, surface runoff and groundwater flow (F_{i1}), from the CO_2 released by the precipitation of calcium carbonate (curved arrow 1), and by upwelling from the deeper ocean (F_{i5}). Upwelling refers to the various processes responsible for the transport of water, dissolved inorganic and organic carbon and nutrients from mainly intermediate-depth waters of the ocean to the surface,

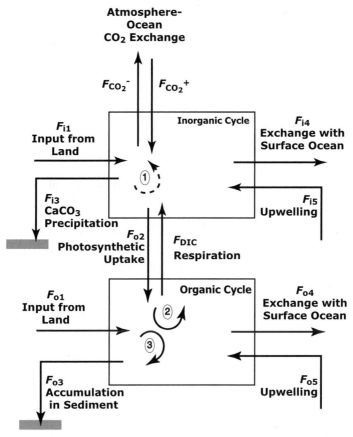

Figure 9.6   Conceptual model of the inorganic and organic carbon cycles in the coastal ocean and their interactions. F_{ij} refer to the fluxes from reservoir i to j (after Mackenzie *et al.*, 1998a). See text for explanation.

in this case to the coastal zone waters. Outflows of inorganic carbon are via net water outflow from the coastal ocean (F_{i4}), accumulation of calcium carbonate formed in the coastal zone, in coastal marine sediments (deposition less dissolution, F_{i3}), and used as CO_2 in photosynthesis (F_{o2}). The organic carbon cycle is linked to the inorganic cycle through biologically driven reduction and oxidation processes, corresponding to primary production (F_{o2}) utilizing CO_2, and respiration and decay (F_{DIC}, where DIC stands for total dissolved inorganic carbon). Analogous to the inorganic carbon cycle, there are inputs of organic carbon from land (dissolved and particulate, F_{o1}) and by transport of dissolved organic carbon from the deeper ocean, referred to as coastal upwelling (F_{o5}). Removal of organic carbon from the coastal zone is through net out-flow to the open ocean (F_{o4}), accumulation in coastal sediments (*in situ* produced and land-derived, F_{o3}), and respiration and decay or remineralization (F_{DIC}). Remineralized organic carbon produced *in situ* (curved arrow 2) or imported from land (curved arrow 3) contributes to F_{DIC}.

The material balance for the organic carbon reservoir in the coastal margin, using the flux notation in Fig. 9.6 for a period $\Delta t = 1$ yr, can be written as:

$$\Delta C_{org} = (F_{o1} + F_{o2} + F_{o5}) - (F_{o4} + F_{o3} + F_{DIC}), \tag{9.1}$$

where C_{org} is the mass of organic carbon in the reservoir, in mol C. Gross photosynthesis (F_{o2}) and respiration and decay (F_{DIC}) are the linkages between the organic and inorganic carbon cycles. The balance of inorganic carbon, [DIC] in mol C, is:

$$\Delta[DIC] = (F_{i1} + F_{i5} + F_{DIC}) - (F_{i3} + F_{i4} + F_{o2}) \pm F_{CO_2} \tag{9.2}$$

Inputs and outputs of carbon produce changes in DIC that equilibrates with atmospheric CO_2. A change in the DIC concentration from an initial value of $[DIC_0]$ to a new equilibrium value after all the input and output fluxes have been accounted for, $[DIC_{eq}]$, is the difference

$$\Delta[DIC] = [DIC_{eq}] - [DIC_0] \tag{9.3}$$

The mass of air-sea exchange of CO_2, flux $F_{CO_2} \times 1$ yr, can be obtained by substitution of F_{DIC} from equation (9.1) for F_{DIC} in (9.2) and also writing the individual terms as fluxes that are shown in Fig. 9.6 as follows:

$$
\begin{aligned}
F_{CO_2} &= (F_{i4} - F_{i1} - F_{i5}) + F_{i3} + (F_{o4} + F_{o3}) - (F_{o1} + F_{o5}) + \Delta C_{org} + \Delta[DIC] \\
&= (F_{DIC\,outflow} - F_{DIC\,inflow}) + (F_{CaCO_3\,ppt} - F_{CaCO_3\,diss}) \\
&\quad + (F_{C_{org}\,outflow} + F_{C_{org}\,sed} - F_{C_{org}\,inflow}) + \Delta C_{org} + \Delta[DIC] \tag{9.4}
\end{aligned}
$$

In the notation of Fig. 9.6 and equation (9.4), F_{CO2} is negative when the net CO_2 flow is from seawater to the atmosphere and the coastal ocean is a CO_2 source; F_{CO_2} is positive when the CO_2 flow is to seawater and the coastal ocean is a CO_2 sink.

2.2 CO_2 Air-Sea Exchange

The individual inorganic and organic carbon fluxes that determine the CO_2 exchange between seawater and the atmosphere are discussed in this section.

The change in DIC and the equilibrium value $[DIC_{eq}]$ depend not only on the atmospheric CO_2 concentration, but also on the shift in the carbonate equilibria because of the net removal of $CaCO_3$ from seawater and a resultant increase in the dissolved CO_2 as a fraction of DIC. A calculation of $[DIC_{eq}]$ that is needed for the CO_2 air-sea flux is based on the changes in total alkalinity and DIC of seawater that are discussed in Chapter 5. A simplified calculation, presented below, is based on the parameter θ that defines a relationship between the net removal of $CaCO_3$ and change in DIC due to production of CO_2 in seawater of normal salinity. The ratio of CO_2 released from seawater of normal salinity to $CaCO_3$ removed by precipitation, between 15°, and 25°C, and the atmospheric CO_2 concentration between the pre-industrial 280 ppmv and the near-present 375 ppmv, is (Table 5.5):

$$\theta = \frac{\Delta[\text{DIC}]}{F_{\text{CaCO}_3}\Delta t} = 0.51 \text{ to } 0.66 \tag{9.5}$$

In the coastal zone, where most of the $CaCO_3$ is produced by biological processes, $(F_{\text{CaCO}_3 \text{ ppt}} - F_{\text{CaCO}_3 \text{ diss}})$ has been called *net ecosystem calcification* or *NEC* in the marine ecological literature.

The balance of the organic carbon inputs and outputs to the coastal ocean is called *net ecosystem production (NEP)* or *net ecosystem metabolism (NEM)*:

$$NEM = F_{o4} + F_{o3} - F_{o1} + \Delta C_{\text{org}} \approx F_{o4} + F_{o3} - F_{o1} \tag{9.6}$$

Net ecosystem metabolism is essentially equivalent to the difference between gross primary production (GPP) and the production of CO_2 by autotrophic and heterotrophic respiration of organic carbon that includes organic carbon from external inputs and *in situ* primary production. *NEM* is a negative quantity when the input of organic carbon to the coastal ocean (F_{o1}) is greater than its removal by outflow to the open ocean (F_{o4}) and net storage in sediments (F_{o3}). In this algebraic notation, the excess of organic carbon input over its removal is a contribution to the CO_2 flux out of the coastal zone.

A relationship between a change in DIC of seawater due to a change in atmospheric CO_2 concentration was developed by Bacastow and Keeling (1973) and approximated by Revelle and Munk (1977) in the form of:

$$R = \frac{C_{\text{atm},t}/C_{\text{atm},0} - 1}{C_{\text{DIC},t}/C_{\text{DIC},0} - 1} \tag{9.7}$$

$$R \approx R_0 + D \times (C_{\text{atm},t}/C_{\text{atm},0} - 1) \tag{9.8}$$

where C_{atm} is CO_2 concentration in the atmosphere and C_{DIC} is the dissolved inorganic carbon concentration [DIC] in surface ocean water, subscript 0 denotes the initial value and subscript t a value at a later time t. Subsequently, R became known as the Revelle factor and its different forms have been discussed by Zeebe and Wolf-Gladrow (2001). Equation (9.8) with constants $R_0 = 9$ and $D = 4$ is Revelle and Munk's (1977) approximation to the curve of R calculated by Bacastow and Keeling (1973) for an average surface ocean water of total alkalinity 2.435×10^{-3} mol-equivalent/liter, temperature 19.59°C, chlorinity 19.24 per mil, and initial pH = 8.271. With $R_0 = 9$, the buffer mechanism of seawater causes a fractional rise of DIC in coastal ocean surface seawater that is one-ninth of the increase of CO_2 in the atmosphere. The change in DIC, as written in equation (9.3), becomes:

$$\Delta[\text{DIC}] = \frac{(C_{\text{atm},t} - C_{\text{atm},0})C_{\text{DIC},0}}{R_0 C_{\text{atm},0} + D \times (C_{\text{atm},t} - C_{\text{atm},0})}. \tag{9.9}$$

A time-dependent change of [DIC], $d[\text{DIC}]/dt$, is obtained by differentiation of equation (9.8) that gives a simpler approximate relationship in equation (9.10a) and a more complete relationship (9.10b) with a second-order term:

$$\frac{d[\text{DIC}]}{dt} = \frac{[\text{DIC}_0]}{R_0 C_{\text{atm},0} + D \times (C_{\text{atm},t} - C_{\text{atm},0})} \times \frac{dC_{\text{atm}}}{dt} \qquad (9.10a)$$

$$\frac{d[\text{DIC}]}{dt} = \left\{ \frac{[\text{DIC}_0]}{R_0 C_{\text{atm},0} + D \times (C_{\text{atm},t} - C_{\text{atm},0})} \right.$$
$$\left. - \frac{D \times [\text{DIC}_0] \times (C_{\text{atm},t} - C_{\text{atm},0})}{[R_0 C_{\text{atm},0} + D \times (C_{\text{atm},t} - C_{\text{atm},0})]^2} \right\} \times \frac{dC_{\text{atm}}}{dt} \qquad (9.10b)$$

The CO_2 air-sea flux in equation (9.4), using the preceding definitions of the other flux terms, can now be written as:

$$F_{CO_2} = (F_{i4} - F_{i1} - F_{i5}) + NEC + NEM + \frac{d[\text{DIC}]}{dt} \qquad (9.11)$$

Table 9.2   Estimates of measured air-sea CO_2 exchange from worldwide shallow-water locations (estuaries, reefs, shelves)

Location/Region	Net CO_2 flux (mol C m^{-2} yr^{-1})	Reference
Global coastal zone		
without estuaries	+1.17	Borges (2005)
including estuaries	−0.38	
Global estuaries	−41	Borges (2005)
European estuaries (Elbe, Ems, Rhine, Scheldt, Tamar, Thames, Gironde, Duoro, Sado)	−36.5 to −277	Frankignoulle et al. (1998)
Oregon, Cape Perpetua	+7.29	Hales et al. (2003)
New Jersey	+0.43 to +0.84	Boehme et al. (1998)
Bermuda, Hog Reef flat	−1.20	Bates et al. (2001)
Mid Atlantic Bight (MAB)	+1.00	DeGrandpre et al. (2002)
South Atlantic Bight (SAB)	−2.50	Cai et al. (2003)
Baltic Sea	+0.90	Thomas and Schneider (1999)
North Sea	+1.35	Kempe and Pegler (1991)
North Sea average	+0.95 to +1.33	
range from South to North	−0.5 to +2	Thomas et al. (2004)
Scheldt Estuarine plume	−1.10 to −1.90	Borges and Frankignoulle (2002)
Gulf of Biscay	+1.75 to +2.88	Frankignoulle and Borges (2001)
Galician Coast	+0.66 to +1.17	Borges and Frankignoulle (2001)
Northern Arabian Sea	−0.46	Goyet et al. (1998)
East China Sea	+2.91	Tsunogai et al. (1999)
East China Sea	+1.20 to +2.80	Wang et al. (2000)
Moorea, French Polynesia	−0.55	Gattuso et al. (1993)

Note: Negative values denote CO_2 flux from ocean to the atmosphere; positive values denote flux from the atmosphere to ocean water. Some of the values are annual integrated flux measurements and others monthly or occasional measurements.

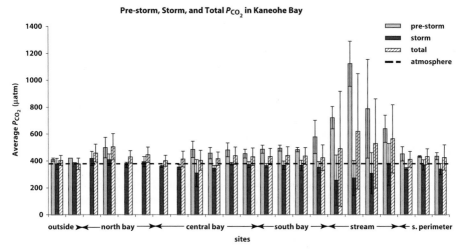

Figure 9.7   Average P_{CO_2}, calculated from total alkalinity and dissolved inorganic carbon bi-weekly measurements, for the surface waters of the major estuary of Kaneohe Bay, Oahu, Hawaii, and the major stream, Kaneohe Stream, that enters it, for the period September 2003 through September 2004. The P_{CO_2} is shown for various regions of the bay: north bay, central bay and south bay, for waters directly outside the bay, and for Kaneohe Stream and its perimeter. Light-grey bars represent pre-storm conditions (the background level; 9/22/03-11/10/03); dark-grey bars are for storm and near post-storm conditions of 12/19/03, 2/4/04, 3/9/04 and 4/20/04; and the crosshatch bars represent the annual average for all data collected from September 2003 through September 2004. The dashed line is the mean P_{CO_2} of the boundary layer of the atmosphere near its base as measured at Cape Kumukahi, Hawaii. Notice that the annual average of the P_{CO_2} measurements relative to the atmospheric value indicates that the bay is a source of CO_2 to the atmosphere. However during and following storms, the P_{CO_2} of bay surface waters is driven down below the atmospheric value because of enhanced phytoplankton productivity due to increased runoff and nutrient loading of the bay, and the bay becomes a net sink for atmospheric CO_2 for a variable but short period of time (data and figure courtesy of Kathryn E. Fagan).

The preceding equation defines the CO_2 flux across the air-sea interface of the coastal ocean as an algebraic sum of four terms: the difference between the outflows and inflows of DIC (Fig. 9.6) into the coastal ocean, dissolved CO_2 produced by precipitation of calcium carbonate in net ecosystem production, net ecosystem metabolism that is the net difference of the inputs and outputs of organic carbon, and the change in the DIC concentration caused by a change in the atmospheric CO_2 concentration.

The relative importance of the individual processes that determine the CO_2 air-sea flux in equation (9.11) is discussed in the next section. Some of the measurements of the present-day CO_2 flux at different locations in coastal and open oceanic sections, and estuaries are summarized in Table 9.2, showing flux values that are highly variable both in their magnitude and direction.

An example of this variability is shown in Fig. 9.7 for the air-sea exchange of CO_2 in the major estuary of Kaneohe Bay, Oahu, Hawaii. Although the estuary, about 14 km long and up to 5 km wide, on an annual basis is a source of CO_2 to the atmosphere from calcification processes involving major coral reef development in the bay, the direction, magnitude, and spatial distribution of the net CO_2 flux vary considerably during the year. The bay on an annual basis is net autotrophic and immediately following storms that deliver large subsidies of nutrients to the estuary from the land, via river and groundwater flows, the bay for some period of time, because of the enhanced organic productivity in phytoplankton blooms resulting from the storm-derived nutrient loading, becomes a sink of atmospheric CO_2. As the productivity declines, the bay returns to its more normal state of being a source of CO_2 to the atmosphere. These observations, particularly characteristic of many proximal coastal ocean ecosystems of the world, support two major conclusions: determining the air-sea exchange CO_2 flux for coastal environments and the integrated global coastal ocean is a difficult task because of the high variability of this flux and must involve time series carbon measurements of

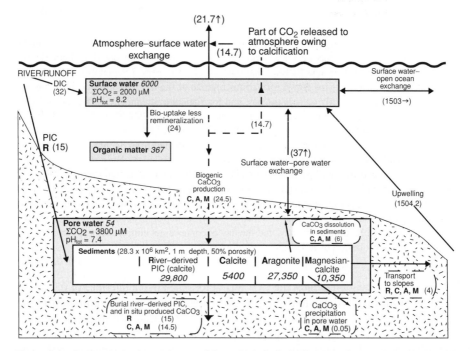

Figure 9.8   Inorganic (carbonate) part of the carbon cycle in the coastal ocean in pre-industrial time (from Andersson *et al.*, 2005, and Ver *et al.*, 1999a). Reservoir masses in *italics* are in units of 10^{12} mol C. Arrows denote carbon fluxes between reservoirs in units of 10^{12} mol C/yr (shown inside parentheses). Two major domains are the water column (surface water and dissolved inorganic carbon) and the pore water-sediment system (pore water, dissolved inorganic carbon, river-derived particulate inorganic carbon (PIC), calcite, aragonite, and 15 mol % magnesian calcite, a mean composition representing the range in composition of magnesian calcites). For two-way arrows the direction of the net flux is shown next to the flux estimate. The dashed lines indicate carbon flux owing to $CaCO_3$ production, equations (9.4), (9.5), and (9.12).

sufficient duration and spatial distribution to determine the flux quantitatively; and the net air-sea CO_2 flux is in part a reflection of both the net ecosystem metabolism (*NEM*) and net ecosystem calcification (*NEC*) rates.

3 Inorganic and Organic Carbon

3.1 A More Detailed Conceptual Model

It is clear from the preceding discussion that the inorganic or carbonate part of the carbon cycle is intimately linked to its organic part through the reduction of CO_2 in photosynthesis by primary producers, and oxidation of organic matter in autotrophic respiration and by microbial activity. Nevertheless, it is instructive to consider the inorganic and organic parts separately, as they are schematically shown in Figs. 9.8 and 9.9. The masses of carbonate and organic carbon in the different reservoirs of the

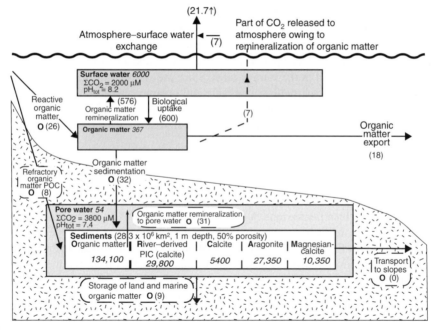

Figure 9.9 Organic (C_{org}) part of the carbon cycle in the coastal ocean in pre-industrial time (from Andersson *et al.*, 2005, and Ver *et al.*, 1999a). Reservoir masses in *italics* are in units of 10^{12} mol C. Arrows denote carbon fluxes between reservoirs in units of 10^{12} mol C/yr (shown inside parentheses). Two major domains are the water column (surface water, dissolved inorganic carbon, and organic matter) and the pore water–sediment system (pore water and organic matter in sediments). The dashed lines indicate carbon flux owing to the net imbalance between GPP and total remineralization of organic matter or net ecosystem metabolism, *NEM*, equations (9.6) and (9.11).

coastal ocean and the interreservoir fluxes have been used in the global carbon cycle models TOTEM (Terrestrial Ocean aTmosphere Ecosystem Model, Mackenzie *et al.*, 2001; Ver *et al.*, 1999) and SOCM (Shallow-water Ocean Carbonate Model, Andersson and Mackenzie, 2004; Andersson *et al.*, 2003). One of the prominent features of the carbonate cycle is the dominant abundance of land-derived detrital calcite and biogenic aragonite in coastal sediments, followed by magnesian calcites and calcite. For the sediment layer 1 m in thickness (Fig. 9.8), total residence time of inorganic carbon calculated with respect to inputs is about 1850 years, with the main removal flux being the burial in sediments of the detrital calcite and *in situ* produced carbonates. Water residence time in the coastal water column is shown as about 4 years, although there are also shorter estimates of 2 to 3 years (Chavez and Toggweiler, 1995).

The organic carbon reservoir of the coastal water column includes dissolved organic carbon (DOC) and particulate organic carbon (POC), in part transported from land and in part formed *in situ* by primary production. Input from land consists of DOC and part of POC classified as a reactive particulate fraction and taken as 50% of total POC in rivers (Fig. 8.4; Smith and Hollibaugh, 1993). The remaining 50% of POC is considered a refractory fraction, at least on a short time scale. Organic carbon that is exported to the open ocean is eventually remineralized, in whole or in part. The uncertainties in the published estimates of the carbonate and organic carbon fluxes in the Holocene and earlier, during the Last Glacial period, are considerable, and a variation within a factor of two is to be expected in the storage rates. Since the end of pre-industrial time, the inputs of dissolved and particulate inorganic and organic carbon to the coastal zone have increased significantly and the projected trend is that of a continued increase on a centurial time scale (Lerman and Mackenzie, 2004; Mackenzie *et al.*, 2004; Fig. 8.4).

3.2   Remineralization of Organic Matter

Degradation of organic matter in sediments by microbial processes generally produces aqueous CO_2 and bicarbonate ions. A number of such reactions involving the reduction of oxygen, nitrate, ferric iron, and sulfate are given in Table 9.3 and their energy yields are also shown in Fig. 9.10. The latter depend on the reaction stoichiometry and the reactants and products. Overall, the energy change decreases from remineralization of organic matter by oxygen through the reduction of nitrate, Fe^{3+} iron, and sulfate to anaerobic decomposition of organic matter that produces methane, CO_2, and hydrogen.

A major benthic process involving the interaction between the inorganic and organic carbon cycles that was initially observed for shallow-water carbonate sediments and then extended to open ocean sediments was that involving the degradation of organic matter by a bacterially mediated oxygen reaction (Emerson and Bender, 1981; Table 9.3, reaction 1) and the consumption of the produced CO_2 in the process of dissolution of calcium carbonate. In an abbreviated and idealized form, comparable to reaction (5.23):

$$CH_2O + O_2 + CaCO_3 = Ca^{2+} + 2HCO_3^- \qquad (9.12)$$

Table 9.3   Organic matter oxidation (1), reduction by organic matter of nitrate, ferric iron, and sulfate (2–4), and anaerobic decomposition to methane and byproducts (5)

Process	Reaction[2]	$\Delta G_R^{\circ(1)}$ (kJ/mol C)
1. Remineralization	$CH_2O + O_2 = CO_2 + H_2O$	−478.4
2. a. Denitrification	$5CH_2O + 4NO_3^- + 4H^+ = 5CO_2 + 2N_2 + 7H_2O$	−486.2
b.	$2CH_2O + 2NO_3^- + 2H^+ = 2CO_2 + N_2O + 3H_2O$	−436.1
c.	$2CH_2O + 2NO_3^- = N_2O + 2HCO_3^- + H_2O$	−391.4
3. a. Fe^{3+} iron reduction	$CH_2O + 4Fe(OH)_3 + 8H^+ = CO_2 + 4Fe^{2+} + 11H_2O$	−412.6
b.	$CH_2O + 4FeOOH + 8H^+ = CO_2 + 4Fe^{2+} + 7H_2O$	−278.3
c.	$CH_2O + 7CO_2 + 4Fe(OH)_3 = 4Fe^{2+} + 8HCO_3^- + 3H_2O$ [3]	−55.0
d.	$2Fe(OH)_3 + H_2 = 2Fe(OH)_2 + 2H_2O$ (kJ/mol Fe)	−31.5
4. a. Sulfate reduction	$3CH_2O + 4H^+ + 2SO_4^{2-} = 3CO_2 + 5H_2O + 2S$	−140.1
b.	$2CH_2O + 2H^+ + SO_4^{2-} = 2CO_2 + 2H_2O + H_2S$	−120.0
c.	$2CH_2O + SO_4^{2-} = 2HCO_3^- + H_2S$	−75.3
5. a. Methano-genesis	$2CH_2O = CO_2 + CH_4$	−69.4
b.	$3CH_2O + H_2O = 2CO_2 + CH_4 + 2H_2$	−47.7
c.	$CH_3COOH = CO_2 + CH_4$	−27.6

Note: Standard Gibbs free energy of the reaction at 25°C in kJ per 1 mol carbon of organic matter, written in a shorthand notation CH_2O for glucose, $C_6H_{12}O_6^{(1)}$.
[1] Standard free energy of formation of glucose ($C_6H_{12}O_6$) $\Delta G_f^{\circ} = -918.78$ kcal/mol computed from the free energy change of the photosynthesis reaction, taken as $\Delta G_R^{\circ} = 2870$ kJ/mol or 478.4 kJ/mol C (Chapter 2).
[2] Standard Gibbs free energy of formation values from Lide (1994) and Drever (1997). Gaseous species: $O_2, CO_2, N_2, N_2O, H_2$, and CH_4. Liquid: H_2O. Aqueous species: $H^+, H_2S, NO_3^-, Fe^{2+}$, and SO_4^{2-}. Solids: ferrihydrite $Fe(OH)_3$, goethite $FeOOH$, ferrous hydroxide $Fe(OH)_2$, and rhombic sulfur S.
[3] This reaction is the sum of 3a and $8CO_2 + 8H_2O = 8H^+ + 8HCO_3^-$. Canfield (1993) gives for this reaction −114 kJ/mol CH_2O.

or with the organic matter based on the C:N:P Redfield ratio of marine phytoplakton:

$$(CH_2O)_{106}(NH_3)_{16}H_3PO_4 + 138\ O_2 + 124\ CaCO_3$$
$$= 16\ H_2O + 16\ NO_3^- + HPO_4^{2-} + 124\ Ca^{2+} + 230\ HCO_3^- \qquad (9.13)$$

This reaction is only possible if the pore waters are undersaturated with a carbonate phase and there are no competing reactions that increase pore water carbonate saturation state, such as ammonia release from the decaying organic matter that may increase the pore water pH. Figure 9.11 shows the increase in total alkalinity versus the increase in DIC for shallowly buried pore waters bathing sediments containing nearly pure calcite, aragonite, and magnesian calcites with up to 20 mol % $MgCO_3$ occurring in the Gulf of Calvi in Corsica. Notice that the slope of this relationship is 1.04, close to the slope predicted in equation (9.13), $2 \times 124/230 = 1.08$ and not 2 as would

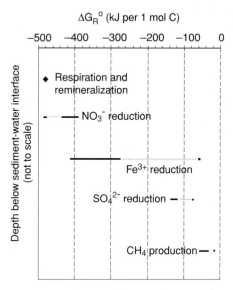

Figure 9.10   Energy change in organic matter decomposition reactions (Table 9.3) shown as a sequence of reduction reactions with increasing depth below the sediment-water interface, from nitrate reduction near the top, followed by ferric iron and sulfate reduction, and methanogenesis deeper in the sediment. Oxidative remineralization of organic matter has the greatest energy yield and methanogenesis the smallest.

be expected if the dissolution were simply related to undersaturation with carbonate minerals: $CaCO_3 = Ca^{2+} + CO_3^{2-}$. The fact that the regression line does not pass through 0 in Fig. 9.11 reflects the fact that dissolved CO_2 has been added to the pore waters from aerobic bacterial respiration of organic matter prior to carbonate dissolution thus lowering the carbonate saturation state of the pore water sufficiently for dissolution to occur. Moulin *et al.* (1985) showed that due to this very early diagenetic reaction in carbonate deposits, principally containing particles of the red alga *Lithothamnium* sp. in Calvi Bay, nearly 75% of the carbonate sediment initially deposited is dissolved.

   Another benthic process that can be important in the dissolution (and precipitation) of carbonate minerals in sediments is that of sulfate reduction (Table 9.3, reactions 4a-c). This is a complex process, but it can be described schematically by the following equation:

$$1/53(CH_2O)_{106}(NH_3)_{16}H_3PO_4 + SO_4^{2-}$$
$$= 2HCO_3^- + HS^- + 16/53\,NH_3 + 1/53\,H_3PO_4 + H^+ \qquad (9.14)$$

This reaction results in the decrease in sulfate and pH and the concomitant increase in alkalinity, hydrogen sulfide, phosphate, ammonia, and DIC, commonly observed in anoxic sediment pore waters with increasing depth. During the early stages in sulfate reduction (range of approximately 2 to 35% decrease in sulfate concentration), the pore

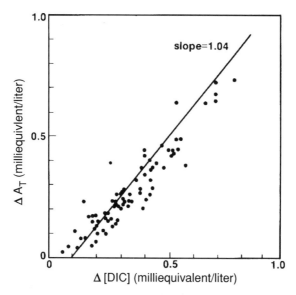

Figure 9.11   Increase in total alkalinity (A_T) and dissolved inorganic carbon (ΣCO_2) or DIC over that observed in overlying seawaters, Gulf of Calvi, Corsica. Note in particular the nearly 1:1 slope of the relationship (Moulin *et al.*, 1985).

water may become undersaturated with respect to carbonate minerals (Fig. 9.12), particularly the highly soluble magnesian calcites and aragonite, and carbonate phases may dissolve. As the extent of sulfate reduction increases further, the pore water becomes supersaturated with respect to carbonate minerals and a carbonate phase may precipitate. The extent of the reaction depends strongly on the organic carbon available and the C/N ratio of the labile carbon. This reaction pathway has been demonstrated both in model calculations and has been observed in pore waters (Mackenzie *et al.*, 1995; Morse *et al.*, 1985). The relative importance of carbonate dissolution in aerobic pore water environments and that of dissolution in anaerobic pore waters is not well known because of the lack of data on pore water composition accompanied by solid phase chemistry and mineralogy, particularly for shallow-water sediments. Suffice it to mention that both processes may occur during early diagenesis and are used in a model, discussed in the subsequent sections of this chapter, to assess changes in shallow-marine pore water chemistry and carbonate saturation state under rising atmospheric CO_2 concentrations and temperature and enhanced burial of organic carbon.

3.3   Carbon Fluxes

For the carbon cycle in the coastal zone near the end of pre-industrial time, taken as the year 1700, the direction and magnitude of the CO_2 exchange between coastal ocean waters and the atmosphere can be estimated from the fluxes shown in Figs. 9.8 and

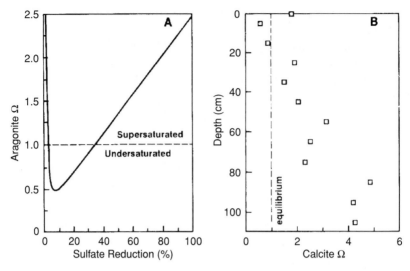

Figure 9.12   A. Model calculation of the saturation state (Ω) of marine pore waters of sediments with respect to aragonite as a function of the percentage of dissolved sulfate reduced in the pore waters starting with a pore water sulfate concentration of 0.028 mol/kg of seawater (Ben-Yaakov, 1973; recalculated by Morse and Mackenzie, 1990). B. The observed saturation state of Mangrove Lake, Bermuda marine pore waters with respect to calcite. Notice the undersaturation with respect to calcite at shallow depths and the supersaturation at deeper depths in the sediment. At the shallower depths in the sediment, calcite, aragonite and the highly soluble magnesian calcites could dissolve, and at deeper depths a carbonate phase(s) could precipitate (Mackenzie et al., 1995).

9.9 and equation (9.4). Below, three estimates of the exchange are given, based on somewhat different approaches.

For dissolved inorganic carbon (DIC), there are two main inputs, from land by rivers (32×10^{12} mol C/yr) and from intermediate ocean depths by coastal upwelling, and exchange with the open surface ocean. DIC is also removed as CO_2 in primary production and calcium carbonate formation and storage in sediments. For organic carbon, there is the riverine input of reactive and particulate organic carbon (the latter considered as at least temporarily refractory) and its removal by export to the ocean and storage in sediments. Taking these fluxes with no change in the organic carbon concentration in the coastal ocean and no increase in the atmospheric CO_2 concentration, the sea-air flux from equation (9.4) is:

$$F_{CO_2} = (1503 - 32 - 1504.2) + (24.5 - 6) + (18 + 9 - 26 - 8) + 0$$
$$= (-33.2 + 18.5 - 7) \times 10^{12} = -21.7 \times 10^{12} \, \text{mol/yr} \qquad (9.15)$$

This value is shown as the sea-to-air flux in Figs. 9.8 and 9.9, which also show the individual contributions to the CO_2 flux from DIC (in units of 10^{12} mol C/yr), $-33.2 + 18.5 = -14.7$, and from the net imbalance between GPP and total remineralization of organic matter, -7. The increase in DIC due to an increase in atmospheric CO_2 on a yearly basis that is given by the term $d[\text{DIC}]/dt$ in equation (9.11) is relatively small in comparison to the other terms.

A more detailed solution, based on the preceding flux values and the CO_2-transfer computation that is described in Chapter 5, gives a slightly higher pre-industrial sea-to-air CO_2 flux:

$$F_{CO_2} = -24.7 \pm 0.1 \times 10^{12} \, \text{mol/yr} \tag{9.16}$$

An approximate value of the sea-air CO_2 exchange can also be obtained from the balance of dissolved CO_2 in the DIC inflow and outflow, CO_2 production in $CaCO_3$ net deposition, and remineralization of organic matter. Then, as explained in Section 2.2 and below, the CO_2 flux at a steady state can be written from equation (9.11) as:

$$F_{CO_2} \approx \alpha_{sw} \cdot (F_{\text{DIC out}} - F_{\text{DIC in}})_{sw} - \alpha_{rw} \cdot F_{\text{DIC in, rw}} - \theta \cdot NEC + NEM \tag{9.17}$$

The DIC input to the coastal ocean, as shown in Fig. 9.8, consists of an upwelling flux of ocean water and riverine inflow, and the DIC outflow of seawater. Dissolved CO_2 represents different fractions of DIC in fresh and ocean water: in river water, taken as pure water at a pH of 7.5, the CO_2 fraction is $\alpha_{rw} = 0.077$ at $15°$ and 0.066 to $25°C$, but it is much smaller in seawater at the same temperatures because of its higher alkalinity, $\alpha_{sw} = 0.0050$ to 0.0038, at a pH of 8.2. Using these values of the CO_2 fractions α_{sw} and α_{rw} with the DIC outflow and inflow values, other fluxes as in equation (9.15), the sea-to-air flux is:

$$F_{CO_2} \approx -20.4 \pm 0.2 \times 10^{12} \, \text{mol/yr} \tag{9.18}$$

The three estimates of the global sea-air CO_2 exchange are comparable, although it should be borne in mind that they are all based on one set of inflow and outflow values in the model of the coastal ocean and sediments.

The calculated CO_2 flux since the beginning of industrial time to the early years of the 21st century and beyond is shown in Fig. 9.13. This calculation, done using the model SOCM, takes into account increases in the inputs of carbon, nitrogen, and phosphorus from land to the coastal zone owing to human activities and the documented historical increase in atmospheric CO_2 concentrations and their projection into the future. The greater delivery of nutrients to the coastal ocean, the rising atmospheric CO_2, and a possibly increasing global temperature of the Earth's surface affect the main processes of the inorganic and organic carbon cycles in the coastal zone: primary production, storage of organic matter in sediments, seawater chemistry, and precipitation, dissolution, and storage of calcium carbonate minerals, all of which control the air-sea exchange of carbon dioxide.

Figure 9.13 also shows Borges's (2005) mean values of the coastal zone air-sea exchange of CO_2 based on synthesis of most of the observational measurements available to date. It should be kept in mind that these data are of variable quality and still do not cover a major proportion of the global coastal zone area. In addition, few time-series data are available. The open circle in Fig. 9.13 is the mean air-sea CO_2 exchange flux

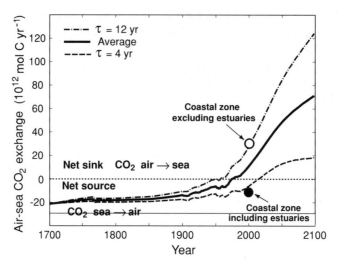

Figure 9.13   Net air-sea CO_2 exchange (10^{12} mol C yr^{-1}) between 1700 and 2100 calculated by *SOCM* adopting the coastal water residence time (τ) of 4 and 12 years. The solid line indicates the average flux of the two scenarios. The data points (Borges, 2005) indicate that the coastal zone including estuaries (solid circle) is a net source of CO_2 to the atmosphere, and excluding estuaries (open circle) it is a net sink of CO_2. The small area of the estuaries ($\sim 1 \times 10^6$ km^2), less than 5% of the global continental shelf ($\sim 26 \times 10^6$ km^2), indicates that the strength of the estuarine source of CO_2 is unusually high; that is, its specific area CO_2 flux is significantly greater than the specific area CO_2 flux of the whole coastal zone that is dominantly shelf area (Andersson *et al.*, 2005).

if one does not include estuaries in the estimate; the filled circle includes the data from estuaries. The area specific global coastal zone air-sea CO_2 flux without including the estuaries is +1.17 mol C m^{-2} yr^{-1} (a sink of atmospheric CO_2) and including estuaries is −0.38 mol C m^{-2} yr^{-1} (a source of atmospheric CO_2). The strong role of the estuaries in the global coastal zone air-sea CO_2 exchange balance is evidenced by their very significant area-specific mean air-sea CO_2 exchange flux of −35.8 mol C m^{-2} yr^{-1}. Thus the model results in Fig. 9.13 agree very well with the observational data of Borges (2005) and show that in the future, the whole global coastal ocean will likely become a sink of atmospheric CO_2 owing to a combination of the rising atmospheric CO_2 with increased biological production that leads to a greater net storage of organic carbon in sediments. The global shelf area is already a sink of atmospheric CO_2. The magnitudes of the biogeochemical mechanisms that ultimately control the air-sea CO_2 exchange are shown in Fig. 9.14, calculated for the 300 years of the immediate past and projected for 300 years of the future Industrial Age.

A striking feature of the projected trends is the reversal of the coastal ocean's role from being a CO_2 source to a CO_2 sink owing to the change in the carbon cycle

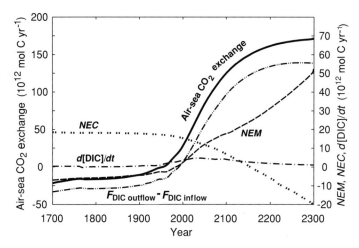

Figure 9.14   Air-sea CO_2 exchange, equation (9.11), owing to net ecosystem metabolism (*NEM*, equation (9.6)), net ecosystem calcification (*NEC*, $CaCO_3$ precipitation less dissolution), atmospheric CO_2 concentration ($d[DIC]/dt$; equation (9.10)), and the imbalance of DIC between export to the open ocean and input from rivers and upwelling ($F_{DIC\ outflow} - F_{DIC\ inflow}$) between 1700 and 2300. Positive values of air-sea exchange are for the CO_2 flux from the atmosphere to seawater, negative values for the flux out of the coastal water (Andersson *et al.*, 2005).

dynamics. The contribution of the atmospheric CO_2 rise is small, as was mentioned earlier. Net ecosystem metabolism (*NEM*) changes from negative to positive, generating an excess of organic carbon in the coastal ocean that is in part stored in sediments and in part exported to the open ocean. Remineralization of organic matter also contributes to DIC in coastal water, increasing the DIC concentration in outflow. Even a more pronounced change takes place in the inorganic carbon cycle, as shown by the decrease in net ecosystem calcification, *NEC*. Excess of $CaCO_3$ mineral precipitation over dissolution declines until the magnitudes of the two processes are reversed and there is a net loss of $CaCO_3$ from sediment (*NEC* < 0) from approximately the middle of the 22nd century to the year 2300. The reason behind this carbonate loss is the dissolution of the relatively soluble Mg-calcites and aragonite, as discussed in the next section, that occurs because of a higher dissolved CO_2 concentration in coastal water due to the increase in atmospheric CO_2 and remineralization of organic carbon in carbonate sediments. Dissolved CO_2 lowers the degree of seawater saturation with respect to the carbonate minerals and promotes their dissolution. Remineralization or decay of organic matter in sediments (Figs. 9.9 and 9.12) generates at the beginning of the modeling period, year 1700, a flux of 31×10^{12} mol C/yr that increases with time (Fig. 9.15), contributing to the carbonate mineral dissolution in sediments according to the reactions of equations (9.13) and (9.14).

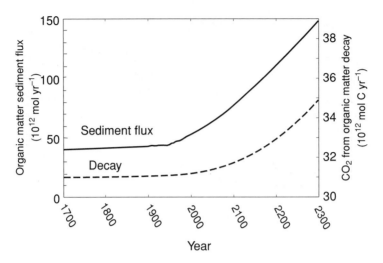

Figure 9.15   Calculated rates of sedimentation and remineralization or decay of organic matter in the coastal ocean (Andersson *et al.*, 2005).

4   Marine Calcifying Organisms and Ecosystems

4.1   Calcium Carbonate Saturation State and Calcification

Increasing atmospheric CO_2 and subsequently decreasing carbonate saturation state of surface ocean water (ocean acidification) may have a negative effect on marine calcifying organisms because their ability to calcify depends in part on the carbonate saturation state. The rate of calcification or the rate at which marine organisms make their $CaCO_3$ skeletons depends on the carbonate saturation state of seawater with respect to the mineral phase formed by the biomineralization processes. Greater supersaturation is usually accompanied by faster calcification rates. This relationship has been demonstrated for such different calcifying organisms as coccolithophorids (Riebesell *et al.*, 2000; Sciandra *et al.*, 2003; Zondervan *et al.*, 2001), foraminifera (Bijma *et al.*, 1999; Spero *et al.*, 1997), coralline algae (Agegian, 1985; Borowitzka, 1981; Gao *et al.*, 1993; Mackenzie and Agegian, 1989; Smith and Roth, 1979), and scleractinian corals (Gattuso *et al.*, 1998; Marubini and Thake, 1999; Marubini *et al.*, 2001, 2003; Reynaud *et al.*, 2003). Similar results were obtained from experiments with typical calcareous communities in incubation chambers and mesocosms, and on the artificial reef of Biosphere 2 (Halley and Yates, 2000; Langdon *et al.*, 2000, 2003; Leclercq *et al.*, 2000, 2002). Although substantial variations have been observed between species and communities, the major results and conclusions of most studies have been similar: the rate of calcification has been observed to decrease as a function of decreasing carbonate saturation state, as shown in Fig. 9.16 for different communities of carbonate organisms. In addition, the saturation state and temperature of seawater also have an effect on the Mg-content of calcite in a species of a

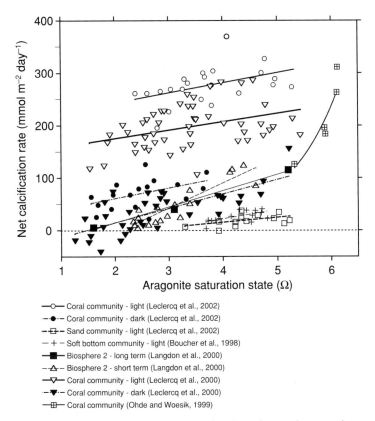

Figure 9.16   Community net calcification rate as a function of aragonite saturation state under different environmental conditions adopted from the references cited in figure. Positive values represent net precipitation and negative values represent net dissolution of carbonate minerals. Solid lines represent the best fit to the data. From Andersson *et al.* (2005).

coralline alga (Fig. 9.17). The degree of saturation of seawater with respect to carbonate minerals was discussed in Chapter 5 and it may be reiterated here that because the Ca^{2+}-ion concentration in seawater does not vary much, the saturation state depends on the CO_3^{2-}-ion, the concentration of which is a function of atmospheric CO_2. Dissolution of CO_2 in seawater decreases the carbonate-ion concentration, as shown in the reaction below, and it consequently decreases the saturation state with respect to the carbonate minerals:

$$CO_2 + H_2O + CO_3^{2-} = 2HCO_3^{-} \qquad (9.19)$$

A reaction between dissolved CO_2 and carbonate minerals (Chapter 5) adds alkalinity to seawater, also buffering the change in the H^{+}-ion concentration that would be caused by adding CO_2 to pure water:

$$CaCO_3 + CO_2 + H_2O = Ca^{2+} + HCO_3^{-} \qquad (9.20)$$

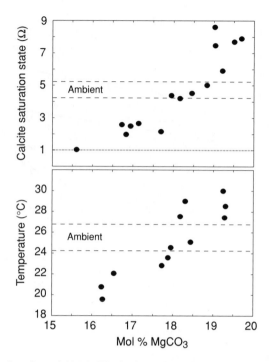

Figure 9.17   Variations in mol % $MgCO_3$ in the red coralline alga *Porolithon gardineri* as a function of calcite saturation state and temperature (after Agegian, 1985, and Mackenzie and Agegian, 1989).

Empirical relationships for the increase of the calcification rates of corals and other calcifying organisms as a function of increasing degree of supersaturation of seawater have been proposed by a number of investigators (Gattuso *et al.*, 1998, 1999; Leclercq *et al.*, 2002; see also Chapter 4).

The degree of supersaturation of surface ocean water with respect to calcite has been decreasing due to the increase in atmospheric CO_2 (Fig. 9.18). In addition, recent observations from open ocean environments have shown a shoaling of the carbonate saturation horizon in several regions of all major ocean basins (Feely *et al.*, 2004). Based on current experimental results and the observed increase in atmospheric CO_2 since the onset of the Industrial Age, one would expect the rate of calcification of marine calcifying organisms to have already declined by 6 to 14% and to decrease further by 11 to 44% owing to a doubling of the atmospheric CO_2 concentration relative to pre-industrial conditions (Buddemeier *et al.*, 2004; Gattuso *et al.*, 1999; Kleypas *et al.*, 1999; Langdon, 2002).

However, no observations of such a decline exist at this time. On the contrary, analyses of drill cores taken from *Porites* colonies along the Great Barrier Reef suggest that the rate of calcification of these corals has increased rather than decreased between 1880 and the later part of the 20th century, a phenomenon attributed to an increase in

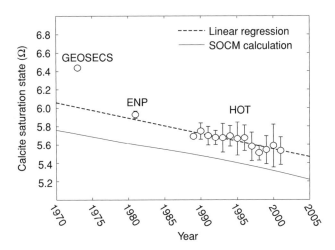

Figure 9.18 Annual surface-water calcite saturation state (open circles) at station ALOHA (HOT, 2004) and single measurements taken at nearby stations during the CO_2 dynamics cruise in 1981 (ENP) (Chen et al., 1986) and the GEOSECS cruise in 1973 (Takahashi et al., 1980). The error bars indicate one standard deviation. Linear regression is based only on the data from HOT. (dashed line: $y = -0.0252t + 56.1284$; $R^2 = 0.82$; $p \ll 0.01$). The dissolved inorganic carbon system and calcite saturation state were calculated at in situ temperature and salinity in each case from either total alkalinity or P_{CO_2} together with total DIC using the program CO2SYS (Lewis and Wallace, 1998) and the carbonate dissociation constants of Mehrbach et al. (1973), refit by Dickson and Millero (1987). From Andersson et al. (2005).

the average annual sea surface temperature during the 20th century (Lough and Barnes, 2000). Similar results showing increased calcification rates toward the present have also been obtained from a coral core taken in French Polynesia (Bessat and Buiges, 2001). It is important to recognize that the rate of calcification is not controlled only by carbonate saturation state and temperature that is discussed in the next section, but also by light, as shown by the data in Fig. 9.16 and Marubini et al. (2001), nutrients, and metabolic and photosynthetic activity of the organism if the organism is autotrophic or dependent on autotrophic symbionts (Ferrier-Pagès et al., 2000; Marubini and Atkinson, 1999; Paasche and Brubak, 1994; Sciandra et al., 2003).

The question that remains to be answered is whether lower carbonate saturation state and decreased rates of calcification will negatively affect corals and other calcifying organisms ecologically. Obvious adaptational advantages given to organisms that calcify include such functions as structural support and protection from predators and desiccation, whereas other hypotheses suggest that calcification may facilitate nutrient and/or bicarbonate uptake through an increase in surface area (Cohen and McConnaughey, 2003; McConnaughey and Whelan, 1997; McConnaughey et al., 2000).

As a result of the lower carbonate saturation state of surface water, the physical strength and structure of the skeletons and shells formed by marine calcifiers may become weaker and more susceptible to environmental stress in general, both natural and anthropogenic. As a consequence, calcareous organisms may not thrive in a world characterized by high CO_2 and low carbonate saturation state and may be at an evolutionary disadvantage in their environment relative to other non-calcifying organisms that occupy the same habitats. For example, it would be anticipated that coral reef habitats might experience a successive transition from being dominated by corals and coralline algae to being dominated by seaweeds and fleshy macro algae. A recent report from the Caribbean goes so far as indicating that the average hard coral cover on reefs in this region has declined from 50% to 10% in the last three decades (Gardner *et al.*, 2003). Although local factors, both natural and anthropogenic, such as disease, storms, temperature stress, predation, over-fishing, sedimentation, eutrophication, and habitat destruction were attributed to the reported decline, the effect of decreasing carbonate saturation state cannot be ruled out as a co-factor. Numerous worldwide observations report similar significant declines in coral reef health, which along with future climate projections has lead to the statement of a global "coral reef crisis" (Buddemeier *et al.*, 2004).

4.2   Surface Water Temperature and Calcification

Different relationships have been reported between the rates of biological calcification and sea surface temperature. A positive linear relationship between the rate of calcification and sea surface temperature for the species of the coral genus *Porites* was postulated by Lough and Barnes (2000) from data from different geographical locations characterized by different annual average sea surface temperatures: the Hawaiian archipelago (Grigg, 1982, 1997), Phuket, Thailand (Scoffin *et al.*, 1992), and the Great Barrier Reef (Lough and Barnes, 2000). In contrast to these conclusions, a negative parabolic relationship between the rate of calcification and temperature was reported for individual coral colonies and coralline algae (Andersson *et al.*, 2005; Clausen and Roth, 1975; Mackenzie and Agegian, 1989; see Chapter 4). The projections on how calcification rates may be affected in the future by increasing temperature significantly depend on such relationships. The results of SOCM suggest that global shallow-water carbonate production could either increase by almost 70% or decrease by approximately 44% by the year 2100 depending on the temperature and saturation state relationships adopted in the model. Similar projections made by McNeil *et al.* (2004), adopting a positive linear temperature and a positive linear saturation-state dependence, suggest that net coral calcification significantly increases between the years 1900 and 2100 by approximately 40% as the positive effect of ocean warming outweighs the negative effect of decreasing carbonate saturation state.

The positive linear relationship observed between coral calcification and the average annual temperature in different geographic locations is most likely due to increased metabolism and/or increased photosynthetic activity by symbiotic zooxanthellae (Buddemeier *et al.*, 2004). In calcifying organisms hosting symbiotic algae, photosynthesis and calcification are strongly coupled. Calcification rates are approximately three to five times higher in light conditions when photosynthesis is active than in the dark when this process is inactive (Fig. 9.16). It has been suggested that primary production in some algae may be limited by CO_2 concentration (Raven, 1993, 1997; Riebesell, 1993). If carbon fixation by symbiotic algae is limited by the availability of CO_2, photosynthesis and consequently calcification may increase in the future owing to increasing concentrations of CO_2 that would then partly counteract the effects of a lower carbonate-mineral saturation state of surface water (Leclercq *et al.*, 2002). However, there is currently no conclusive evidence confirming that the primary production of coral reef ecosystems is limited by the availability of CO_2 and its potential to compensate for a decrease in calcification rate.

Present-day corals and other calcifying organisms have evolved in their environment for thousands of years under relatively constant, but naturally changing, conditions of temperature and salinity, allowing them to adapt and acclimate to the prevailing regional climate conditions. We are now living in a world in which the rate of accumulation of CO_2 in the atmosphere from human activities and, possibly, the rate of temperature rise during the past 30 years are outside the conditions that characterized the Earth's surface for most of the past 740,000 years of the data available for the period before the Industrial Age (Chapter 10, Section 4.1). Human activities have driven the Earth's system away from the environmental conditions that bracket the interglacial-glacial dynamics (Chapter 10) and it is difficult to make projections for the future. However, the response of marine calcifying organisms to increasing sea surface temperature probably depends to some extent on the rate of temperature change. Certainly in a warmer world and one that is warming quickly, it is likely that there will be more bleaching events affecting corals. In addition, if the future temperature change in the tropics were smaller than the average as predicted (Houghton *et al.*, 2001), the direct effect of temperature on the marine calcification rates of shallow, warm water benthic communities (Fig. 9.19)

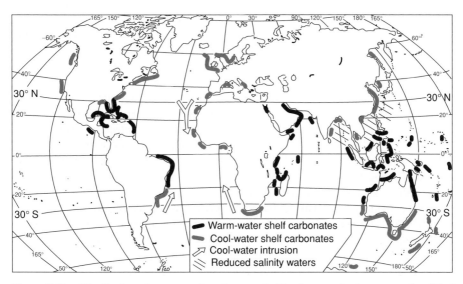

Figure 9.19   Distribution of warm- and cool-water shelf carbonates in the present (modified from Nelson, 1988; by permission of Elsevier, publishers of *Sedimentary Geology*).

of hermatypic corals, green and red algae, benthic foraminifera, and bivalves might be relatively small. It might be the case that the cooler water benthic carbonate communities of bryozoans, mollusks, echinoderms, and coralline algae of predominantly the high and temperate latitudes might be more directly affected by temperature change than those in the tropics. In addition, as CO_2 continues to invade the ocean in the future, the carbonate saturation state of surface waters at high latitudes could drop below that of saturation with respect to aragonite and even calcite (Fig. 5.6). This along with rising sea surface temperatures could have a significant impact on the calcification rates of these cool-water carbonate communities and even their skeletal mineralogy and chemical composition.

5   Present and Future of the Coastal Ocean Carbon System

5.1   Carbonate Sediments and Pore Water

Carbonate sediments in the coastal ocean are mixtures of calcite, aragonite, and magnesian calcites of variable composition, ranging up to more than 20 mol % $MgCO_3$. The carbonate sediment composition shown in Fig. 9.8 consists of 48% calcite, of detrital origin and produced *in situ*, 38% aragonite, and 15% magnesian calcite ($Ca_{0.85}Mg_{0.15}CO_3$). However, among the minerals produced in the coastal zone, aragonite is the most abundant (63%), followed by magnesian calcite (24%) and calcite (13%). As discussed in Chapters 4 and 5, the mineral solubility increases from calcite to aragonite and Mg-calcite with 15 mol % and higher $MgCO_3$. In general, in undersaturated solutions, inorganic aragonite dissolves faster than calcite and the latter dissolves faster than 15 mol % Mg-calcite. However, the dissolution kinetics of biogenic materials depends on a variety of factors, such as skeletal microstructure and not simply mineralogy, and hence is quite complex (Chapter 4).

During early diagenetic reactions on the seafloor, dissolution of carbonate minerals generally follows a sequence based on mineral thermodynamic stability, progressively leading to removal of the more soluble phases until the most stable phases remain. Such selective dissolution of metastable carbonate minerals has been observed both in experimental and natural environments by many investigators (Balzer and Wefer, 1981; Chave, 1962; Halley and Yates, 2000; Leclercq *et al.*, 2002; Morse and Mackenzie, 1990; Neumann, 1965; Schmalz and Chave, 1963; Wollast *et al.*, 1980). According to SOCM calculations, global surface waters of the coastal ocean in the year 2300 would still be supersaturated with respect to calcite, aragonite and a 15 mol % Mg calcite, but undersaturated with respect to higher magnesian calcite compositions (Fig. 9.20A). Thus the calcites with a higher Mg-content could dissolve in contact with surface waters upon death of the organisms forming them. More significant perhaps is the forecast that because reactive organic matter loading of coastal sediments is anticipated to increase in the future, the enhanced decomposition of organic carbon in the pore waters of these sediments will result in a more substantial decrease in carbonate saturation state of the pore waters than overlying surface waters due to the

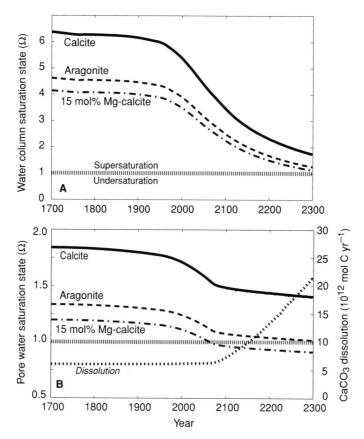

Figure 9.20   Saturation state of the coastal water column and pore water with respect to calcite, aragonite and 15 mol % magnesian calcite at 25°C and S = 35 psu. **A**. Surface water. **B**. Pore water and total carbonate dissolution between 1700 and 2300 (from Andersson *et al.*, 2005).

reactions of equations (9.13) and (9.14). Consequently, magnesian calcite compositions that were once in metastable equilibrium with the pore water or with which the pore water was once supersaturated would have a greater tendency to dissolve. The global pore waters of coastal sediments may even become undersaturated with respect to a 15 mol % Mg calcite and nearly undersaturated with respect to aragonite by the year 2300 (Fig. 9.20B).

The dissolution curve in Fig. 9.20B represents model-calculated values for all the carbonate minerals, starting with the dissolution flux from sediments of 6×10^{12} mol C yr^{-1} at the beginning of industrial time that is about 25% of the total $CaCO_3$ precipitation rate (Fig. 9.8). In coastal ocean seawater that is supersaturated with respect to calcite, aragonite, and only some of the magnesian calcite compositions, dissolution may affect mainly the higher-Mg-calcite compositions, with about 20 mol % or higher

$MgCO_3$, and it may be taking place in microenvironments around the carbonate mineral particles that are close to the organic matter remineralized by bacterial reactions. In sediments, the continuous decomposition of organic matter by microbes, releasing CO_2 and maintaining a low carbonate saturation state within the sediment pore water, is the major factor controlling the extent of calcium carbonate dissolved (Fig. 9.20B), as discussed in Section 3.1.

5.2   Carbonate System Balance and Industrial CO_2

The partitioning of industrial CO_2 between the atmosphere and surface ocean water was discussed in Chapter 5. For the surface ocean water to be buffered against increasing acidity and a lower pH by the rising atmospheric CO_2, dissolution of $CaCO_3$ minerals in the global surface ocean essentially has to balance the net invasion of anthropogenic CO_2, about 160×10^{12} mol C/yr or 2 Gt C/yr. Even if dissolution of $CaCO_3$ in the surface sediment layer were equal to the current rates of $CaCO_3$ production of about 95×10^{12} mol C/yr (Milliman, 1993), the surface water would be still only partially buffered. However, dissolution rates of this magnitude are very unlikely since most of the $CaCO_3$ produced in pelagic environments, about 70×10^{12} mol C/yr, sinks and dissolves at greater depths. Thus because the physical exchange of water between the shallow-water ocean environment and the open ocean is much faster than the turnover time of the ocean, substantial dissolution of metastable carbonate minerals in the shallow-ocean region is necessary to produce sufficient alkalinity to counteract any changes in the surface water chemistry owing to increasing atmospheric CO_2. Current estimates of $CaCO_3$ dissolution in the global coastal ocean range from 6.7×10^{12} mol C/yr (Milliman, 1993) to 10×10^{12} mol C/yr (Langdon, 2002), the higher estimate corresponding to about 7% of the average net anthropogenic invasion of CO_2 into the global ocean during the 1980s, at the rate of 160×10^{12} mol C/yr (Sarmiento and Gruber, 2002). Direct measurements of carbonate dissolution range from 0 to 13.7 mmol $CaCO_3$ m^{-2} day^{-1}, which corresponds to 0 to 140×10^{12} mol C/yr, if extrapolated to the entire shallow-water ocean environment (Langdon, 2002). Dissolution of this magnitude is more than 5 times the amount of $CaCO_3$ produced annually within the global coastal region (24.5×10^{12} mol C/yr, Fig. 9.8; Milliman, 1993) and it implies a substantial loss of calcium carbonate minerals from reef structures and sediments. Continued increase in atmospheric CO_2 and the lowering of the saturation state of surface water would result in a relatively faster dissolution of magnesian calcites with their higher Mg-content, thereby changing somewhat the mineral composition of the sediment toward a higher fraction of low-Mg calcites and pure calcite.

As can be surmised from the discussion above, the observed changes in the carbonate saturation state of the world's oceans, including its surface coastal waters and the model predictions of saturation state for the future, do not bode well for reefs and other carbonate ecosystems. In addition, because the effects of temperature on individual calcifying organism and communities is still not well established, it is difficult to conclude that these organisms will acclimate rapidly enough, particularly if the future sea surface temperature changes are in the upper range of predictions for the future

(Chapter 11). Many corals and other marine calcifying organisms are currently living near the upper threshold of their temperature range. Thus the problem of changing carbonate saturation state of the world's coastal and open ocean waters, ocean acidification, and the consequences of increasing temperature are of concern to the ecological well-being of the coastal ocean and its carbonate communities.

We are now in a position in this book to discuss in more detail the behavior and evolution of the natural carbon cycle through geologic time (Chapter 10). This in turn sets the stage for further discussion of the human influences on the carbon cycle during the past 200 to 300 years of the industrialized era known as the Anthropocene and on into the future (Chapter 11).

Chapter 10

Natural Global Carbon Cycle through Time

In this chapter we review the evolution of the long-term behavior of the global carbon cycle through geologic time and some events of biological evolution and climatic change relevant to that behavior. A discussion of glacial-interglacial changes in the carbon cycle follows to provide background of carbon system behavior prior to major human interference in the carbon cycle. We utilize and emphasize portions of the previous chapters to present a coherent story of trends in the carbon cycle and events that took place and led to major reorganizations of the cycle. This material provides background for discussion in Chapter 11 of the modern carbon cycle and its drivers of nutrient N and P since major human interference in the cycle, a period of time that has become known as the Anthropocene.

1   The Hadean to Archean

Figure 10.1 illustrates schematically some important trends and events in biogeochemical and physical features of the Earth's surface environment during the Hadean and Precambrian. As suggested by Watson and Harrison (2005) and others even earlier, in the Hadean, roughly 200 million years after the Earth formed, the Earth had already settled into a pattern of crust formation, erosion, and sedimentary recycling similar to that recorded in the preserved rocks from the known future eras of plate tectonics. The hydrosphere had begun to develop in contact with an atmosphere devoid of O_2 and enriched in CO_2 and possibly containing organic gases, as discussed in Chapter 2. It also may have contained substantial quantities of hydrogen, leading to the efficient production of organic compounds (Tian *et al.*, 2005). Once temperatures of the Hadean Earth

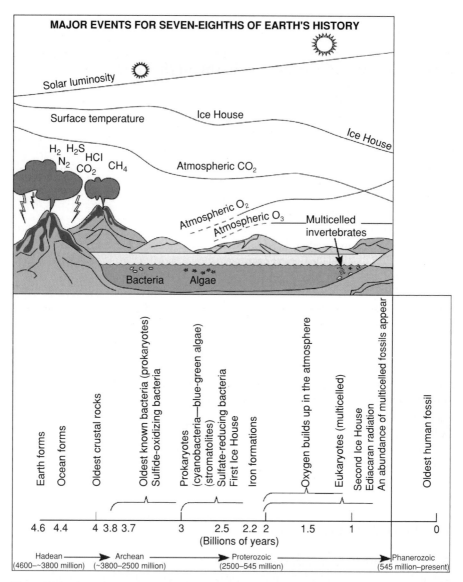

Figure 10.1   Some important atmosphere, ocean and biological trends and events in the history of Earth's surface environment during seven-eighths of geologic time. As free O_2 accumulated in the atmosphere, ozone (O_3) built up in the stratosphere and organisms became less exposed to intense solar ultraviolet radiation (modified from Broecker 1985). Geologic time scale is given in Figs. 1.1 and 7.1.

had fallen to the critical temperature of H_2O, water vapor could begin to condense into rain containing substantial concentrations of the acid gases of carbon dioxide, hydrogen chloride, and sulfur and nitrogen species dissolved in it and creating strongly acidic rain. Much of the H_2O originally present in the primordial atmosphere as water vapor would have condensed to liquid form by the time the planet had cooled to about 200° to 300°C. Thus the major primordial oceans appeared in a hot, chemically reducing environment. Their composition was set by reaction of hot acidic rains with the early, mainly basalt-like crust of the planet. It is likely that nearly 80% of the crust formed within 200 million years of the solidification of the mantle. The reaction of hot, acidic rains with the early crust led to the formation of dissolved and solid weathering products that were transported to the primordial ocean basins by rivers and groundwaters. The solid products of weathering, made of metal-aluminosilicate minerals, were sedimented owing to gravity and accumulated in the oceans as sedimentary deposits of sand, silt, and clay. The dissolved constituents were stored briefly in the ocean and subsequently entered into inorganic reactions removing them in part from the ocean, the removal rate depending on the residence time of the dissolved constituent in the ocean. Although some investigators (Chapter 2) believe the early oceans were similar in composition to a soda lake enriched in dissolved sodium and carbonate, it would be difficult to maintain such a composition of the bulk of the oceans for too long a period of time because of the necessity to rid the atmosphere of the substantial amounts of HCl that must have been present in the primordial atmosphere as attested to by the composition of the acid volatiles. This HCl would have reacted with the early crust, as in the reaction between sodium-feldspar (Na-aluminosilicate) and hydrochloric acid, producing a clay mineral (kaolinite), dissolved or solid silica, and aqueous metal and chloride ions:

$$2NaAlSi_3O_8 + 2HCl + 9H_2O = Al_2Si_2O_5(OH)_4 + 2Na^+ + 2Cl^- + 4H_4SiO_4$$
(10.1)

Reactions such as (10.1) and those with K, Ca, and Mg-bearing silicates led to an early chloride or NaCl ocean with a salinity that was perhaps nearly twice that of present-day seawater of 35 (Chapter 2). The conclusion of a higher salinity ocean in the Hadean-earliest Archean stems from the fact that there are no bedded evaporites in the oldest Archean sediments and the Na, Ca, SO$_4$, and Cl salts of these deposits may have resided in the primordial oceans in dissolved form. The pattern of evaporite occurrence in the geologic record suggests that there was no continuous recycling of these rock types (Chapter 7).

The carbon chemistry of the Hadean-earliest Archean ocean was significantly different from that of the later part of the Precambrian and of course that of the modern oceans. At the time when the Hadean oceans were at a temperature of close to 100°C, their water was moderately acidic with a pH of about 5.6, if atmospheric P_{CO_2} were as high as 10 atm, 6.1 if it were 1 atm, and 6.7 if it were 0.1 atm (Fig. 2.5). Both total alkalinity and the DIC concentrations of Hadean and probably early Archean seawater were higher than those of the rest of geologic time. Total alkalinity may have approached 30 milliequivalent/liter and DIC 130 mmol/liter in the Hadean ocean falling to concentration levels in the Archean more similar to the modern values of several mmol/liter. By the end of the Hadean, the oceans had likely attained surface water temperatures below 100°C, allowing the earliest organisms to evolve and proliferate (Brock, 1973). The appearance of a record of the earliest presence of life on the planet is found in ^{13}C-depleted graphite carbon found in 3.9 billion-year-old rocks and at this time or shortly before, carbon began to accumulate in the sedimentary organic carbon reservoir.

The exact nature of the first organism is not known but certainly early in Earth's history the Archaea and Bacteria arose, represented by the extreme halophiles and thermophiles and the methanogens (see Chapter 2, Fig. 2.6). This event was followed closely by the appearance of the Bacteria represented by the green and purple sulfur bacteria that oxidize reduced sulfur to sulfate (modern example: genus *Chlorobium*). These bacteria were responsible for the first additions of oxidized sulfur to the waters of the Archean Earth. The antiquity of bacterial sulfate reduction is necessarily coupled to the emergence of the photosynthesizing sulfur bacteria because of the need for a sulfate reservoir with sufficient quantities of sulfate that would be used by the sulfate respirers (modern example: genus *Desulfovibrio*). The sulfate-reducing bacteria take dissolved sulfate from water and reduce it to sulfide; in the process, as they oxidize organic matter back to CO_2, H_2O, and nutrients, elemental sulfur and the mineral pyrite (FeS_2) may be formed. Dissimilatory sulfate reduction can be conceived of as an adaptive reversal of the pathway of sulfide oxidation by the photosynthetic sulfur bacteria in which H_2S and CO_2 are restored to the environment (Holser *et al.*, 1988). On the early Earth devoid of oxygen, the Archaea and Bacteria probably flourished in the harsh environmental conditions of the time. In the modern world, both the sulfur-oxidizing and sulfate-reducing bacteria live in habitats that are anaerobic such as fetid marine muds, as do the methanogenic bacteria. These organisms are also found associated with hydrothermal springs on the modern sea floor, an environment in which they may have originally evolved. As free oxygen arose in the Precambrian, the bacteria living at the surface and that could not tolerate oxygen in the environment retreated to anaerobic environments. Photosynthesising Cyanobacteria ("blue-green algae") responsible for early oxygen production and eventually for free O_2 in the atmosphere probably evolved more than 3.0 Ga ago. Photosynthetic microbial mats preserved in cherts have been recognized from the 3.14 Ga old Buck Reef Chert of South Africa, indicating that photosynthetic organisms had evolved by this time and were living in a stratified ocean supersaturated in dissolved silica (Tice and Lowe, 2004). Thus certainly by 3.0 billion years ago, light-powered, biologically driven, coupled carbon and sulfur cycles had emerged.

Precipitation of a carbonate mineral from the early oceans probably started concomitantly with the earliest biological evolutionary events and perhaps before. The presence of calcite limestones in the scant Archean record of the rocks is some evidence that the mineral precipitated was calcite. Sumner (1997) concluded that Archean deep subtidal seawater was probably supersaturated with respect to calcite and that *in situ* inorganic precipitation of calcite on the seafloor was responsible for formation of the Archean limestones. In addition, a layered carbonate structure termed stromatolite is first seen in limestones of the Archean and becomes a pervasive type of carbonate rock in the early Proterozoic. These carbonate rock stromatolite structures appear to be the result of inorganic precipitation of calcium carbonate from seawater (Grotzinger and Rothman, 1996). Thus it is distinctly possible that the oceans were at least saturated, and perhaps supersaturated, with respect to calcite throughout their volume. The formation of the stromatolite structures could also have involved oxygen-producing Cyanobacteria that through their photosynthetic mechanism would draw down aqueous CO_2 during organic productivity and increase the carbonate saturation state of seawater leading to the precipitation of calcium carbonate. Stromatolite structures become more common in late Archean and Proterozoic rocks, perhaps indicating the proliferation and expansion of the Cyanobacteria into the environment as the oxygen they produced went to oxidize first reduced gases in the atmosphere and reduced substances on land and in the ocean, but eventually began to build up in the atmosphere as free O_2. With the production of free atmospheric O_2, the initial ozone shield of the planet could start to develop due to the net reaction $3O_2 = 2O_3$. The development of the ozone layer would enable the Cyanobacteria to expand into the environment from areas in which they were sheltered from the intense UV radiation on the Archean Earth, thus generating more O_2 and enhanced deposition of organic carbon on the early Earth. Although the causes are still strongly debated, it appears that atmospheric O_2 concentrations arose fairly rapidly about 2.2 to 2.6 Ga ago (Petsch, 2004; Rye and Holland, 1998).

If seawater in the Hadean and early Archean were saturated with respect to calcite and probably even supersaturated, at the high DIC and total alkalinity concentrations mentioned earlier, there would have been little dissolved calcium remaining in seawater. Both the storage of carbon in the inorganic carbon reservoir of limestones and in the organic carbon reservoir of sedimentary organic matter led to the fall of atmospheric CO_2 concentrations during the Precambrian. The decrease in concentrations was controlled to a significant degree by the balance between the weathering of silicate minerals on land that consumed atmospheric CO_2 and the production of CO_2 during the subduction of carbonates on the seafloor and their reaction with silicate minerals according to:

$$CaCO_3 + SiO_2 = CO_2 + CaSiO_3 \qquad (10.2)$$

and the subsequent venting of the CO_2 to the atmosphere by volcanic processes. In addition, the decomposition of sedimentary organic matter in subducting sediments and the process of equation (10.2) in deep sedimentary basins led to the release of CO_2 to the atmosphere (see also Chapter 2).

2   The Archean to Proterozoic

As atmospheric CO_2 concentrations fell during the Precambrian, they began to exhibit oscillations. These oscillations extend into the Phanerozoic and appear to be correlated with climatic change. Extended times of relatively high CO_2 concentrations are warm periods of time and have been called Greenhouses or Hot Houses, whereas extended periods of low atmospheric CO_2 concentrations are cold periods of time and have been called Ice Houses.

It is likely that atmospheric CO_2 was relatively low during the continental-scale glaciation of 2.5 Ga ago, a time known as the First Ice House when atmospheric CO_2 and temperatures were low enough for a continental glaciation to occur. What caused the Earth to be driven into this Ice House state? This is a difficult question to answer but certainly it involves the carbon and water cycles of Earth. One possibility is that the cause was related to the intensity of plate tectonic motions and that just prior to the First Ice House, plate tectonic activity was ebbing and sea level falling because of decreasing accretion rates of ancient mid-ocean ridge systems. This resulted in less CO_2 being emitted to the atmosphere because of metamorphism and volcanism and more being consumed in weathering the continents in existence at the time, lowering atmospheric CO_2 levels. The planet cooled because of a less effective greenhouse and the atmosphere became drier. Because solar luminosity was about 20% lower than at present during the First Ice House, the effective radiating temperature of Earth at that stage was colder by several degrees Celsius than that of today. Recovery from this Ice House to a more equable climate probably involved an increase in plate tectonic activity leading to increased accretion rates of mid-ocean ridges, their increase in volume, and a rise in global sea level, flooding the continental margins and interiors. This led to an imbalance between the production of CO_2 by volcanism and consumption of CO_2 by weathering and a rise in atmospheric CO_2. As the planet began to warm, more water vapor evaporated into the atmosphere, giving rise to an increase in atmospheric water vapor, thus strengthening the greenhouse effect, a positive feedback to the initial warming. With higher temperatures and rainfall, weathering on land would increase and moderate the rise of atmospheric CO_2, a negative feedback to the warming.

The Proterozoic Eon from 2.5 to 0.542 Ga ago marks the transition in time between the dominantly anoxic world of the Archean and that of the oxygen-rich world of the Phanerozoic (Figs. 1.1 and 7.1). During the early part of the Proterozoic of 2.2 to 1.6 Ga ago, large volumes of sedimentary rock rich in iron associated with silica, carbonate, and sulfide, known as the banded iron formations (or BIFs, Fig. 7.2), were deposited probably during an extended Hot House when atmospheric CO_2 concentrations were relatively high and there was some free O_2 in the atmosphere. The banded iron formations are unique sedimentary deposits, never to be found in abundance again in younger rocks. The origin of these deposits is still controversial and there are at least 11 hypotheses for their origin. It is likely that the accumulation of the iron formations required low oxygen concentrations so that much of the global ocean of the time was anoxic, particularly the deeper waters of the ocean where Fe^{2+}, H_2S, and HS^- could accumulate. Also, without silica-secreting organisms like diatoms and radiolarians, the oceans at this time were probably supersaturated with respect to amorphous silica (opal-A, of solubility approximately 2×10^{-3} mol Si/kg in seawater at $25°C$). The upwelling of these deep ocean waters rich in dissolved iron, silica, reduced sulfur, and DIC into the surface ocean environment containing some dissolved oxygen could have led to the precipitation of the variety of iron minerals as laminations and bands found in the iron formations: iron oxide, iron carbonate, iron sulfide, and iron silicate. The mineral composition formed would depend on the redox and pH conditions of the seawater at the site of iron mineral deposition and the concentrations of reduced iron and sulfur, dissolved silica, and DIC in the seawater.

The cause(s) of the end of the banded iron formations, like their origin(s), is not well known, but it may have resulted from progressive oxidation of deep ocean interior waters or, as Poulton *et al.* (2004) suggest, continued ocean anoxia after final deposition of the banded iron formations and the transition to highly sulfidic bottom waters. Sulfate concentrations in seawater prior to 2.2 Ga ago were very low, on the order of less than 1 mmol/liter and perhaps less than 200 μmol/liter. Calculations by Kah *et al.* (2004) based on the sulfur isotopic composition of sulfate found in marine carbonates indicate that low sulfate levels of 1.5 to 4.5 mmol/liter, or 5 to 15% of the modern value of 28 mmol/kg, persisted in the world's oceans for more than 1 billion years after the initial oxidation of the Earth's near-surface environment 2.2 billion years ago. The sulfidic bottom ocean waters could have resulted from the rise in atmospheric O_2 approximately 2.2 Ga ago leading to enhanced sulfide weathering on land and an increased flux of sulfate to the ocean. This increased sulfate flux resulted in increased rates of sulfate reduction in the ocean and hence higher dissolved sulfide concentrations and consequently removal of dissolved Fe^{2+} as ferrous sulfide (pyrite, FeS_2) from the ocean, the source of iron for the banded iron formations. The sulfidic and reducing ocean conditions could have persisted into the Neoproterozoic until a second major rise in atmospheric O_2 occurred due to enhanced eukaryotic diversification between 0.8 and 0.58 Ga ago that might have led to a greater oxygenation of waters throughout the ocean basins. These changes would also have major additional implications for the global carbon cycle. Low concentrations of sulfate in seawater and high sulfidic levels forming by the reduction of sulfate imply higher DIC and total alkalinity concentrations, perhaps lower Ca^{2+} concentrations because of the supersaturation of seawater with respect to $CaCO_3$ at a higher total alkalinity. The sulfidic and suboxic to anoxic seawater might have had a lower pH, and higher CH_4, NH_4^+, and PO_4^{3-} concentrations and also have been in contact with higher reduced sulfur gas and methane levels in the atmosphere. Thus the seawater and atmospheric conditions, which began in the Hadean-Archean, may have persisted for a long time in the Precambrian reflecting the protracted nature of the oxygenation of Earth's atmosphere-ocean-sediment system. A persistently sulfidic Proterozoic atmosphere may have delayed the establishment of eukaryotic life on land due to toxic levels of H_2S in the atmosphere, as was hypothesized by Kump *et al.* (2005).

One recent suggestion for the slow development of an oxygenated ocean is that of Falkowski *et al.* (2000): a geochemical limitation or "bottleneck" that was imposed on the ocean-atmosphere system by feedbacks between organic carbon burial, net oxygen evolution, and the beginnings of a global nitrogen cycle that involved the important redox reactions of nitrogen fixation, nitrification, and denitrification. The argument is this: under anoxic conditions ammonium (dissolved NH_3 rather than molecular nitrogen, N_2) would be abundant in the oceans and as oxygen rose, the processes of nitrification and subsequent denitrification would have rapidly removed fixed nitrogen from the oceans. If nitrogen were the limiting or controlling nutrient, its consequent scarcity in the ocean due to denitrification would impose a strong constraint on a continuous rise in atmospheric O_2, in essence because of depriving the marine oxygenic photoautotrophs of an essential nutrient. This in turn would slow oceanic productivity and the rate of burial of organic carbon in sediments and thus the rate of accumulation of O_2 in the atmosphere. In other words, the oceanic system went through an extended nitrogen-limited phase during which time export production of organic carbon was limited and hence the buildup of atmospheric O_2 was slow. Falkowski and colleagues envision that the delay in development of a fully oxygenated ocean may have taken only several hundred million years since the appearance of oxygenic photosynthesis, but conceivably it could have taken longer to establish a global nitrogen cycle similar to that of the modern system. Although there is no evidence at present to support this hypothesis of a nitrogen cycle greatly different from the present, such slower evolution would accord with the recent ideas on a sulfidic interior ocean for much of the Proterozoic.

The emergence of eukaryotes (Fig. 10.1) is a milestone in the history of the organic world, although its timing is still questionable. The earliest forms of eukaryotes were single-celled organisms and were followed by the emergence of multi-cellular algae. The eukaryotes were more efficient than the Cyanobacteria at generating oxygen and their appearance led to a more rapid and important buildup of oxygen in the atmosphere and probably increased accumulation of organic carbon on the sea floor. Eventually these organisms developed into two broad groups; one evolved into the plants and the other into the animals.

The modern carbon cycle may not have taken root until well into the Neoproterozoic. The potential rise of Ca^{2+} concentrations at the close of the Proterozoic could have been one condition necessary for the Big Bang of Organic Evolution at the Precambrian-Cambrian boundary, the appearance of carbonate and calcium-phosphatic shelled organisms of nearly all major extant phyla. Whatever the case, as iron formation deposition abated and the system of ocean-atmosphere began to be more fully oxygenated, the ocean carbon cycle took on an appearance more like that of the modern oceans, where organic carbon is produced in the surface ocean and much of the dead sinking organic matter is oxidized in its transit to the sea floor by bacteria requiring oxygen for their metabolism. Most of the mass of the reduced organic and oxidized inorganic carbon reservoirs found in sedimentary rocks had been established by the end of the Proterozoic or even earlier.

Near the end of the Proterozoic, the first metazoans appeared. The earliest metazoan fossils, collectively called the Ediacaran fauna, were first found more than 50 years ago in Neoproterozoic rocks from the Ediacara Hills in Southern Australia. These organisms seem to have been floating, gelatinous forms with flat, quilted and intricately intertwined textures, and were probably large protoctists and animals. They apparently lived mainly in the subtidal zone of the sandy beaches that characterized parts of the Neoproterozoic sea. Similar soft-body forms have been found in similar age rocks in England, Greenland, Siberia, Southwest Africa, and 20 other locations. The Ediacarans may have been ancestors to animals preserved in the Phanerozoic Early Cambrian sediments (the Burgess Shale popularized by Steven Jay Gould, 1991) or they may have been a false start in organic evolution.

Earth's environment was substantially cooler 800 to 550 million years ago and the Second Ice House occurred at the end of the Precambrian. At this time ice sheets may have reached the equator. This period of time is now known as the Snowball Earth, when the greenhouse effect was weak and the Earth was so cold that a substantial area of the surface waters of the ocean was most likely frozen. It appears, as suggested previously for the earlier Precambrian glaciation of 2.5 Ga ago, that tectonic changes could have triggered the progressive transition from the previous Hot House climate to that of the Ice House during Neoproterozoic time, culminating in the Snowball Earth climatic conditions (Hoffman *et al.*, 1988; Donnadieu *et al.*, 2004).

3 The Phanerozoic

At the beginning of the Phanerozoic, some 540 Ma ago, plate tectonic activity was relatively less intense, global atmospheric CO_2 concentrations and sea level possibly low, and the climate was cool. The planet was still in the climatic conditions of the Second Ice House. At the beginning of this Era, there was also an organic evolutionary event (the Big Bang of organic evolution) with the appearance of shelled forms of life representing most of the existing phyla on Earth today. For the first time, the rocks contain the fossilized remains of well-preserved, hard body parts. It is not our intention to discuss the subject of organic evolution throughout the Phanerozoic, as this is a detailed and complex subject and one that has been well treated in other books. Suffice it to state that as Phanerozoic time progressed, biodiversity as represented by the number of families preserved as fossils in the rock record slowly increased, but this progression was interrupted occasionally by major, massive extinction events occurring 510, 440, 365, 245, 208, and 65 Ma ago. Each of these events was likely to involve major reorganization of the carbon cycle (see below).

During the Phanerozoic there is a distinct pattern of oscillation of atmospheric CO_2. This pattern was recognized initially in model calculations of the global carbon cycle (Berner and Kothavala, 2001; Berner *et al.*, 1983) and then later supported by proxy records of atmospheric CO_2 found in paleosol (fossilized soils) iron oxyhydroxides and $CaCO_3$, marine sedimentary carbon, the fossil stomatal record of plants, and the boron isotopic composition of carbonate fossils (Fig. 10.2A). Atmospheric CO_2 concentrations rose steeply to relatively high levels, perhaps reaching 7000 ppmv, approximately 500 Ma ago and then fell to low concentrations 250 to 350 Ma ago. Concentrations then rose again into the Middle Mesozoic, about 100 to 150 Ma ago, and perhaps reached levels of 2500 ppmv and then fell again somewhat erratically to the levels of the Pleistocene and Holocene Epochs of 180 to 280 ppmv. The modeled Phanerozoic long-term oscillations are a result of the interplay between the outgassing of CO_2 and weathering changes consuming CO_2, for example, during uplift of mountain ranges and the fall of sea level. The large downward trend in atmospheric CO_2 during the Paleozoic probably reflects to some extent the appearance of vascular land plants that accelerated the rates of weathering and introduced a new sink of carbon in bacterially resistant organic matter that accumulated in marine and nonmarine sediments. Later, in the Mesozoic Era, the overall downward trend from the Cretaceous on probably reflects the appearance of the angiosperms with their numerous roots and root hairs, and later on the C_4 grasses, and their role in accelerating weathering rates at a time when global sea level was falling and the land area for weathering was increasing.

Geologists have long recognized that relatively long periods of time spanning tens of millions of years existed in the Phanerozoic when the climate was warm and that these

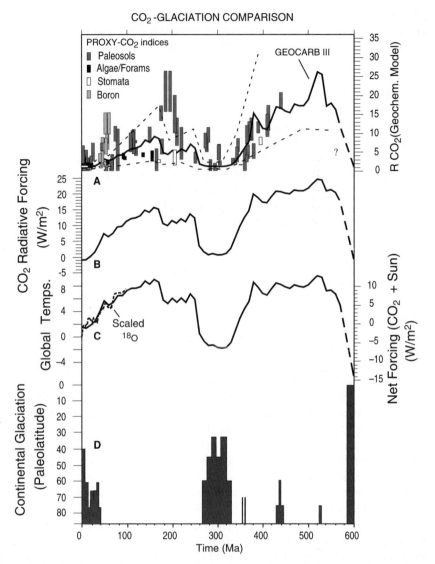

Figure 10.2   A. The model GEOCARB calculated atmospheric CO_2 concentrations ($RCO_2 =$ times pre-industrial concentration of 300 ppmv) through the Phanerozoic compared with the proxy information for CO_2 from paleosols, algae and foraminifera, leaf stomatal density and the boron isotopic composition of seawater. B. Calculated net radiative forcing for CO_2 gas assuming the CO_2 concentrations in (A) and corrected for changes in the Sun's luminosity over time. C. On the left-hand axis, $\delta^{18}O$-derived paleotemperatures for low-latitude, tropical seawater from Veizer *et al.*, 2000. The deep-sea $\delta^{18}O$ curve derived from fossil foraminifera and belemnites for the past 100 million years has been scaled to global temperature variations. On the right-hand axis, the net forcing from both CO_2 and changes in the Sun's luminosity. D. Times of major continental-scale glaciations in the Phanerozoic (Crowley and Berner, 2001; figure by permission of R. A. Berner).

extended periods of time were interspersed with shorter periods of cold and continental-scale glaciations. During the warmth of the Cretaceous, 65 to 145 Ma ago, dinosaurs lived in what is now Alaska and ocean bottom temperatures, which are now about 2°C, were a warmer 15°C. In the Phanerozoic there are at least two Hot House periods (545 to 360 Ma and 240 to 30 Ma ago) and two Ice House periods (360–240 Ma ago and the last one beginning in the Cenozoic 30 Ma ago) (Fig. 10.2C). Geological evidence shows that major continental-scale ice sheets were present during late Neoproterozoic time and during the two Ice Houses of the Phanerozoic (Fig. 10.2D). The glaciation at about 440 Ma ago, which appears to have occurred at a time of high radiative net forcing (Fig. 10.2B), may have been the result of the location of the land mass of Gondwanaland, on which the glaciation occurred, that lay essentially tangential to the South Pole and also to a small negative excursion in atmospheric CO_2 at this time. Thus it appears that the first-order agreement between the continental glaciation record and atmospheric CO_2 levels for the Phanerozoic supports the conclusion that atmospheric CO_2 plays an important role in long-term climatic change.

This conclusion was recently challenged by Shaviv and Veizer (2003; Veizer, 2005). They point out that there is a major discrepancy between, on the one hand, the low-latitude record of sea surface temperatures as deduced from the $\delta^{18}O$ record of calcite and aragonite fossil shells (Chapter 7), high levels of atmospheric CO_2, and the net radiative forcing based on those levels, and, on the other hand, changes in the Sun's output of energy during the mid-Mesozoic 120 to 220 Ma ago (Fig. 10.2B, C). The low-latitude $\delta^{18}O$ record appears to be at variance with other climate records that show high-latitude warming and an absence of continental-scale glaciation (Fig. 10.2C, D). Veizer and co-authors believe changes in the Sun's energy field and the generation of cosmic rays modulate climate through the Phanerozoic and may even be responsible to some extent for modern climatic change. Connections between the climate, variations in solar luminosity, and cosmic ray flux were proposed earlier (Svensmark, 1998). The cosmic ray flux from the Sun depends on solar activity and the intensity of the flux depends on the intensity of the solar and terrestrial magnetic fields that act as a shield for the Earth against cosmic rays. The highly energetic cosmic ray particles, of solar or galactic origin, when they strike the Earth's atmosphere, can potentially generate cloud condensation nuclei that can lead to the formation of clouds and act as a mirror reflecting solar energy back to space. Variation in the cosmic ray flux could lead to more or less cooling of the Earth by modifying its albedo and thus be a principal driver of climatic change over time. The tie between climate and the atmospheric greenhouse gas CO_2 is obvious (Chapter 3); the higher the atmospheric CO_2 levels, the warmer the global temperature, and the lower the levels, the cooler the climate. Thus some of the cooler periods of Phanerozoic time as deduced from the $\delta^{18}O$ record that do not follow the trend between atmospheric CO_2 and climate may be a result of changing cosmic ray flux. However, the correlation between the paleoclimatic record and reconstructed cosmic ray paleoflux has been criticized by a group of 11 scientists (Rahmstorf et al., 2004).

The explanation of the discrepancies or agreement between the modeling results shown in Fig. 10.2 with the proxy data for paleo-CO_2 from pedogenic calcium carbonate and goethite, the distribution of the stomatal pores of fossil C_3 plants, the $\delta^{11}B$ of planktonic foraminifera, and the $\delta^{18}O$ carbonate record probably lies in the future.

However, it should be kept in mind that the $\delta^{18}O$ record may be biased by both diagenesis and changes in seawater and atmospheric CO_2 composition over time (Chapter 7). It certainly is not possible at present to conclude unequivocally that the principal driver for multimillion-year climatic change through the Phanerozoic was changes in the cosmic ray flux, but it also should be kept in mind that the model and proxy paleo-CO_2 trends for this period of time are not well constrained and show a broad envelope of upper and lower bounds of variation, as shown in Figs. 10.2A and 3.7. As the old cliché goes, "the truth may lie somewhere in between" with atmospheric CO_2 being the principal driver for climatic changes through geologic time as modified by the comic ray flux, particularly on shorter time scales.

As atmospheric CO_2 concentrations were varying through the Phanerozoic so was seawater composition (Chapter 7). Several authors have tried to document variations in the CO_2-carbonic acid system and carbonate saturation of seawater for portions of the Phanerozoic (e.g., Arvidson et al., 2006; Lasaga et al., 1985; Locklair and Lerman, 2005; Tyrell and Zeebe, 2004). Figure 10.3 shows the variations in the various seawater CO_2-carbonic acid parameters and the atmospheric CO_2 concentration curve as calculated from the model MAGic (Arvidson et al., 2006). The dynamical model MAGic describes the elemental cycling of the sedimentary materials involving the elements sodium, potassium, calcium, magnesium, chloride, carbon, oxygen, iron, sulfur, silicon, and phosphorus through much of the Phanerozoic. The model incorporates the basic reactions that control atmospheric CO_2 and O_2 concentrations, continental and seafloor weathering of silicate and carbonate rocks, net ecosystem productivity, basalt-seawater exchange reactions, precipitation and diagenesis of chemical sediments and authigenic silicates, oxidation-reduction reactions involving carbon, sulfur and iron, and subduction-decarbonation reactions. The major coupled reservoirs in the model are: shallow and deep cratonic silicate and carbonate rocks and sediments, seawater, atmosphere, oceanic sediments and basalt, and the shallow mantle. We discuss here only the changes in seawater carbon chemistry as derived from this model in light of the atmospheric CO_2 and climatic changes described above.

As atmospheric CO_2 (Fig. 10.2A) and the concentrations of calcium, magnesium and sulfate (Fig. 7.10) varied through the Phanerozoic, so did the carbon chemistry of seawater (Fig. 10.3). Generally, during times of high atmospheric CO_2 (Hot Houses), the carbonic acid content and the total alkalinity of seawater were relatively high and the pH low. DIC (not shown in Fig. 10.3), dominated by HCO_3^-, followed the same pattern as total alkalinity. At intervening times of low atmospheric CO_2 (Ice Houses), the converse was true. Despite the lower pH, the saturation state of seawater with respect to calcite was generally high during Hot House climatic conditions, reflecting the relatively high total alkalinity and calcium concentrations of seawater at this time. The converse was true during Ice House conditions. The saturation state of seawater with respect to dolomite during the Phanerozoic exhibits peaks and troughs that at times accord with changes in the calcite saturation and at times are offset due to changes in the magnesium concentration in seawater. The calculations derived from MAGic reinforce the conclusions from the evaporite fluid inclusion data and other modeling results that seawater chemistry was not constant during the Phanerozoic but varied considerably

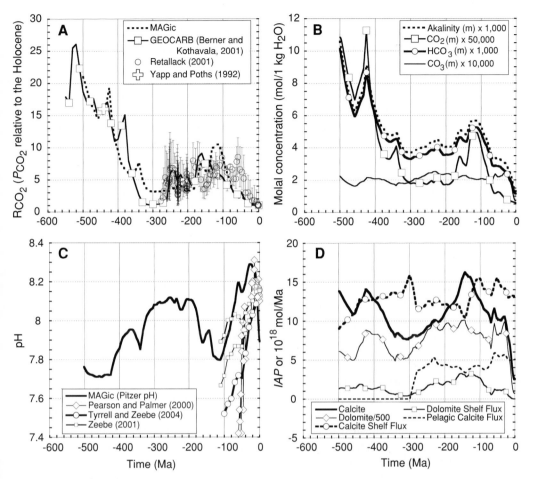

Figure 10.3   Phanerozoic generalized, time-averaged trends in: A. Atmospheric CO_2 as calculated from the GEOCARB and MAGic models and compared with the proxy paleosol data of
Yapp and Poths (1992) and the stomatal index data of Retallack (2001); B. CO_2-carbonic acid
system parameters of seawater; m is molal; C. Paleo-pH compared with the pH proxy data from
boron isotopes (Pearson and Palmer, 2002) and the models of Tyrrell and Zeebe (2004) and Zeebe
(2001); and D. Changes in the carbonate saturation state of seawater with respect to calcite and
dolomite, the calcite and dolomite accumulation fluxes on the shelf and the pelagic calcite flux to
the deep sea. The trends in B, C, and D were calculated using the model MAGic (after Arvidson
et al., 2006, by permission of R. A. Arvidson).

and such variations extend to the CO_2-carbonic acid system and carbonate saturation
state. Figure 10.3 also shows the fluxes of carbonate related to accumulation of calcite
and dolomite on the shelf and pelagic calcite in the deep sea during the Phanerozoic.
Unless abiotic deposition of calcite occurred in the deep sea during pre-Mesozoic time,

there was no deep-sea carbonate deposition during the Paleozoic (Chapter 7): there is no evidence of deep-sea carbonates in the sedimentary rock record or from Paleozoic ophiolite complexes, although the record is biased because all pre-200 Ma deep-sea sediments have been lost to subduction; also, the major pelagic calcifying organisms contributing to post-Paleozoic carbonate accumulation in the deep sea, the planktonic foraminifera, Coccolithophoridae, and Pteropoda, did not appear until the Mesozoic. Both the calcite and dolomite shelf carbonate flux records during the Phanerozoic show peaks and troughs that represent changing seawater carbon chemistry and global first-order changes in sea level. The dolomite shelf flux is particularly interesting since it exhibits a cyclicity that accords with the rise and fall of global atmospheric CO_2 and sea level, as well as seawater total alkalinity and Mg/Ca and SO_4/Ca ratios during the Phanerozoic (Chapter 7). This confirms that the preservation of dolomite in the Phanerozoic carbonate rock record follows a cyclical pattern to some extent (Fig. 7.6) and that this pattern is related to both changes in seawater chemistry and the availability of shallow-water environments for dolomite formation.

The general pattern of cyclicity in atmospheric CO_2 levels and ocean chemistry during Phanerozoic time was interrupted abruptly at certain times, likely resulting in reorganizations of the global carbon cycle and effects on biological evolution. An example of one type of reorganization involves events that take place at times of oceanic anoxia in shelf and abyssal marine environments, such as during the mid-Cretaceous and Late Devonian. For those times, Kump et al. (2005) envision a situation similar to that at the beginning of the development of modern anoxic basins that exist today, such as the Black Sea. In these basins, mildly acidic and sulfidic deep waters are separated from the atmosphere by an oxygenated surface layer at the base of which is a zone of dissolved reduced sulfur concentrations increasing with increasing depth, the sulfide chemocline. The flux of H_2S across the chemocline is regulated by the photosynthetic and chemosynthetic green, purple, and colorless bacteria that exist at the chemocline. If in the past, H_2S concentrations below the chemocline increased beyond a critical threshold during anoxic events, then the sulfide chemocline could have risen to the surface, leading to high fluxes of H_2S to the atmosphere and possibly toxic levels of H_2S in the atmosphere. The surface ocean would then be populated by the bacteria mentioned above. In addition, if this occurred, the abundance of hydroxyl radical (OH*) used in scavenging the enhanced flux of H_2S to the atmosphere from the ocean would be reduced to very low levels. The depletion of OH* would also lead to higher concentrations of CH_4 in the atmosphere, perhaps exceeding 100 ppmv, because the OH* radical is also responsible for the destruction of atmospheric CH_4. Furthermore, the ozone shield may have been partially or totally destroyed because of the reaction of H_2S in the stratosphere with atomic O, leading to depletion of atomic O in the stratosphere which is the most important reactant in the formation of stratospheric ozone. Such a series of catastrophic events could act as a "kill mechanism" (Kump et al., 2005), leading to the mass extinctions and reorganization of the global carbon and other biogeochemical cycles at the end of the Permian, Late Devonian, and in the Cenomanian-Turonian of the Cretaceous. However, as the Earth's history shows, if such events occurred, they did not lead to permanent changes.

4   Pleistocene to Holocene Environmental Change

4.1   The Records of Change

The last Ice House began approximately 30 Ma ago and, during the past 1.8 Ma of the Pleistocene-Holocene record of environmental change, the planet has experienced oscillations in warmth and cold, the latter being associated with the advance of large continental ice sheets, such as those of Greenland and Antarctica today. It is likely that, by 1 to 2 Ma ago, as in the continental glaciation of 300 Ma ago, atmospheric levels of carbon dioxide had fallen enough for changes in solar insolation to affect climate significantly. Analyses of air trapped in bubbles of ice cores and of the water and aerosols contained in the ice have provided us to date with one of the most detailed and informative records of changes in the environment of the atmosphere, ocean, and land during the past 420,000 years, with some information dating back 740,000 years (EPICA Community Members, 2004). Figure 10.4 shows the trends with time of several environmental variables as recorded in the 3500-meter ice core recovered from drilling the glacial ice at Vostok, Antarctica. The 420,000-year record includes the present interglaciation (interglacial stage) plus four previous ones and four continental glaciations (glacial stages). The temperature record from the ice core that reflects the past temperatures of the environment at Vostok exhibits a strong 100,000-year cyclicity, as predicted by the Milankovitch theory of three main orbital periodicities (Imbrie *et al.*, 1984; Milankovitch, 1941): changes in the ellipticity of the earth's orbit, with a period of 100,000 years; the obliquity or tilt of the Earth's axis to the plane of its solar orbit, about 41,000 years; and the wobble or precession of the axis, with a period of about 23,000 years. The records of temperature, atmospheric carbon dioxide, and methane roughly track each other in the data from the Vostok ice core for the past 420,000 years. The latter two variables reflect global atmospheric concentrations of CO_2 and CH_4. The oxygen isotopic record ($\delta^{18}O$) is that of atmospheric O_2 trapped in ice. The latter reflects changes in the oxygen isotope ratio of seawater that was the source of water vapor deposited as snow and consolidated into ice at Vostok. The $\delta^{18}O$ of H_2O in atmospheric precipitation of the cold high latitudes is much lower, -30 to -50 ‰ on the SMOW scale, whereas tropical surface ocean water is $\delta^{18}O \approx 0$ ‰ (Box 6.1; Broecker, 1995, p. 14; Moser and Stichler, 1980). To some extent these values are a record of the air temperature over the ice at the time of formation, and the $\delta^{18}O$ and δD of seawater reflect global temperatures. The dust record shows that the precipitation over Vostok contained more dust particles at the climax of each glacial stage, probably reflecting a more dusty and windy atmosphere during continental glaciations. Carbon dioxide during this whole time interval oscillated in approximately 100,000-year cycles by about 100 ppmv from 180 to 280 ppmv, and CH_4 by about 0.35 ppmv, and both follow the ice core temperature record obtained from the Greenland Ice Core Project (GRIP) since the last interglaciation, the period of time for which the ice cores overlap in age. The 100,000-year $\delta^{18}O$ cycle observed in the ice cores is also seen in the oxygen isotopic composition of foraminifera collected from deep-sea calcareous oozes that serves as a proxy for ice sheet volume and for ocean water surface temperatures. This record goes

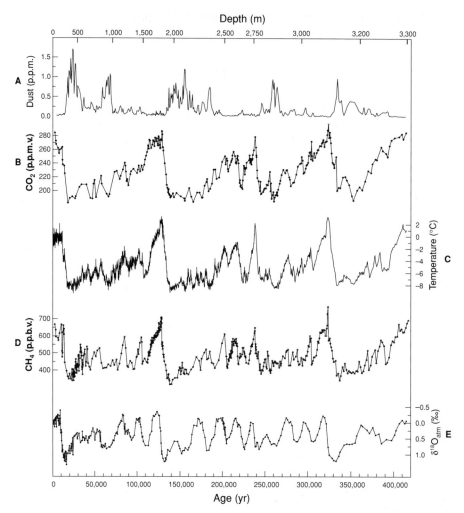

Figure 10.4   Trends in atmospheric CO_2 and CH_4 concentrations, regional air temperatures, dust loading, and $\delta^{18}O$ as recorded in the ice drilled at Vostok, Antarctica. The oxygen isotopic record represents changes in the oxygen isotopic composition of atmospheric O_2 that is related to ice volume, and to some extent it reflects the air temperature over the ice at the time of formation of the ice. Temperature is the deviation from the modern annual average surface temperature at Vostok of $-55.5°C$ (after Petit *et al.*, 1999).

back nearly 740,000 years. In addition, the sedimentary record of major, continental glacial advances and retreats found in Pleistocene glacial deposits of North America, Europe, and Asia also exhibits a 100,000-year cyclicity. Thus the 100,000-year pattern is seen not only in ice core environmental parameters but also in other records from both the terrestrial and marine realms.

What mechanisms control the atmospheric CO_2 oscillations on the 100,000-year and shorter time scales of the Pleistocene? The answer to this question is not well known at present and at least a dozen hypotheses have been proposed to account for the 80 to 100 parts per million changes in atmospheric CO_2. Table 10.1 shows some of the major processes and mechanisms that could affect atmospheric CO_2 on the time scale of interglacial-glacial change. It is certainly the case that whatever mechanisms caused rises in CO_2 from a glacial termination were not necessarily the same mechanisms that caused a decline in CO_2 going into a glacial stage. However, aside from this problem of CO_2 oscillation, and probably even more provoking, is the question of why do atmospheric CO_2 and CH_4 concentrations lie within such a narrow band of variation during glacial-interglacial cycles; for CO_2 about 100 ppmv and for CH_4 about 350 ppbv? As can be seen in Fig. 10.5, the climate system appears to have operated within a relatively constrained domain of atmospheric CO_2 (and CH_4) and temperature for the past 420,000 years. On millennial time scales, changes in atmospheric CO_2 recorded in ice cores are highly correlated with changes in temperature. High-resolution analyses of the ice core records show that there are periods of time when temperature can change relatively sharply without discernible changes in atmospheric CO_2 concentrations, but the converse does not seem to be true.

It is obvious from Figs. 10.4 and 10.5 that in terms of temperature and both atmospheric CO_2 and CH_4 concentrations, the carbon cycle has been well regulated during glacial-interglacial time. This implies that there are mechanisms within the cycle involving carbon exchanges between the atmosphere, ocean, land, and sediments that involve feedbacks between the cycle and climate and the cryosphere. The feedbacks for CO_2 and CH_4 encompass both the ocean and terrestrial ecosystems. The oceans in recent, pre-industrial time were a source of CO_2 to the atmosphere because of heterotrophic respiration and calcification processes (Chapter 9). Based on the historical time scale of glacial-interglacial temperature variations, prior to the human perturbation of the carbon cycle, the Earth was within a few thousand or less years of another glacial cycle, if the periodic pattern of the last 420,000 years were to repeat itself. So what would have happened if there were no human interference in the Earth's environmental systems in the future? To enter the glacial cycle, it would probably require a change in the Milankovitch forcings that should entail a series of events that lead to a reduction of CO_2 and CH_4 in the atmosphere and cooling of the planet. First to consider are the processes in the oceans that could remove CO_2 from the atmosphere.

4.2   Processes Controlling CO_2 in the Ocean

The distribution of CO_2 between the atmosphere, ocean water, and sediments is a dynamic process that is more complex than the individual processes of the gas and mineral solubilities and formation of organic matter that were described in Chapters 5 and 6. The three main controlling processes interacting among themselves are: the solubility pump, the biological pump, and the carbonate pump. The solubility pump simply involves the uptake of CO_2 due to P_{CO_2} differences between the atmosphere and ocean waters and its strength depends on the intensity of the thermohaline circulation and on latitudinal and seasonal changes in ocean ventilation. The ocean takes up CO_2 mainly in the colder temperate and high latitudes and the ocean outgases CO_2 generally in the warmer tropics. For example, the strength of the solubility pump at the equator is

Table 10.1   Summary of some major processes that could affect atmospheric carbon dioxide on the interglacial-glacial time scale

Processes (and controlling factors)	Example references
Coastal and oceanic domains	
Growth of the coastal zone, exchange with open ocean (ocean and ice volumes)	Imbrie *et al.* (1984); Fairbanks (1989)
Water-atmosphere exchange, vertical water exchange, coastal upwelling, thermohaline circulation (temperature, salinity, wind)	Broecker (1982; 1995); Schmitz, Jr. (1995); Manabe and Stouffer (1994); Oaillard and Labeyrie (1994); Yu *et al.* (1996); Fichefet *et al.* (1994); François *et al.* (1997); Smith *et al.* (1999); Kheshgi *et al.* (1999)
Precipitation and dissolution of $CaCO_3$ (temperature, alkalinity, input from land, CO_2 in the system, coral reef and other carbonate-secreting organisms)	Berger (1982); Milliman (1993); Walker and Opdyke (1995); Keir (1995); Nozaki and Oba (1995); Farrell and Prell (1989); Archer and Maier-Reimer (1994); Morse and Mackenzie (1990); Sigman *et al.* (1998); Kleypas *et al.* (1999); Buddemeier *et al.* (2004)
Bioproduction, biological pump, remineralization, storage (nutrient availability, temperature)	References in Rich (1998); Boyle (1988a, b); Anderson and Sarmiento (1994); Broecker and Henderson (1998); Farrell *et al.* (1995); Kumar *et al.* (1995); Sigman and Boyle (2000)
Input of organic matter from land, its remineralization, storage, export to open ocean (land erosion and weathering, in situ processes)	Smith and Hollibaugh (1993); Wollast (1998); Kump and Alley (1994); Milliman and Syvitsky (1994)
CO_2 from oceanic volcanism and spreading zones	Des Marais (1985)
Land domain	
Deglaciated land area (temperature, elevation, ice volume)	Imbrie *et al.* (1984); Fairbanks (1989)
Net primary production (temperature, water, CO_2, nutrient N and P)	Woodwell *et al.* (1998); Sundquist and Broecker (1985)
Changes in biomes (e.g., C_3 and C_4 plants; forests and grasslands) and land-area vegetation cover (temperature, water, zonal latitudinal variation)	Crowley (1991; 1995); Crowley and Baum (1997); Friedlingstein *et al.* (1995); Bird *et al.* (1994); Prentice and Fung (1990); Van Campo *et al.* (1993); Cole and Monger (1994)
Formation and remineralization of humus, recycling of nutrients (temperature, mean lifetime of phytomass, burial of organic matter)	Harrison *et al.* (1995); Houghton *et al.* (1985)
Exchange with the atmosphere (temperature, respiration, denitrification)	McGuire *et al.* (1995); Delgado *et al.* (1994)
Weathering of silicates and carbonates, release of P (temperature, water, state of the postglacial regolith)	Brady (1991); Brady and Carroll (1994); Probst (1992); Munhoven and Francois (1996); Munhoven (1997); Lovelock and Kump (1994); Tiessen (1995); Van Cappellen and Ingall (1996); Gaillardet *et al.* (1999); Mortatti and Probst (2003)
CO_2 from continental volcanism	References in Dawson (1992)

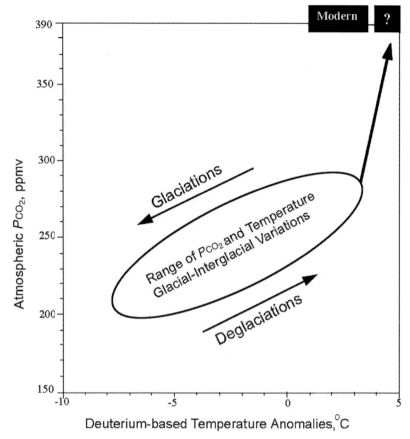

Figure 10.5   Overall range (space within the oval) of atmospheric CO_2 concentrations and deuterium-based isotopic temperature anomalies, obtained by using hydrogen isotopes in water melted from ice, as recorded in the Vostok ice core, Antarctica. Temperature anomaly is the deviation from the modern annual average surface temperature at Vostok of $-55.5°C$. The general directions of change in atmospheric CO_2 and temperature during transitions from glaciations to interglaciations and vice versa are shown as arrows. The arrow in the upper right of the diagram represents atmospheric CO_2 and temperature changes since the Industrial Revolution of 1850 with an indication of the direction of these changes for the future. It should be pointed out that the general changes in CH_4 concentrations in the Vostok core (Figure 10.4) broadly follow those of CO_2 and thus a similar relationship would be observed for CH_4 versus temperature. One can conclude from this that climatic variations in the past 420,000 years have operated within a relatively constrained domain, and we are out of that domain in the modern era (modified from Falkowski *et al.*, 2000, original figure by permission of P. Falkowski).

equivalent to an outgassing of CO_2 of about 0.09 Gt C degree latitude^{-1} yr^{-1}, whereas at N30° and S30°, it is equivalent to approximately 0.02 Gt C degree latitude^{-1} yr^{-1} of uptake (Le Quéré and Metzl, 2004). Carbon dioxide is more soluble in cold waters and sequestration of atmospheric CO_2 in the ocean interior depends on the formation of cold, dense seawater masses principally at the high latitudes of the North Atlantic and around Antarctica in the Southern Ocean confluence. These waters in their transit to the deep become part of the lower limb of the ocean conveyor belt circulation that moves water at depth in the ocean from the North Atlantic into the interior of the deep Indian and Pacific Oceans to upwell and return again to the North Atlantic at intermediate water depths of about 1500 m and shallower as the upper limb of the conveyor belt circulation. A single water molecule that sinks to depth in the ocean takes between 1000 and 2000 years to complete the trip; thus, CO_2 is effectively prevented by this pump from re-equilibrating with the atmosphere on a time scale of hundreds of years. Because of the solubility pump alone, the concentration of carbon at depth in the ocean is 5% higher than at the surface of the ocean because the CO_2 at depth equilibrated with the atmosphere at cold temperatures.

The biological pump involves the process of photosynthesis that lowers the P_{CO_2} in the surface waters of the ocean and thereby promotes the uptake of atmospheric CO_2. Presently, the amount of carbon exported to the deep ocean interior by this pump, where much of it is oxidized by heterotrophic respiration, is not well known but it is approximately 17% of the organic carbon produced each year in net primary production in the ocean of approximately 63 Gt C/yr, or 11 Gt C/yr. It is the process of the biological pump that maintains atmospheric CO_2 at levels of 150 to 200 ppmv lower than they would be if all the phytoplankton in the ocean died (Falkowski *et al.*, 2000).

The final pump is the carbonate pump in which DIC is removed from the surface waters of the ocean by calcifying organisms such as foraminifera, coccolithophorids, and pteropods. The inorganic carbon produced sinks into the deep ocean where much of the production dissolves below the lysocline adding DIC to the deep ocean interior. About 0.84 Gt C/yr as biogenic $CaCO_3$ is exported from the euphotic zone of the open ocean to its interior depths; however because of dissolution in transit to the sea floor, only 25%, or 0.21 Gt C/yr, accumulates on the ocean floor. In the coastal ocean, 0.29 Gt C/yr as biogenic carbonate minerals is produced by calcifying organisms, of this amount 0.17 Gt C/yr accumulate in coastal sediments. The carbonate pump actually leads to the evasion of CO_2 from the ocean to the atmosphere during the calcification process (Chapters 5 and 9).

Changes in the intensity of any of the three pumps described above, as glacial cooling began, could result in changes in atmospheric CO_2 concentrations. Based on the physical or solubility pump alone, it would be expected that as the climate began to cool, more CO_2 would be taken up by the colder waters at the sea surface. A 10°C-temperature decrease significantly increases the CO_2 solubility in seawater: for a temperature decrease from 20° to 10° or from 15° to 5°C, the CO_2 solubility increases by 35 to 39% (K'_0, at the salinity S between 35 and 37, Box 5.1). However, a salinity increase from 35 to 36 or 37, corresponding to a lower sea level at the Last Glacial Maximum, would decrease the CO_2 solubility due to the "salting out effect" only by about 1%, which is much smaller than the opposite effect of the lower temperature. This is where changes in the nature of the biological pump come into play, but it too seems not to be strong enough, no matter how it operates, to explain all of the CO_2 drawdown during a glaciation. As a whole, the atmosphere, land vegetation, and soil humus at the LGM contained 1,300 to 1,400 Gt C less than their carbon content at the end of pre-industrial time, as shown by the reservoir sizes in Fig. 10.7. If all this carbon was stored in the ocean, the dissolved carbon content of ocean water at the LGM would be about 40% higher than at the end of pre-industrial time.

The role of biological and physical processes in changing atmospheric CO_2 concentrations can be briefly summarized by the following range of views on this issue. Broecker and Henderson (1998) concluded that the timing of the rise in atmospheric CO_2 seen at the termination of the past two glacial stages, coupled with the timing of North Atlantic circulation changes relative to the melting of the ice sheets, eliminates any scenarios to explain the rise in CO_2 from the glacial terminations which call on sea level change, North Atlantic nutrient redistribution, or North Atlantic cooling. They argue that the changes in atmospheric CO_2 are tied to changing conditions in the Southern Ocean. It is likely that in some way the fertilization of the Southern Ocean, which is a high-nutrient, low-chlorophyll (HNLC) concentration region of the world's oceans, and productivity enhanced by iron in the dust flux are in part causes for the drawdown of CO_2 during glaciations. This conclusion would accord with the observation from ice cores that the dust flux was low during interglaciations and increased into glacial maxima because of a windier global atmosphere (Fig. 10.4). The increased glacial dust flux may have caused increased nitrogen fixation by planktonic, photosynthetic Cyanobacteria such as the genus *Tricodesmium* ("sea-saw-dust"), allowing for a greater drawdown of CO_2. After the dust flux had dropped to its interglacial stage levels, the fixed nitrogen would be redistributed throughout the ocean due to the respiration of the sinking organic matter of the Cyanobacteria and the relatively long residence time of nitrate in the ocean of several thousand years. Thus major productivity changes would then be global rather that confined to the Southern Ocean, as the NO_3^- mixed through the ocean, concomitant with a slow increase in atmospheric CO_2 over time scales of thousands of years as the dust load was substantially reduced and the system relaxed from the perturbation on productivity.

On the other hand, Kohfeld *et al.* (2005) have argued from analyses of sedimentary records of marine productivity at the peak and the middle of the last glacial cycle that increased iron fertilization of marine phytoplankton in the Southern Ocean, increased ocean nutrient content or its utilization, or shifts in the C:N:P ratio of the dominant plankton types could not account for the 80 to 100 ppmv drawdown of CO_2 during the last glacial stage. They estimate that less than half of the glacial-interglacial variations in CO_2 can be attributed to changes in the strength or other attributes of the biological pump. They also inferred that certain physical processes, such as reduced ventilation of CO_2-rich deep water due to increased sea ice formation or changes in the location or mechanisms of deep water formation, must play an important role in glacial-interglacial variations in atmospheric CO_2.

4.3   Sea Level Rise and the Coastal Zone

Up until this point in the discussion, most of the comments related to controls on the rise and fall of atmospheric CO_2 during the time of the Vostok ice core record have dealt with changes in physical and biogeochemical processes operating in the vast expanse of the open ocean. Yet one must not lose sight of the fact that sea level may undergo rise and fall fluctuations totaling 180 m or so during glacial maxima to total melting of the continental ice sheets. Thus it is possible that processes and mechanisms related to the fall and rise of sea level and the exposure and flooding, respectively, of the continental shelves could play some role in regulating glacial-interglacial fluctuations in atmospheric CO_2. The conclusion of Broecker and Henderson (1998) that the nearly synchronous change of atmospheric CO_2 and Southern Hemisphere temperature, as seen in the Vostok record, apparently preceded the melting of the Northern Hemisphere ice sheets for the past two glacial terminations has led many investigators to discard the coastal oceans as a possible modulator of atmospheric CO_2 fluctuations. This is despite the fact that since the beginning of the rise of atmospheric CO_2 following the Last Glacial Maximum (LGM), the oceans have lost alkalinity due to the accumulation of carbonate sediments on reefs, banks, and shelves of flooded land and that this accumulation must have been accompanied by release of CO_2 to the atmosphere (Chapters 5 and 9). This release rate is significantly more than that required to account for all the additional CO_2 in the atmosphere of approximately 200 Gt of carbon since the last glaciation.

It can be shown that as atmospheric CO_2 and temperature rose since the LGM, the total amount of DIC stored in coastal ocean waters rose and the pH and carbonate saturation state of these waters declined (Fig. 10.6), mainly due to uptake of atmospheric CO_2 by the expanding area of coastal ocean. With the Industrial Revolution, these trends were exacerbated by human activities and will continue to be into the future (Chapter 11). Notice also in Fig. 10.6 that since the LGM up until major human interference in the global carbon cycle, accumulation of calcium carbonate has been growing in the area previously exposed during the LGM low sea level stand. In addition, the coastal ocean was becoming progressively more net heterotrophic during this period of time. Both of these trends in these process mechanisms led to increasing release of CO_2 to the atmosphere from the growing coastal ocean from about 2×10^{12} mol C/yr (0.024 Gt C/yr) in the LGM to 20×10^{12} mol C/yr (0.22 Gt C/yr) in pre-industrial time (Table 10.3). CO_2 releases from a surface ocean layer and coastal ocean are given in Tables 10.2 and 10.3 for the LGM and the end of pre-industrial time. This rate of CO_2 emission is certainly significant in terms of the average rate of accumulation of carbon in the atmosphere since the LGM of about 0.8×10^{12} mol C/yr or 0.01 Gt C/yr and cannot be neglected in terms of the growth of atmospheric CO_2 since the LGM. Despite this significant flux, it should be pointed out that Köhler et al. (2005), using the coupled ocean-atmosphere-biosphere model BICYCLE forced by proxy data, concluded that only about 5% of the rise in atmospheric CO_2 of 80 ppmv during the transition from the LGM at 29,000 to 19,000 years ago to the beginning of the Holocene around 11,000 years before present could be accounted for by coral reef growth. Whatever the case, it is clear that the coastal ocean was a modulator of atmospheric CO_2 change during glacial-interglacial transitions and vice versa. Indeed the coastal ocean and its interactions with the open ocean and atmosphere could be the "canary in the cage" for understanding Pleistocene atmospheric CO_2 fluctuations and for the homeostasis of the natural system.

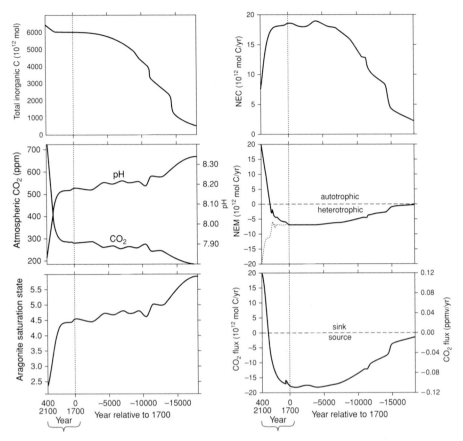

Figure 10.6   SOCM calculation for coastal ocean (left column) total dissolved inorganic carbon (DIC), surface water pH, atmospheric P_{CO_2}, and aragonite saturation state and (right column) net ecosystem calcification (*NEC*), net ecosystem production (*NEP*) and the CO_2 flux resulting from these processes since the Last Glacial Maximum (LGM) with projections into the future. Atmospheric CO_2 is from ice cores and Mauna Loa, Hawaii, record. Depending on the carbon material balance involved in the calculation of *NEC* and *NEP*, coastal ocean waters can act as a net source or net sink of atmospheric CO_2 (Chapters 5 and 9). Since 1700 and particularly since the Industrial Revolution of 1850, the trends in these variables are changing dramatically (see discussion in Chapter 9).

4.4   Role of the Land in the CO_2 Control

Aside from the ocean, terrestrial ecosystems also exchange CO_2 rapidly with the atmosphere and can regulate atmospheric CO_2 concentrations, but unlike the oceans there is no physical or dissolution pump for plants on land. CO_2 is removed from the atmosphere through photosynthesis and stored in organic matter. It is returned to the atmosphere via a number of respiratory pathways that operate on various time scales:

(1) autotrophic respiration by the plants themselves; (2) heterotrophic respiration, in which plant-derived organic matter is oxidized primarily by soil microbes; and (3) disturbances, like fire, in which large amounts of organic matter are oxidized in very short periods of time. On a global basis, terrestrial carbon storage primarily occurs in forests. The sum of carbon in living terrestrial biomass and soils in pre-industrial time is approximately four to five times greater than the CO_2 in the atmosphere, but the turnover time of terrestrial carbon is on the order of decades. Direct determination of changes in terrestrial carbon storage has proven extremely difficult to measure even today. The contribution of terrestrial ecosystems to carbon storage in modern times is inferred from changes in the concentrations of atmospheric gases, especially CO_2 and O_2, their isotopic composition, inventories of land-use change, and models. However, most of the models require accurate knowledge of the oceanic uptake of CO_2. To determine how the changing climatic conditions of the glacial-interglacial past have affected the types and distribution of phytomass on land, their carbon storage and loss, and their response and feedback to climatic change, models of the distribution of biomes for the Last Glacial Maximum and previous interglaciation have been developed and compared with proxy data. These data include, for example, pollen from lacustrine sediment cores and ocean $\delta^{13}C$ from foraminiferal calcite in deep-sea cores.

The most striking changes seen in the LGM continental vegetation distributions as distinguished from modern time are the poleward shifts and compression of the northern midlatitude biome belts, a major reduction in the area of taiga, a major redistribution of the tundra with much of its present area covered by ice, fragmentation of the African rain forest, and expanded tropical rain forest on the exposed continental shelves of the low sea level stand during LGM time (Crowley, 1995; Prentice and Sykes, 1995). Carbon storage calculations indicate that the continental biosphere stored about 200 to 400 Gt less organic carbon at the LGM than in pre-industrial time, and added to these numbers is a deficiency of an additional 600 to 1200 Gt C in soil humus (Ajtay et al., 1979; Friedlingstein et al., 1995; Prentice and Fung, 1990). The peat-land area was less extensive during the LGM primarily because of ice coverage, whereas the present-day estimate of peat is about 150 Gt C (Fig. 1.5). On balance, an increase in organic carbon storage on land or its deficit at the LGM was 600 to 1600 Gt C. An increase of this magnitude would accord reasonably well with the documented increase of ^{13}C of ocean water DIC, as observed in the $\delta^{13}C$ signature changes in calcite foraminifera of the deep sea during the LGM to present interglacial transition. Thus it is likely that the terrestrial biosphere was a net sink of several hundred Gt C released from the ocean during the climatic transition from the LGM to pre-agricultural time, despite the fact that the rising sea level decreased somewhat the land available for growth after it was covered by the rising sea level.

The changes in the main reservoirs and fluxes between the LGM and the end of pre-industrial time or the beginning of the Anthropocene are shown in the cycle diagram in Fig. 10.7. The data for this figure are summarized in Tables 10.2 and 10.3. It may be noted that the land reservoir of organic carbon consisting of the vegetation and soil humus is, as explained above, smaller at the LGM than at pre-industrial time. Also smaller at the LGM are the carbon transport fluxes from land to the oceanic coastal

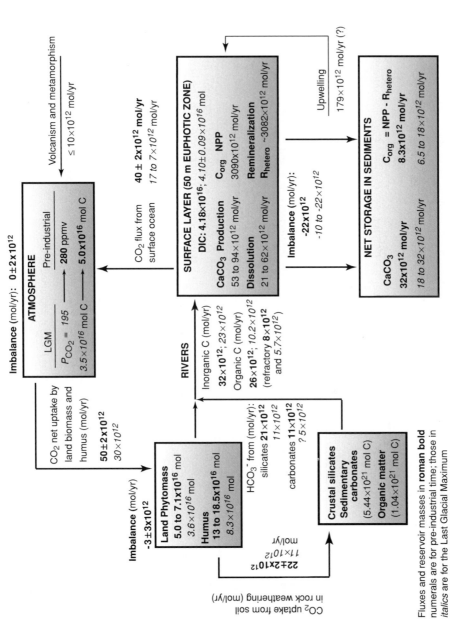

Figure 10.7 Carbon in the atmosphere, and land and oceanic reservoirs at the LGM and in pre-industrial time (modified from Lerman and Mackenzie, 2005).

Table 10.2　Conditions in pre-industrial time and at the Last Glacial Maximum used in estimation of CO_2 release from a surface ocean layer

Time	Temperature (°C)	River flow (10^{16} kg/yr)	Atmospheric CO_2 mass (10^{16} mol)	CO_2 uptake in weathering	CO_2 net land storage	River input to the ocean		$CaCO_3$ net sediment storage (y)	C_{org} net sediment storage (z)
						C_{inorg} (v)	C_{org} (w)		
				(10^{12} mol/yr)					
LGM	9	2.64	3.48	*11*	*30*	*23*	*10.2*	18 to 32	6.5 ± to 18
Pre-industrial	15	3.74	5.00	*21*	*50*	*32*	*26*	32.1	8.3

Note: Pre-industrial and LGM data in italics are from Lerman and Mackenzie (2005). Parameters v, w, y, and z defined in Chapter 5.

Table 10.3　CO_2 flux from ocean to atmosphere due to $CaCO_3$ precipitation and imbalance between production and respiration of C_{org} (Fig. 10.7)

Initial atmospheric CO_2 (ppmv)	Ocean surface temperature (°C)	Initial alkalinity ($A_{T,0}$) (10^{-3} mol/kg)	Initial DIC_0 (10^{-3} mol/kg)	C_{org} net storage rate (10^{12} mol/yr)	$CaCO_3$ net storage rate (10^{12} mol/yr)	CO_2 flux from surface ocean (10^{12} mol/yr)	CO_2 flux from coastal zone (10^{12} mol/yr)
195 (LGM conditions)	25	2.808	2.190	6.5 to 18	18	15.7 to 5.7	~5 to 0.6
	5	2.582	2.255	6.5 to 18	18	17.3 to 7.0	
195 (Pre-industrial conditions)	25	2.808	2.190	8.3	32.1	35.4	—
	5	2.582	2.255	8.3	32.1	39.3	
280 (Pre-industrial conditions)	25	2.808	2.293	8.3	32.1	38.0	21 ± 5
	5	2.582	2.332	8.3	32.1	42.0	

Note: Rates of $CaCO_3$ net storage (production less dissolution) are for a 50 m surface ocean layer, approximating the euphotic zone (water mass $M_w = 1.854 \times 10^{19}$ kg; Fig. 5.10). Input from land to the euphotic zone and storage rates from Table 10.2. For the LGM, the higher flux value corresponds to the lower storage rate of C_{org}. Computation of the fluxes is explained in Chapter 5.

zone. This is accounted for by a combination of such factors as a smaller land surface area, smaller river volume discharge at a lower temperature, and slower rates of organic carbon oxidation and mineral-CO_2 weathering reactions. The source of CO_2 for mineral weathering is believed to be its production in soil by remineralization of organic matter (Chapter 8). The weathering reactions produce DIC and the bicarbonate-ion, HCO_3^-, in river water from the CO_2 reacting with silicate minerals as well as carbonates. The calculated sea-to-air CO_2 flux is smaller at the LGM owing to a lower rate of $CaCO_3$ production and net storage in sediments and, conversely, a higher rate of organic matter storage. Significantly, the ocean is not balanced with respect to its carbon content because the inputs by rivers are smaller than the sum of carbon storage in sediment and flow to the atmosphere. This imbalance reflects a shift in the formation and storage of $CaCO_3$ to the growing coastal zone since the LGM (Milliman, 1993).

The increase in the terrestrial biosphere sink strength was most likely due in part to climate-induced biome shifts and in part to enhanced growth of existing LGM vegetation due to rising atmospheric CO_2, temperature, and changes in nutrient cycling on land. Terrestrial net primary production (NPP) is not saturated by even present-day atmospheric CO_2 concentrations of about 380 ppmv. Consequently, as atmospheric CO_2 increased during the LGM to the present climatic transition, terrestrial plants and consequently soil humus were a potential sink for atmospheric carbon simply due to fertilization by the rising atmospheric CO_2 levels. The activity of the principal carbon-fixing enzyme in C_3 plants (Rubisco, Chapter 6) increases with increasing CO_2 concentrations, saturating between 800 and 1000 ppmv CO_2, well above the 180 to 280 ppmv CO_2 concentrations of the LGM and the pre-industrial modern world, respectively. Increased carbon fixation by plants as a function of rising CO_2 is shown in Fig. 6.6.

Now we turn briefly to the glacial-interglacial changes in atmospheric methane, CH_4. Although methane is thermodynamically unstable with respect to carbon dioxide even at very low concentrations of oxygen (Chapter 2), it occurs metastably in the atmosphere, where the flux into the atmosphere is mainly from bacterial production, and is oxidized by the hydroxyl radical to CO and then CO_2. As a glaciation begins, changes in atmospheric CH_4 concentrations appear to be more related to processes in terrestrial ecosystems and in wetlands, rather than to processes in the open ocean. As the planet cools and ice spreads across the continents, the strength of the natural sources of CH_4 to the atmosphere would weaken. The high-latitude tundra region which is an important source of CH_4 to the atmosphere at present would be covered by ice and the colder temperatures would lead to a decrease in methane emissions produced by methanogenesis in tropical wetlands. Methanotrophy, a process that consumes methane in soils and is the major natural biological sink of CH_4, would also probably slow but its magnitude today represents only about 10 to 20% of the natural wetland flux to the atmosphere. The oceanic flux of CH_4 to the atmosphere is poorly known but is presently of the order of 10% of the natural wetland flux. Thus the drawdown in CH_4 during a glaciation is most likely related to the weakening of the methane fluxes to the atmosphere. Its increase is most likely mainly due to the warming and wetting of tropical wetlands and swamps. The shorter term changes in atmospheric CH_4, as seen in the Vostok ice core record, are not always in step with the changes in temperature and atmospheric CO_2, suggesting that the mechanisms controlling the CH_4 changes are not all the same as those for CO_2, which is not surprising considering the different natural sources and sinks of these gases.

In summary, atmospheric CO_2 and CH_4 concentrations during the past 420,000 years were locked into a similar pattern of the major 100,000-year and to some extent shorter time scale fluctuations as that of the air temperature at the Antarctic Station Vostok and the global mean temperature, as recorded in the $\delta^{18}O$ record found in the calcite of foraminifera of the deep sea. More recent data from the European Project for Ice Coring in Antarctica (EPICA, 2004), Dome Concordia ice core, show that the cycles of CO_2, CH_4, and temperature extend back in time to 650,000 years before present (Siegenthaler et al., 2005). However, prior to 430,000 years before present, the range in CO_2 variations is almost 30% smaller (260 to 180 ppmv) than that of the last four glacial cycles as seen in the Vostok and Dome C ice cores (290 to 180 ppmv). This lock-step is remarkable and its meaning is still not well understood. Certainly the processes and mechanisms described above played a major role in regulating this pattern of change. The last interglacial was slightly warmer than modern pre-industrial time and sea level was higher than its present-day level by approximately 5 m (Broecker, 2002). Atmospheric CO_2 and CH_4 concentrations were very similar to modern pre-industrial time, perhaps a little higher at the most climatically equable time of the last interglaciation. The preceding interglaciations show a similar pattern of change and the climatic maxima preceding the LGM show to a first approximation similar degrees of atmospheric levels of CO_2 and CH_4 and temperature change as found for the LGM. The asymmetry in the pattern of change of CO_2, CH_4, and temperature as one goes into and out of glaciations is also remarkable: the relatively smooth and rapid transitions from glacial maxima into interglaciations (despite the climatic reversal associated with the Younger Dryas cooling event 13,000 to 11,500 years ago) and the longer, by a factor of about four times, very irregular transitions from the interglaciations to the glacial maxima. Changing conditions in the ocean related to the ocean processes and mechanisms discussed above seem to be responsible for most of the atmospheric CO_2 change from the LGM to the immediate pre-industrial world that resulted in not less than 600 Gt C being released from the ocean, with 200 Gt C accumulating in the atmosphere.

4.5 Overview

The resolution of the mechanisms controlling the rise and fall of atmospheric CO_2 during the Pleistocene interglaciations and glaciations, respectively, lies in the future, as does the remarkable consistency in atmospheric CO_2 and CH_4 upper and lower bounds during the glacial-interglacial periods. One can certainly conclude that during glacial-interglacial time, climate and environmental variables were constrained within upper and lower bounds for both glaciations and interglaciations and that patterns of change in climate and environmental variables were very consistent from glaciation to glaciation. This implies a tightly coupled system, one that is self regulating and homeostatic, and an external or internal driver of change that is remarkably consistent, such as the Milankovitch periodic orbital forcings. It is conceivable that once the system is slightly moved by this forcing, a whole cascade of events follows that amplify the changes in the system.

Broecker (2002) has pointed out three types of forcing that can be potentially responsible for the general pattern of climatic change observed during much of Pleistocene-Holocene time. These include: (1) changes in seasonality associated with the Earth's orbital cycles, the Milankovich forcings; (2) reorganization of the ocean's thermohaline conveyor belt circulation pattern that is associated with catastrophic inputs of fresh water to the Northern Atlantic, and (3) fluctuations in the Sun's energy emission associated with the appearance and disappearance of sunspots, that is sunspot activity. With respect to the latter, this is a weak forcing that one also must consider in conjuction with possible changes in the cosmic ray flux. Some of the larger and abrupt changes in climate seen in the ice core record and in the stratigraphic and isotopic records of deep-sea sediments might indeed have been paced by the distribution of solar energy as related to the changes in the Earth's orbital parameters. In addition, it is clear that reorganization of the ocean's thermohaline circulation was also a driver of climatic change. The latter would certainly result in major modifications of equatorial to poleward heat and moisture transport patterns. Nevertheless, these three forcings appear too weak in and of themselves to account for the intensity of climatic change during glacial-interglacial times. There were probably amplifiers or positive feedbacks to the initial forcing and once the climate was slightly changed by one of the forcings, the changes to follow were amplified. The potential amplifiers include: (1) feedbacks in the biogeochemical cycles of the natural greenhouse gases of H_2O, CO_2, CH_4, and N_2O and resulting changes in atmospheric composition; (2) continental dust and sea salt aerosol loading of the atmosphere; (3) sea ice coverage; (4) water vapor content of the atmosphere; (5) cloud cover; (6) ice sheets and their geographical extent, and (7) changes in net production and storage rates of carbonate and organic carbon in the ocean. One intriguing feature of these amplifiers is that they were certainly stronger during times of glaciation than during times of interglaciation. Whatever the ultimate conclusion to the story of glacial-interglacial climatic and environmental change, we still have a long way to go to understand it fully and complete the writing of the story.

The final chapter of the carbon story in the Earth's geobiosphere is the following Chapter 11 that deals with the period of human industrial and agricultural perturbations of the last two to three centuries that are also known as the time of the Anthropocene. In this chapter we will see that the human species has become a geologic force in the surface environment of the Earth, leading to major reorganizations of the global carbon cycle and the important nutrient element cycles of nitrogen and phosphorus, on which it depends and with which it is coupled and interacts, with potential consequences for humankind and natural and managed ecosystems.

Chapter 11

The Carbon Cycle in the Anthropocene

This last chapter of the book examines the global carbon cycle and its links to those of nitrogen and phosphorus in the Earth's outer shell that has been continuously perturbed by human activities for some time. The carbon cycle is linked to other cycles, particularly nutrient N and P, mainly by the biological formation of organic matter that requires N, P, and other elements for the functioning of living cells. The human industrial perturbation started about 200 years ago, the beginnings of the agricultural perturbation are much older, and both may be expected to intensify in the future. Understandably, questions of the future can be answered only from reasonable projections of the past and present, and such answers should be viewed as indications of the trends rather than accurate estimates of where the global environment would be in the next centuries.

1   Characteristics of the Anthropocene

As the Earth recovered from the Last Glacial Maximum (LGM), the planet's surface warmed 5° to 7°C; the atmosphere became less windy and dusty and more moist; atmospheric trace greenhouse gases of CO_2, CH_4, and N_2O rose in their concentrations, sea level rose 120 m because of the melting of ice sheets, the oceanic coastal area or the continental shelf expanded about threefold, the terrestrial biosphere stored progressively more organic carbon in its vegetation and soils, and ultimately the human species became a force modifying its environment. At that time, *Homo sapiens* Linnaeus was a well established species, not yet very populous, but with the ability of making fire (and unknowingly contributing to atmospheric carbon dioxide).

The term Anthropocene, as mentioned in Chapter 1, was originally proposed for the period since the late 1700s or the Industrial Age, the record of which appears in the ice cores. At that time, the atmospheric change emanated primarily from the coal-burning industries of Europe, but on a smaller scale it started much earlier with agricultural practices of crop cultivation and other land-use activities accompanying them, such as deforestation and displacement of the natural vegetation. The beginnings of agricultural practices, sometimes also referred to as one of the "agricultural revolutions," go back about 5000 or more years in the Middle East (Mesopotamia), and also in China and Meso-America (e.g., Harlan, 1971; Richerson *et al.*, 2001). One of the changes in the carbon cycle produced by agriculture is a faster recycling and shorter residence time of carbon on land where annual crops and grasses replace the longer-living trees and shrubs. Interestingly, the development of agriculture as a cultivation of land and domestication of animals followed a shift from the more humid to dry conditions in parts of the Middle East and Africa about 7000 years ago, and it grew through the warmer and cooler periods since the year 1100 that lasted several decades to a century each (Eddy, 1976).

Among the many changes in the environment that can be directly attributed to humans, the global phenomenon of carbon dioxide emissions from the burning of fossil fuels (coal, followed by petroleum, and later by natural gas) and land-use practices is a consequence of the world population growth and its need to consume energy and food.

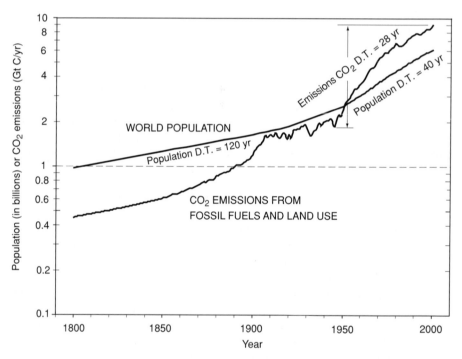

Figure 11.1   Growth of world population and carbon dioxide emissions from the burning of fossil fuels and changing land use patterns from 1800 to 2000. The growth rate of world population accelerates in the late 20th century, as evidenced by its shorter doubling time (D.T.). Emissions have been growing faster than the population during the 200 years. The trend of CO_2 emissions also closely parallels that of rising atmospheric CO_2 concentrations for much of this period of time. Population data 1800–1950 are from United Nations (1999), 1950-2000 from U. S. Census Bureau. Fossil fuel emissions are from Marland et al. (2003), and land-use emissions are from Houghton's (1995) and Houghton et al. (1998), projected to 2000.

Figure 11.1 shows the growth of world population in 200 years, from 1800 to 2000, and the growth of CO_2 emissions from fossil fuel burning and land-use practices. In this period, the global population increased 6-fold, but the industrial and land emissions increased 20-fold. The logarithmic scale of the figure shows that the population growth rate became faster in 1900 and again in 1950, following the end of World War II. However, the CO_2 emissions were growing faster than the population even in the 19th century and their growth accelerated further in the 20th.

It was discussed in the preceding chapters that the net increase in atmospheric CO_2 concentration by one-third, from pre-industrial 280 ppmv to the present 380 ppmv, represents only about 43% of the emissions, the remainder being distributed between the ocean and land.

2 Major Perturbations of the Carbon Cycle: 1850 to the Early 21st Century

The coupling of the carbon, nitrogen, and phosphorus cycles is rooted in these elements' occurrences as the major constituents of organic matter (Chapters 2, 6) where the atomic abundance ratios of C:N:P are characteristic of the aquatic and land primary producers, and the ratios are called the Redfield ratios. Because the dissolved forms of N and P that can be utilized by primary producers (nitrate, ammonium, and phosphate ions) are less abundant than CO_2 in the atmosphere or DIC in waters that supply dissolved CO_2, nitrogen and phosphorus play a somewhat variable role as a limiting or controlling nutrient under different environmental conditions. On a global, long time scale phosphorus is more likely to be the controlling nutrient in ocean and continental waters because of its geochemical behavior, whereas nitrogen is limiting in some marine systems on the shorter time scale and in local and regional environments (Schlesinger, 1997; Tyrrell, 1999).

Figure 11.2 shows a conceptual multi-reservoir diagram with rates of transfer, the fluxes, for the coupled, global biogeochemical cycles of carbon (C), nitrogen (N) and phosphorus (P) in a portion of the outer shell of the Earth, its surface system of land, atmosphere, coastal ocean, open ocean, and shallow sediment column. This rather unique representation of the biogeochemical cycles of C, N, and P shows simultaneously all the major processes and fluxes involving C, N, and P transport and exchange in the surface system prior to major human interference in these biogeochemical cycles. In addition, it includes a separate coastal ocean reservoir as distinct from the open ocean. The fluxes of C, N, and P shown in the diagram are based on those estimated from geologic and historical data and documented in Ver et al. (1999; Ver, 1998) and Mackenzie et al. (2001). The Earth's surface environment in the diagram is comprised of four major domains and includes 12 reservoirs:

I.   The atmosphere;
II.   Terrestrial domain that consists of the five reservoirs of living biota, humus, inorganic soil, soil water, and shallow groundwater;

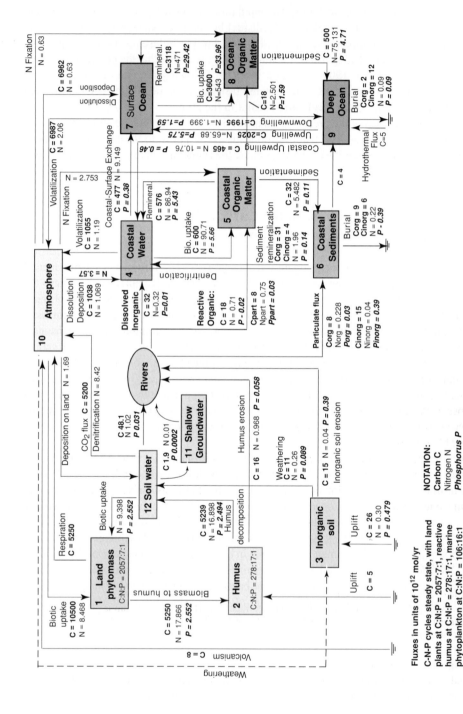

Fluxes in units of 10¹² mol/yr

C-N-P cycles steady state, with land
plants at C:N:P = 2057:7:1, reactive
humus at C:N:P = 278:17:1, marine
phytoplankton at C:N:P = 106:16:1

NOTATION:
Carbon C
Nitrogen N
Phosphorus P

III.   Coastal domain, of three reservoirs: coastal water, organic matter (including
             living biomass and dead organic matter), and sediments; and
IV.   Open ocean domain, with its four reservoirs of surface water, deep water,
             organic matter (living and dead organic matter), and deep-sea sediments.

The rivers are a main purveyor of materials to the coastal ocean and hence through it
to the open ocean. The instantaneous residence time of water in the rivers is only 20 days,
calculated with respect to precipitation over land, and hence rivers are not treated as a
reservoir in the diagram. The atmosphere is a medium of important exchange between
the land and the ocean and the recipient of much of the emissions of C and N from
human activities to the surface environment of the planet.

The model shown in Fig. 11.2 has been employed under the name of TOTEM to
calculate the rise in atmospheric CO_2 over the last 300 years of the Anthropocene due to
fossil fuel and land-use emissions of CO_2. The calculations are in very good agreement
with the historical observational record from globally distributed air sampling stations,
such as Mauna Loa, Hawaii, and from ice core data, and they clearly show that the CO_2
rise in the last 300 years was principally driven by anthropogenic emissions of carbon.
Because of these and other agreements between calculations and observational data, the
model was also used to make projections for the carbon cycle and its driver nutrients of
N and P into the early decades of the 21st century. This has enabled further assessment
of the changes for the future and the partitioning of the anthropogenic fluxes into the
various surface reservoirs through Anthropocene time.

Figure 11.3 shows the major perturbations of the C-N-P system of Fig. 11 from the
year 1850 with projections to 2040 due to the emissions of carbon, nitrogen, and sulfur,
fertilizer application and sewage discharge of N and P, and the rising temperature. These
perturbations include the application of nitrogen- and phosphorus-bearing fertilizers to
croplands, carbon, nitrogen and sulfur emissions to the atmosphere from burning of fossil fuels of coal, oil and gas and land use activities, municipal sewage and wastewater nitrogen and phosphorus disposal, and the temperature variations of the last 150 years. The
past and projected emissions of CO_2 from fossil fuel and land-use activities have been so
far the two main human forcings on the global carbon cycle. Because of the uncertainties
in changing land use patterns, upper and lower bounds of CO_2 emissions from land-use

---

←

Figure 11.2   Conceptual framework of the reservoirs, processes, and fluxes of the coupled global
biogeochemical cycles of carbon (C), nitrogen (N), and phosphorus (P) in the surface system of
the Earth's outer shell of land, atmosphere, coastal ocean, open ocean, and sediments prior to
major human interference in these cycles. The CO_2 weathering flux, shown by a dashed line, may
be coming from the soil humus reservoir (see Figs. 1.4, 10.7). The diagram is constructed for a
mean C:N:P ratio in land plants of 2057:17:1 and for a mean C:N:P ratio in reactive soil humus
of 287:17:1. The original diagram, as published in Ver et al. (1999), is slightly different because
of the use of different mean C:N:P ratios for land plants and reactive soil humus. These mean
ratios are not well known even today for the land biosphere because of their large variability. The
fluxes are shown in units of 10^{12} mol of the element per year; to obtain annual fluxes in grams of
the element, multiply the individual element fluxes by 12.01 for C, 14.01 for N, and 30.97 for P.

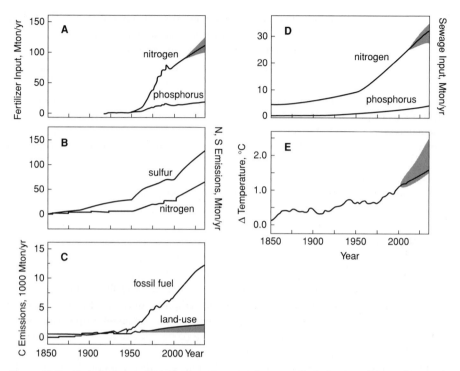

Figure 11.3   Perturbations of the Earth system as relevant to the behavior of the carbon cycle and those of N and P from 1850 projected to 2040: A. Application of N- and P-bearing fertilizers to croplands; B. Atmospheric NO_x and SO_x emissions mainly from fossil fuel combustion; C. Atmospheric CO_2 emissions from fossil fuel burning and land-use activities; D. Changes in global mean temperature. The shaded areas shown in A and D are the ranges of values of the projections based on low and high fertility rates of the UN Population Division (2001); in C the range of projected CO_2 emissions from land-use change (see text); and in E the future global mean temperature changes from Nakićenović et al. (2000).

change are shown in Fig. 11.3. The upper bound is based on projections from Houghton (1995) and Houghton et al. (1998); the lower bound is based on emission estimates of about 1.1 ± 0.5 Gt C/yr in 1995 (T. M. L. Wigley, personal communication, 1999). A summary of the anthropogenic inputs of C, N, and P is given in Table 11.1.

Important industrial production of inorganic nutrient fertilizers was not established until after the end of World War II (Fig. 11.3A). With the increased production of nitrogen fertilizers through the Haber-Bosch process, discovered in the early 1900s, and increased mining of phosphate rock, the consumption of inorganic fertilizers in agriculture increased exponentially. Between 1950 and 1988, there was a nearly sustained increase in global annual consumption of N and P fertilizers from 3 to 79.6 Mt N and from 2.4 to 14.5 Mt P, yielding approximate average annual growth rates of 7%

Table 11.1   Summary of C, N, and P perturbations in the period from 1700 to 2000 (from data in Mackenzie *et al.*, 2001, and Ver *et al.*, 1999)

Perturbation	Time period	C (10^{12} mol)	C Gt	N (10^{12} mol)	N Gt	P (10^{12} mol)	P Gt
Emissions (fossil fuel and industrial)	1850–2000	23,333	280	90.6	1.27	—	—
Land-use change	1700–2000	16,667	200	451.0	6.3	72.8	2.26
Chemical fertilizers	1920–2000	—		156.6	2.2	18.0	0.56
Sewage	1700–2000	774	9.3	132.7	1.9	8.4	0.26
Detergent	1945–2000	—	—	—	—	1.1	0.03

Note: 1 gigaton or Gt = 10^9 ton = 10^{15} grams.

and 4%, respectively. World phosphate consumption in 2002 was 13 Mt P/yr, slightly below its peak of 1988, and nitrogen fertilizer consumption was at a peak of about 85 Mt N/yr.

Significant anthropogenic remobilization of nitrogenous combustion products from the burning of fossil fuels began very shortly after the Industrial Revolution in 1850 (Fig. 11.3). Combustion of coal, oil, and gas in both stationary and mobile sources accounted for approximately 50% of the total natural and anthropogenic early 21st century emissions of NO_x (mainly NO and NO_2) from the land to the atmosphere. In the year 2000, these emissions were equivalent to nearly 33 Mt N/yr (Fig. 11.3B). The fate of anthropogenic combustion-nitrogen in the atmosphere to a significant extent is dictated by the short lifetime of the NO_x gases in the atmosphere, where they react photochemically to form HNO_3 and other products. The mechanisms for transfer to the terrestrial or near-shore surface ocean, and less importantly to the open ocean, reservoirs include wet and dry deposition and sedimentation of large particles. In 1860, the annual rate of release of nitrogen from the anthropogenic source was 0.4 Mt (Dignon and Hameed, 1989). Of this amount, about 60% was returned to the terrestrial realm while the rest was deposited onto coastal marine surface waters. By 2000, the anthropogenic emission of N combustion products had exponentially risen to 33 Mt N/yr.

Unlike fossil fuel burning or the agricultural use of inorganic fertilizer, changes in land-use patterns do not add C, N, and P materials directly from geologic sources. These changes mainly alter the rates of remineralization, weathering, denitrification, and biological uptake. Thus organic nitrogen and phosphorus storage may be shifted from soil humus to terrestrial and oceanic phytomass, inorganic N and P are transferred from land to the coastal margin, and N to the atmosphere. This includes transfer of organic and inorganic material from land by surface water runoff to the coastal ocean owing to soil erosion and mineral dissolution. The increased availability of nutrients from remineralization of humus, coupled with rising atmospheric CO_2 and warming temperatures, can enhance primary production and storage of organic carbon in the terrestrial phytomass, thus increasing the drawdown of atmospheric CO_2. When humus material with an average C:N:P ratio of 140:6.6:1 to 278:17:1 is remineralized, the remineralized N and P can ideally support the growth of plant material with an average C:N:P ratio of 510:4:1 to 2057:17:1. The additional required carbon for plant growth is taken from the atmosphere.

The mean global temperature increase shown in Fig. 11.3E is that derived from historical data and projected into the future for the IPCC IS92a emissions scenario (Nakićenović et al., 2000; see also Fig. 11.11D). Mean global temperature by 2040 is projected to rise about 1.4°C from its 1850 value. The temperature change in terms of the carbon cycle can have an impact on terrestrial photosynthesis and respiration and the water cycle that feeds back to terrestrial plant productivity. In addition, as sea surface temperatures continue to rise, ocean uptake of CO_2 will be slowed and circulation patterns of the ocean may change, as discussed later in this chapter.

3   Partitioning of the Carbon, Nitrogen, and Phosphorus Fluxes

Figures 11.4 and 11.6 show the partitioning of the human-induced fluxes of carbon, nitrogen, and phosphorus on land from 1850 to the beginning of the 21st century and calculated to the year 2040. These mass balance diagrams represent the balance of the anthropogenic sources and sinks of the elements. The natural background fluxes (Fig. 11.2), which are not shown, define the starting point of the system at the quasi-steady state condition prior to major human interference in the biogeochemical cycles of C, N, and P by the perturbations shown in Fig. 11.3: fossil fuel and land-use emissions, uses of fertilizers, and organic waste discharges. Let us start with the fluxes involved with the cycling of carbon. Figure 11.4 shows the partitioning of anthropogenic carbon from the two major inputs of fossil fuel and land-use emissions of CO_2 among the atmosphere, ocean, land, and coastal marine sediment reservoirs. Until about 1950 anthropogenic CO_2 emissions grew slowly and were dominated by land-use emissions in the 19th century. As we entered the 20th century and first coal, then oil, and then gas became progressively more important as fuel sources, anthropogenic fossil fuel CO_2 emissions steadily increased. Since 1950, following the end of World War II, consumption of fossil fuels became much greater as a result of the growing human population and its demand for resources to fuel the expanding global economy of the postwar years. Projections under one of the commonly used scenarios of fossil fuel emission ('business-as-usual', Intergovernmental Panel on Climate Change scenario IS92a) for the year 2040 indicate that about 12 Gt C/yr may be emitted to the atmosphere from the combustion of fossil fuels and 2.5 Gt C/yr from changing land use, giving a total anthropogenic emission of 14.5 Gt C/yr. Where does all this anthropogenic carbon go? The question has been difficult to answer in terms of the relative strengths of the individual sinks, but over the years there has been some resolution of the problem (e.g., Mackenzie et al., 2001; Sabine et al., 2004; Sarmiento et al., 1992; Ver et al., 1999). The sinks for the anthropogenic CO_2 emissions since 1850 are shown in Fig. 11.4 as positive values above the zero line. The sinks are the atmosphere, ocean, terrestrial biota, and, less importantly, the accumulation of organic matter in coastal marine sediments.

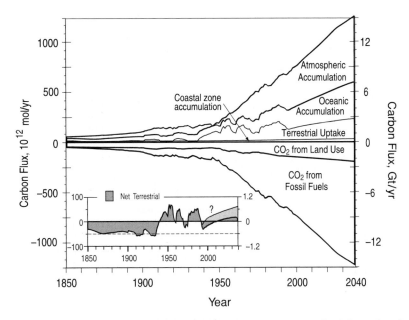

Figure 11.4   Partitioning of the human-induced perturbation fluxes on land for carbon for the period 1850 to 2000 and projected to the year 2040 under a business-as-usual scenario in units of 10^{12} mol C/yr and Gt C/yr. The anthropogenic sources of the element are plotted below the zero line on the (−) side and resulting accumulations and enhanced export fluxes above the zero line on the (+) side of the diagram. Note that the insert shows the net terrestrial organic carbon flux with a range of projections from 2000 that depend on whether or not changes in land use will result in more or less emissions of CO_2 to the atmosphere. It should be emphasized that the diagram is for the balancing of the anthropogenically derived fluxes; in the background there are large exchanges of material that go on naturally (Fig. 11.2) (after Mackenzie *et al.*, 2002).

Between the years 1700 and 2000, a cumulative amount of carbon from fossil fuel combustion and changing land use patterns of approximately 480 Gt C were emitted to the atmosphere (Table 11.1; Fig. 11.5). Of this amount, approximately 200 Gt C accumulated in the atmosphere, 140 Gt C were taken up on land, and 130 Gt C were directly absorbed by the oceans. During this time, increased carbon burial (mainly organic) in coastal marine sediments, equivalent to a cumulative amount of 10 Gt of organic C, was also a small sink of anthropogenic CO_2. Because of the release of carbon during this 300-year period due to land-use activities, increased river discharges of organic and inorganic carbon, and sewage discharge of organic carbon to the ocean, the land actually lost a cumulative net amount of carbon of approximately 76 Gt from year 1700 to 2000.

In concert with the CO_2 emissions from the land for the period 1850 to 2000, there were major anthropogenic inputs of the nutrients N and P to the land environment. For N, the major inputs were from atmospheric deposition, leaching of fertilizer, and land use remobilization of N from humus and the living terrestrial phytomass. These fluxes

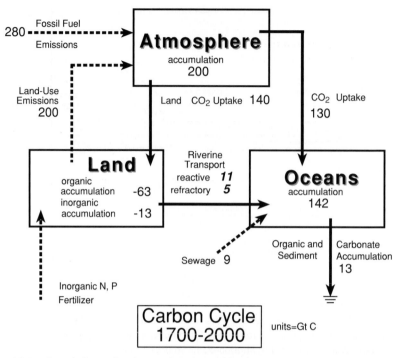

Figure 11.5   Cumulative carbon reservoir gains and losses and cumulative fluxes in Gt C for the period 1700 to 2000 for the land-atmosphere-ocean system. Dashed arrows indicate human perturbations. Note that during this 300-year period of time the atmosphere, ocean and sediments have gained carbon and the land has lost it (− value). This diagram does not include the natural fluxes of C shown in Fig. 11.2 that are assumed to be nearly constant in time but represents only a balance of the anthropogenic fluxes, calculated from model TOTEM; uncertainties in the major uptake and accumulation values are ±10 Gt C (updated from Ver *et al.*, 1999).

were balanced by N accumulations in the biomass and humus, export of dissolved and particulate N to the coastal oceans, and loss of N via denitrification and emission of gaseous N forms. It can be seen from Fig. 11.6A that the anthropogenic mobilization of N in the Earth's surface environment, similarly to the case of carbon, became more important following World War II in the decade of the 1950s. In the year 2000, the total N mobilized on the Earth's surface by human activities was about 150 Mt N/yr. This rate was more than three times greater than that soon after the end of World War II (1950, 45 Mt N/yr) when much of the mobilized N was associated with land use activities (about 90%). As the 20th century progressed, proportionately more of the anthropogenic inputs came from fertilizers, land use, and the combustion of fossil fuels, and these inputs were redistributed into the environment by several processes. Increased application of nitrogenous fertilizers to croplands led to an increase in the export flux of nitrogen to the atmosphere including the emissions of N_2 and N_2O gases by the process of denitrification. In addition, the application of N fertilizers to

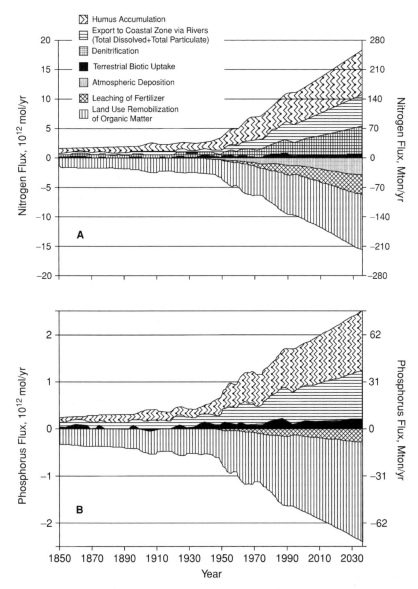

Figure 11.6   Partitioning of the human-induced perturbation fluxes on land for (A) nitrogen and (B) phosphorus for the period 1850 to 2000 and projected to the year 2040 under a business-as-usual scenario, in units of 10^{12} mol (N or P)/yr and Mt (N or P)/yr. The anthropogenic sources of the elements are plotted below the zero line (negative values) and resulting accumulations and enhanced export fluxes above the zero line on the (+) side. The remobilization fluxes for N and P from land-use activities are plotted separately from the terrestrial biotic uptake to highlight their opposing effects on the terrestrial organic reservoirs. The diagrams balance only the antropogenically derived fluxes; in the background there are large exchanges of material that go on naturally (Fig. 11.2) (after Mackenzie *et al.*, 2002b).

the landscape resulted in the leaching of some of the nitrogen into aquatic systems as dissolved NO_3^- and NH_4^+ and an increase in the export of dissolved inorganic nitrogen (DIN), dissolved organic nitrogen (DON), and its particulate forms to the coastal ocean. Finally, fertilizer N remaining in the soil water, N from atmospheric deposition, and N remobilized by land use activities via the degradation of living phytomass and humus have led to the enhanced fertilization of the terrestrial phytomass.

The case for phosphorus shown in Fig. 11.6B is similar to that for N with one important difference: there is no major global flux of P as a gaseous species, although there is evidence suggesting that on a local and regional scale, the emission of phosphine (PH_3) and, perhaps, polyphosphate gas might be of some importance. Dust transport from the land to the ocean is a moderately important flux for getting P to the ocean but the reactivity of this P, mainly bound in iron oxyhydroxides and particulate organic matter, in the surface ocean is not well known. On a global scale, humans have significantly perturbed the P cycle by the mobilization of nutrient P through land use activities and the application of phosphate fertilizer to croplands. For the past 150 years, changing land use practices have increased the mobilization of P into the environment from 6 Mt P/yr in 1850 to 28 Mt P/yr in 2000. Land use activities have also increased loss of P from land due to erosion from 3 to 22 Mt P/yr between 1850 and 2000. Additionally, the increased application of phosphate fertilizers, particularly since the end of World War II, provided a greatly enhanced source of anthropogenic P in the environment. At this time, the total perturbation on the P cycle was about 20 Mt P/yr, 96% of which came from the remobilization and erosional loss of P owing to land-use activities (Mackenzie et al., 2001). The perturbation from phosphate fertilizer leaching into the soil water was small at about 0.7 Mt P/yr (4% of the total). By the year 2000, land use activities accounted for 92% of the total perturbation, about 50 Mt P/yr released to the environment. Phosphorus leaching from fertilizer application accounted for about 8% or about 5 Mt P/yr. This leaching flux represents about 38% of the total of 13 Mt of fertilizer applied annually in 2002 to croplands, golf courses, home gardens, and other small lots. Phosphorus leaching from these sources subsequently enters aquatic systems as dissolved inorganic and organic P (DIP, DOP) and as particulate P (PIP and POP). A portion of this P is exported to the coastal marine realm by river and groundwater discharges, while the rest is used in the fertilization of new phytomass on land.

The net cumulative losses from land over a 300-year period were approximately 2.3 Gt N and 1.6 Gt P (Figs. 11.7 and 11.8). As mentioned previously, during the past recent centuries, there has been a net overall loss of organic carbon from the terrestrial biota and soil humus reservoirs. The pattern of loss of N and P from these reservoirs is similar to that for organic carbon (Fig. 11.9). From 1850 to about 1950, organic C, as well as N and P, were generally lost from both living phytomass and soils on land. In Fig. 11.9, notice the sharp reversal in the carbon mass of the land phytomass reservoir near the middle of the 20th century. This reversal is also seen less pronouncedly in the biotic reservoirs of N and P. From 1950 to 2000, the terrestrial biota gained about 300 Mt C/yr equivalent to 3 Mt N/yr and 2 Mt P/yr. This time of reversal appears to be the start of the enhanced storage of the elements C, N, and P on land due to the combined negative feedback effects to atmospheric CO_2 that include: (1) the fertilization of

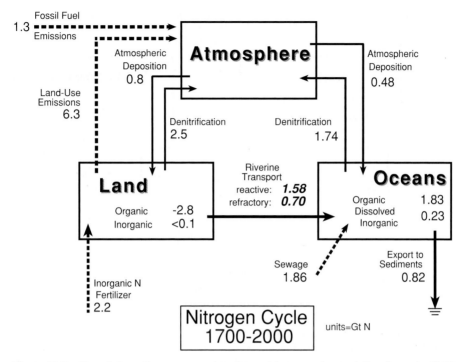

Figure 11.7 Cumulative nitrogen reservoir gains and losses and cumulative fluxes in Gt N for the period 1700 to 2000 for the land-atmosphere-ocean system. Dashed arrows indicate human perturbations. Atmospheric deposition of N on the ocean surface occurs mainly in the coastal ocean environment, where some of this N may be important in the production of organic matter. Note that during this 300-year period of time that the ocean and sediments have gained N and the land has lost it (negative value). The enhanced denitrification due to the anthropogenic perturbation in the oceans and on land implies that the atmosphere is gaining a small quantity of N against its background natural N content because nitrogen fixation has not been augmented significantly by the anthropogenic perturbation. This diagram does not include the natural fluxes of N shown in Fig. 11.2 that are assumed to be nearly constant in time but represents only a balance of the anthropogenic fluxes added to the natural background.

the terrestrial biosphere by rising atmospheric CO_2, (2) release of N and P from the humus and chemical fertilizers on land and deposition of atmospheric N over land, and (3) the re-growth of forests in previously disturbed areas. The temperature rise of approximately $0.7°$ to $0.8°C$ over the past 150 years and possibly slight changes in global precipitation may also have played a role in the enhanced growth of the forests and consequent increased storage of C, N, and P in terrestrial ecosystems. The humus reservoir, except for a slight reversal in C between 1950 and 1980, also shows sustained net losses in masses of C, N, and P throughout much of the entire period from 1850 to 2000, averaging about 170 Mt C/yr, 15 Mt N/yr, and 4 Mt P/yr during the 150 years since the Industrial Revolution.

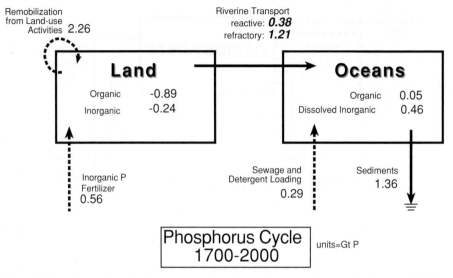

Figure 11.8   Cumulative phosphorus reservoir gains and losses and cumulative fluxes in Gt P for the period 1700 to 2000 for the land-ocean system. Dashed arrows indicate human perturbations. The atmosphere is not a significant medium of transport of anthropogenic P except as dust, some of which may be reactive when it falls on the ocean from the land. Note that during this 300-yr period of time that the ocean and sediments have gained phosphorus and the land has lost it (negative value). This diagram does not include the natural fluxes of P shown in Fig. 11.2 that are assumed to be nearly constant in time but represents only a balance of the anthropogenic fluxes added to the natural background.

It may be the case that a major fraction of the remobilized organic matter was not remineralized but was redistributed on land and sequestered in artificial water bodies, such as agricultural ponds and dammed reservoirs, and in natural lacustrine sediments and river floodplains. This redistribution of carbon may constitute a sink for atmospheric CO_2 of the order of 1 Gt C/yr (Smith *et al.*, 2001; Stallard, 1998). Some of the remineralized N and P was transported to the coastal ocean and other aquatic environments via river and groundwater discharges, stimulating new production and leading to enhanced burial of organic matter (Fig. 11.9). The enhanced burial of organic matter from new production in aquatic environments also constitutes a sink, albeit small, for anthropogenic CO_2, of the order of 0.1 to 0.2 Gt C/yr, equivalent to only 1.4 to 2.8% of the fossil fuel flux in 2004.

Figure 11.10 illustrates the results of the partitioning of the anthropogenic fluxes of C, N, and P and some of the effects of this transfer on the coastal ocean from 1850 to 2040 (see also Fig. 8.4 and Chapter 9). The figure shows the historical and projected future changes in the river fluxes of dissolved and particulate inorganic and organic C, N, and P, the fluxes of atmospheric deposition and denitrification of N, and accumulation of C in organic matter in coastal marine sediments. It is this transfer of C, N, and P materials from the land to the ocean, and also of C and N to the atmosphere, that

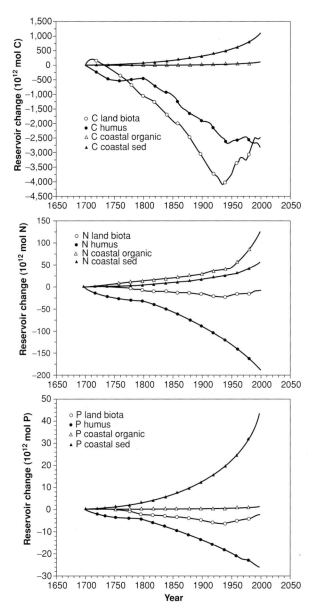

Figure 11.9   Changes in the carbon, nitrogen and phosphorus masses of land plants, humus, coastal organic matter and coastal sediments computed for the period 1700 to 2000, normalized to 0 in 1700. Notice in particular the decline of all three elements in the terrestrial phytomass and humus from 1800 to the mid-20th century and the reversal thereafter, perhaps due to the fertilization of the land plants by rising CO_2, excess nutrient application to the landscape and some re-growth of temperate forests. Also note that the coastal ocean organic pool and sediments are major reservoirs for the loss of materials from the land, as well as the ocean waters for the loss of carbon from changing land-use activities (after Lerman *et al.*, 2004).

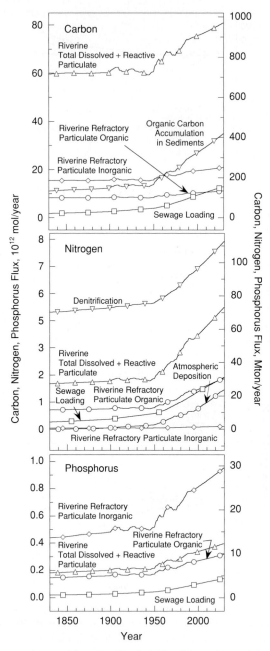

Figure 11.10   Past, present, and future projections for river carbon, nitrogen and phosphorus fluxes from land to the sea in 10^{12} mol/yr and Mt/yr. Because the global coastal ocean is the main recipient of these fluxes from land, the accumulation fluxes of organic carbon in coastal marine sediments, atmospheric deposition of N on the coastal ocean sea surface, and the changes in denitrification fluxes in the coastal ocean are also shown (after Mackenzie *et al.*, 2002).

accounts for much of the decline in the masses of these elements on land for most of industrial time as seen in Fig. 11.9. It can be seen in Fig. 11.10 that the river fluxes of C, N, and P all increase in the dissolved inorganic and organic phases from about 1850 projected to 2040. For example, for carbon, the total river flux (organic + inorganic) increases by about 35% during this period of time. The increased river fluxes are mainly due to changes in land-use practices, including deforestation, conversion of forest to grassland, pastureland, and urban centers, and re-growth of forests, and application of fertilizers to croplands and the subsequent transport of C, N and P into aquatic systems.

Inputs of nutrient N and P to the coastal zone that support new primary production are from the land by river and groundwater flows, from the open ocean by coastal upwelling and onwelling, and to a lesser extent by atmospheric deposition of nitrogen. New primary production depends on the availability of nutrients from these external inputs, without consideration of internal recycling of nutrients. Thus any changes in the supply of nutrients to the coastal zone owing to changes in the magnitude of these source fluxes are likely to affect the cycling pathways and balances of carbon and the nutrient elements in the coastal ocean. In particular, input of nutrients from the open ocean by coastal upwelling is quantitatively greater than the combined inputs from land and the atmosphere. This makes it likely that there could be significant effects on coastal primary production not only because of increased loading of nutrients via river inputs but because of changes in ocean circulation. For example, because of global warming, the oceans could become more strongly stratified owing to freshening of polar oceanic waters and warming of the ocean in the tropical zone. This could lead to a reduction in the intensity of the oceanic thermohaline circulation (ocean circulation owing to differences in density of water masses, also popularly known as the "conveyor belt") of the ocean and hence the rate at which nutrient-rich waters upwell into coastal environments (see also Fig. 11.2).

Total anthropogenic CO_2 emissions to the atmosphere by the year 2040 could be 45% higher than in 2000. The sinks for these emissions, mainly in the atmosphere, ocean, and uptake by terrestrial ecosystems may grow by about 50%, 50%, and 33%, respectively, from the year 2000 to 2040. The change in terrestrial net storage of carbon on land will depend significantly on changing land-use patterns, the resulting CO_2 emissions from these changing patterns, and the future effect of fertilization on land (Fig. 11.4, insert). Such projections to the future can only be considered as tentative estimates that may change if, for example, the continued rise of global temperature would increase the respiration of organic matter on land that would in turn override whatever fertilization effect is going on (Woodwell et al., 1995).

As carbon emissions grow into the 21st century so will nitrogen and phosphorus emissions and inputs from land (Fig. 11.3), estimated to grow by about 45% from year 2000 to 2040. The sinks of these fluxes on land will be in humus accumulation, terrestrial biotic uptake, export of dissolved and particulate N and P to the coastal ocean, and return of N to the atmosphere by denitrification. Denitrification removes N from the pool available for primary production and is strongest on land. In the year 2000, this process accounted for 73% of N removed from land by the combined processes of denitrification and riverine export of N to the coastal ocean. As a result of all these changes on land, the fluxes of C, N, and P materials to the coastal ocean will grow into the future, accounting for changes in the rates of primary production, organic carbon burial in coastal marine sediments, denitrification rates in coastal waters and sediments, transport of atmospheric fixed N to the coastal ocean, and uptake of anthropogenic CO_2 in the coastal and open-ocean surface waters.

4 The Fundamental Carbon Problem of the Future

We conclude this chapter with a look at the future of the global carbon cycle in relationship to the problem of the global warming of the planet (Chapter 3), a subject that very much has been written about. The potential of global warming of the planet due to human activities—mainly the burning of the land phytomass, coal, oil, and gas, venting of CO_2 from cement production and other sources, and venting of the hydrocarbon gases to the atmosphere—is a contemporary environmental problem and one that is probably more significant in terms of the future health of our planet than that of the release of synthetic chlorofluorocarbon (CFC) gases and stratospheric ozone depletion, although both issues are intertwined. In terms of the global carbon cycle, in the case of the industrial CO_2 emissions, humans are mainly increasing the rate of oxidation of organic carbon long buried in the organic carbon reservoir (Fig. 1.4) and, as a consequence, to a lesser extent the rate of decarbonation of $CaCO_3$ by its dissolution from the inorganic carbon reservoir. In the case of land-use emissions, carbon once sequestered on land in phytomass is being released back to the atmosphere. These anthropogenic carbon releases are being stored in the sinks of the atmosphere, ocean, and in recent decades, the temperate forests of the world because of the fertilization processes mentioned previously. Approximately 43% of the total carbon released from 1700 to 2000 has stayed in the atmosphere and 57% was partitioned between the ocean and terrestrial uptake (Fig. 11.5). The amount of CO_2 that has accumulated in the atmosphere could be responsible for as much as 60% of the nearly 1°C global mean surface temperature increase since 1700 or the 0.7° to 0.8°C increase since 1850. This much we know, but the fundamental problem of the enhancement of the greenhouse effect and global warming due to emissions of CO_2 and other greenhouse gases to the atmosphere is not what has happened in the past, but what will happen in the future. We know that the Earth's environment in terms of atmospheric CO_2 and CH_4 concentrations has left the domain that defined it for much of the past 420,000 (Fig. 10.4) and as more recent ice core data show, even 740,000 years. However, because of human activities, we could enter a super-interglaciation and global mean temperatures could rise well above those seen in the ice core record or other historical records for hundreds of thousands of years in the past.

Future emissions of CO_2 to the atmosphere will depend on energy use, the projections of which for the future are based on the perceptions of population growth and distribution, the need for energy to supply the growing world and individual national economies, and the types of energy used for that growth. The future of changing land-use patterns will play a variable and subordinate role to that of fossil fuel emissions. Obviously, burning forested land will release CO_2 to the atmosphere, but forest re-growth in a course of decades on abandoned lands and fertilization processes may sequester CO_2 from the atmosphere. Figure 11.11 shows the various emission scenarios for both anthropogenic CO_2 and SO_2 developed by the Intergovernmental Panel on Climate Change (IPCC) and summarized in the volume Climate Change 2001: The Scientific Basis (IPCC, 2001). The development of the scenarios, other than the business-as-usual IS92A scenario, was begun in 1996 by the IPCC. The scenarios consist of four different narrative storylines and were developed so as to cover a wide range of demographic, economic, and technological driving forces of future anthropogenic greenhouse gas and sulfur emissions. For example, the A1 storyline and scenario family describe a world of very rapid economic growth, global population that peaks in the mid-21st century and declines thereafter, and the rapid introduction of new and efficient technologies. The B1 storyline and scenario family are similar to the A1, but have economic structures that rapidly change toward a service and information economy, with major reductions in the use of materials and the introduction of clean and resource-efficient technologies, as these are perceived at present.

To be noted in particular in Fig. 11.11 is the wide range in CO_2 and SO_2 emission projections for the year 2100 from approximately 5.5 to 29 Gt C/yr and 0.02 to 0.15 Gt S/yr. The projections of the anthropogenic emissions of the greenhouse gases CH_4 and N_2O to the year 2100 also have a wide range of estimates depending on the emission scenario adopted. One positive aspect of the modeling of climate to date with coupled ocean-atmosphere climate models is that the instrumental record of temperature from 1860 to 2000 can be best fitted by models in which both natural and anthropogenic forcings are used in the model simulations. Such forcings include the natural changes in solar and volcanic activity, human-induced changes in atmospheric greenhouse gases, changes in tropospheric ozone and stratospheric ozone, and the effects of aerosols that manifest themselves in particle concentration, size, and number of cloud droplets, and cloud lifetime. This suggests, but does not prove, that the climate models might be able to provide us with some reasonable picture of the future of the climate.

Figure 11.11 also shows the projected global change in atmospheric CO_2 concentrations, mean temperature, and sea level rise that clearly show a significant dependence on the emission scenarios. Estimates for the end of the 21st century vary considerably for CO_2 concentration from 540 to 970 ppmv, temperature change from 1.4° to 5.8°C, and sea level rise from 20 to 88 cm. An important "bottom line" is that even if the global composition of the atmosphere were fixed as we write this book, global mean temperature and sea level would continue to rise because of the thermal inertia of the ocean. Wigley (2005) has shown in model calculations that under hypothetical scenarios of both constant atmospheric composition or constant emissions established in the year 2000, temperature and sea level would continue to rise into future centuries. Depending on the climate sensitivity and the degree of aerosol forcing adopted in the different model scenarios, the global mean temperature rise by the year 2100 could exceed 1°C under a more stringent constant composition commitment and

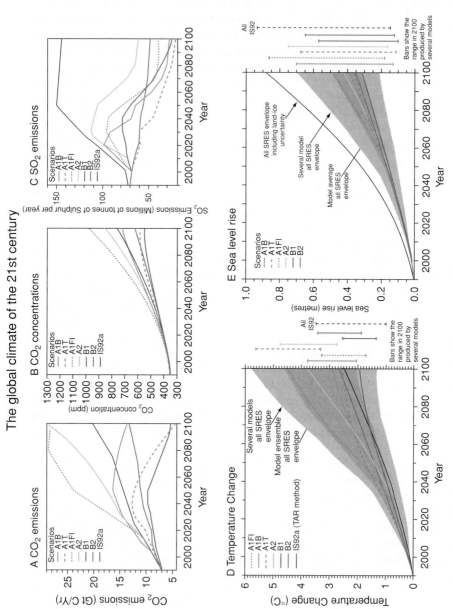

Figure 11.11 Anthropogenic emissions of CO_2 and SO_2 to the atmosphere for the six Intergovernmental Panel on Climate Change Emissions Scenarios (Nakićenović *et al.*, 2000) and carbon system and climate model projections of atmospheric CO_2 concentrations, global mean temperature change and global sea level rise for the 21st century (from IPCC, 2001)

$2°$ to $6°C$ under a constant emission commitment. For sea level rise under the constant composition commitment, the rate of rise could be 10 cm per century and under the constant emission commitment 25 cm per century, both estimates are smaller than those shown in Fig 11.11E. These calculations reinforce what is now common knowledge that, in order to stabilize global mean temperatures soon, the world community needs to commit itself to reducing the emissions of greenhouse gases to the atmosphere to well below present levels.

We will consider the oceanic carbon cycle first in terms of the potential warming of the Earth in the future and the interactions and feedbacks between the carbon cycle, atmospheric CO_2, and climate. As the Earth warms and sea surface temperatures continue to increase, the physical solubility pump will weaken simply because warmer seawater can hold less dissolved CO_2 than colder seawater (Chapter 5). In addition, coupled climate-ocean model simulations show that global warming will weaken the transport of dissolved CO_2 to the deep ocean due to increased density statification of the ocean water column. In the high latitude of the North Atlantic where future warming will be the greatest (by the end of the 21st century, temperatures perhaps as much as $8°C$ higher than late 20th century temperatures), this will be due to both the warming of the sea surface and the melting of sea ice, glaciers, and the Greenland ice sheet, and consequent freshening of high latitude waters. In the lower latitudes, the increased stratification will be mainly a result of warming sea surface temperatures. Approximately 23% of the total, global, sequestration of anthropogenic CO_2 in the ocean has been due to the sinking of high-latitude, North Atlantic waters in the formation of the North Atlantic Deep Water (Sabine *et al.*, 2004). Most models of the effect of climatic change on the intensity of the thermohaline circulation of the ocean, the conveyor belt circulation, show weakening of the circulation in the Northern Atlantic from 1850 to 2100. Because the strength of the circulation varies between models at their initial conditions from 10 to 30 Sv (1 sverdrup or 1 Sv = 10^6 m^3/s), corresponding to the circulation time between 1000 and 2000 years, and the models differ somewhat in their structure, the estimated changes in the thermohaline circulation by the year 2100 are from close to 0 to -15 Sv. If a weakening were to occur, it would result in the reduction in the rate of CO_2 removal from the atmosphere and sequestration in the deep interior ocean. It is too early to say whether an irreversible shut-down of the thermohaline circulation is likely or not or what would be the climatic threshold for this to happen. Needless to say, the magnitude of the potential feedbacks discussed above is critically dependent on how ocean circulation and mixing will respond to the climatic forcing.

Now, what might be the response of the biogeochemistry of the ocean to rising CO_2 and temperature? Coupled climate-biogeochemical models suggest that the biological pump discussed in Chapter 10 tends to counteract the smaller uptake of CO_2 by the solubility pump. However, the efficiency of the biological pump in the future must increase from its present strength, if this pump is to absorb additional CO_2 in the 21st century and beyond. This can be accomplished by any or all of four processes: (1) enhanced utilization of nutrients in the upper ocean, such as the excess NO_3^- in the Southern Ocean; (2) addition of one or more nutrients that now limit primary production, such as iron, to the HNLC areas of the ocean; (3) change in the elemental ratios of C:N:P in marine organisms; and (4) less organic carbon remineralization in the sinking flux of carbon through the oceanic water column (Falkowski *et al.*, 2000). Despite these biogeochemical mechanisms that exist in principle, there is no evidence that future changes in the biological pump would be large enough to counteract significantly the projected CO_2 emissions in the coming century. One of the major concerns in the biogeochemistry of the oceans as CO_2 increases and temperature rises is the effect on ocean water pH and carbonate saturation state, discussed in Chapter 9 for the coastal ocean and that apply to the whole ocean system. As CO_2 enters surface seawater, it titrates the CO_3^{2-} ion in seawater converting it to HCO_3^- and leading to an increase in DIC, lower pH, and lower saturation state of seawater with respect to carbonate minerals. The lowered saturation state has been shown experimentally to lead to decreased calcification rates in individual marine organisms such as scleractinian corals, coralline algae, foraminifera, and coccolithophorids and in communities of calcifying organisms. Acidification of the whole ocean because of the invasion of CO_2 has led to changes in the production, dissolution, and accumulation of biogenic carbonate, and there is already evidence in the dissolution of sinking carbonate shells for the upward expansion or shoaling of the zone of $CaCO_3$ undersaturation in the oceans by 50 to 200 m (Feely *et al.*, 2004; Orr *et al.*, 2005). The limitations of our knowledge of what determines the abundance and distribution of key groups of marine organisms impact our ability to predict specific marine ecosystem responses, even if the changes in atmospheric CO_2 and the resulting changes in the upper-ocean pH and carbonate saturation state are now well known.

As far as the response of the land biosphere to rising atmospheric CO_2 and global warming is concerned, terrestrial plants may become a less effective sink of atmospheric CO_2 in the future simply because as CO_2 increases, net primary production increases by a smaller factor. In addition, increased temperature will probably lead to higher microbial heterotrophic respiration rates of organic matter in soil, which may counteract and even exceed the enhancement of net primary production by the fertilization mechanisms of rising atmospheric CO_2 and nutrient N and P release (Woodwell *et al.*, 1998). It is likely that the continuous increased pressure in the developing world to industrialize, increase food and fiber production by converting forests to agricultural use, and to expand urban centers will increase the flux of carbon to the atmosphere while simultaneously reducing the land area available for active sinks in forest fertilization or re-growth. In the temperate regions of the Northern Hemisphere, the abandonment of agricultural land and re-growth of forests, along with fertilization processes, probably was a terrestrial CO_2 sink, although variable in time, during the past couple of decades. As in the case of marine ecosystems, the direction of the change in the feedbacks to rising atmospheric CO_2 and hence climate change involving the land biosphere are known with reasonable certainty, but the exact magnitudes of the changes are not.

In the longer scheme of events, at about the year 2300 in a business-as-usual scenario, atmospheric CO_2 concentrations are anticipated to start slowly to decrease (e.g., Archer *et al.*, 1998). At this time, the global reservoirs of conventional fossil fuels of approximately 5000 Gt C would be close to exhaustion and anthropogenic CO_2 emissions would have been on the decline for several decades. On a timescale of several centuries beyond the year 2300, the ocean would absorb a substantial fraction of the anthropogenic CO_2 that has accumulated in the atmosphere. The extent of CO_2 up-take by the ocean would depend on the rate of formation of deep and intermediate waters and the mixing time of the oceans, but it would likely result in a shoaling of the carbonate saturation horizons in all major ocean basins, as mentioned above, owing to increasing dissolution of deep-sea calcareous oozes and decreasing rain rates of biogenic carbonate hard parts produced in the surface waters of the coastal and open ocean.

The global carbon cycle is coupled to physical, climatological, and biogeochemical processes that involve not only it, but also the nutrient element cycles of N and P and the other bioessential elements, such as iron. Today we have considerable information about specific processes and other aspects of the carbon cycle, but many of the couplings and feedbacks to other element cycles and the physical climate system are still poorly understood. As we progress further away from the environmental conditions that characterized the pre-industrial Earth's geobiosphere, we certainly will test the limits of our understanding of how the system will respond. As we continue in the new domain of the Anthropocene characterized by the human geologic force, the debate about distinguishing human effects on the carbon cycle and climate from natural variability will hopefully abate in the face of a better understanding of the climate and the biogeochemical cycles of carbon and other elements. However, our ability to predict the future will always have a component of uncertainty; as this is the very nature of science. It is this uncertainty that makes it difficult for policy makers dealing with environmental issues, such as global warming, to formulate policy and make decisions as to how to deal with the problem. However, this uncertainty should not be confused with a lack of knowledge, nor should it necessarily be a reason to postpone prudent policy decisions based on the best scientific information available at the time.

Bibliographic References

Abbe, C., 1911, Meteorology, *Encycl. Brit.*, 11th Ed., v. 18, Cambridge Univ. Press, Cambridge, pp. 264–291.

Abramowitz, M., and Stegun, I. A., 1972, *Handbook of Mathematical Functions with Formulas, Graphs and Mathematical Tables*, 10th printing, U. S. Government Printing Office, Washington, D. C., xiv + 1046 pp.

Agegian, C. R., 1985, The biogeochemical ecology of *Porolithon gardineri* (Foslie), Ph. D. Dissertation, Univ. of Hawaii, Honolulu, xix + 178 pp.

Ahrens, L. H., 1952, The use of ionization potentials. Part I. Ionic radii of the elements, *Geochim. Cosmochim. Acta* **2**:155–169.

Ainsworth, E. A., Davey, P. A., Bernacchi, C. J., Dermody, O. C., Heaton, E. A., Moore, D. J., Morgan, P. B., Naidu, S. L., Yoora, H.-S., Zhu, X.-G., Curtis, P. S., and Long, S. P., 2002, A meta-analysis of elevated $[CO_2]$ effects on soybean *(Glycine max)* physiology, growth, and yield, *Global Change Biol.* **8**:695–709.

Ajtay, G. L., Ketner, P., and Duvigneaud, P., 1979, Terrestrial primary production and phytomass, in: *The Global Carbon Cycle* (B. Bolin, E. T. Degens, S. Kempe, and P. Ketner, eds.), SCOPE 13 (Scientific Committee On Problems of the Environment), Unwin Brothers, Gresham Press, Kingston-on-Thames, U.K., pp. 129–181.

Albarède, F., 1995, *Introduction to Geochemical Modeling*, Cambridge Univ. Press, Cambridge, xx + 543 pp.

Albright, J. N., 1971, Vaterite stability, *Am. Mineral.* **56**:620–624.

Alcock, N. W., 1990, *Bonding and Structure: Structural Principles in Inorganic and Organic Chemistry*, Ellis Horwood, New York, 321 pp.

Allègre, C. J., Manhés, G., and Göpel, C., 1995, The age of the Earth, *Geochim. Cosmochim. Acta* **59**:1445–1456.

Allen, L. H., Jr., 1994, Carbon dioxide increase: direct impacts on crops and indirect effects mediated through anticipated climatic changes, in: *Physiology and Determination of Crop Yield* (K. J. Boote, J. M. Bennett, T. R. Sinclair, and G. M. Paulsen, eds.), American Society of Agronomy, Madison, Wisc., pp. 425–459.

Allen, L. H., Jr., and Amthor, J. S., 1995, Plant physiological response to elevated CO_2, temperature, air pollution, and UV-B radiation, in: *Biotic Feedbacks in the Global Climatic System* (G. M. Woodwell and F. T. Mackenzie, eds.), Oxford Univ. Press, New York, pp. 51–84.

Ambrose, D., 1994, Critical constants, boiling points, and melting points of selected compounds, in: *Handbook of Chemistry and Physics* (D. R. Lide, ed.-in-chief), 75th Ed., CRC Press, Boca Raton, Fla., pp. 6-54–6-64.

Anderson, D. M., and Archer, D., 2002, Glacial-interglacial stability of ocean pH inferred from foraminifer dissolution rates, *Nature* **416**:70–73.

Anderson, L. A., and Sarmiento, J. L., 1994, Redfield ratios of remineralization determined by nutrient data analysis, *Global Biogeochem. Cycles* **8**:65–80.

Andersson, A. J., and Mackenzie, F. T., 2004, Shallow-water oceans: a source or sink of atmospheric CO_2? *Frontiers in Ecol. Env.* **2**(7):348–353.

Andersson, A. J., Mackenzie, F. T., and Ver, L. M. B., 2003, Solution of shallow-water carbonates: an insignificant buffer against rising atmospheric CO_2, *Geology* **31**:513–516.

Andersson, A. J., Mackenzie, F. T., and Lerman, A., 2005, Coastal ocean and carbonate systems in the high CO_2 world of the Anthropocene, *Am. J. Sci.* **305**:875–918.

Anthony, S. S., Peterson, F. L., Mackenzie, F. T., and Hamlin, S. N., 1989, Geohydrology of the Laura fresh-water lens, Majuro atoll: a hydrogeochemical approach, *Geol. Soc. America Bull.* **101**:1066–1075.

Archer, D., and Meier-Reimer, E., 1994, Effect of deep-sea sedimentary calcite preservation on atmospheric CO_2 concentration, *Nature* **367**:260–263.

Archer, D., Kheshgi, H., and Maier-Reimer, E., 1998, Dynamics of fossil fuel CO_2 neutralization by marine $CaCO_3$, *Global Biogeochem. Cycles* **12**:259–276.

Arrhenius, S., 1896, On the influence of carbonic acid in the air upon the temperature of the ground, *Phil. Mag.*, 5th ser., **41**:237–276.

Arthur, M. A., Dean, W. E., and Claypool, G. E., 1985, Anomalous ^{13}C enrichment in modern marine organic carbon, *Nature* **315**:216–218.

Arvidson, R. S., and Mackenzie, F. T., 1997, Tentative kinetic model for dolomite precipitation rate and its application to dolomite distribution, *Aquatic Geochem.* **2**:273–298.

Arvidson, R. S., and Mackenzie, F. T., 1999, The dolomite problem: control of precipitation kinetics by temperature and saturation state, *Am. J. Sci.* **299**:257–288.

Arvidson, R. S., and Mackenzie, F. T., 2000, Temperature dependence of mineral precipitation rates along the $CaCO_3$-$MgCO_3$ join, *Aquatic Geochem.* **6**:249–256.

Arvidson, R. S., Mackenzie, F. T., and Guidry, M. W., 2000, Ocean/atmosphere history and carbonate precipitation rates: a solution to the dolomite problem, in: *Marine Authigenesis: from Global to Microbial* (C. R. Glenn, L. Prevot-Lucas, and J. Lucas, eds.), Soc. Econ. Paleontologists Mineralogists Spec. Pub. No. 65, Tulsa, Oklahoma, pp. 1–5.

Arvidson, R. S., Mackenzie, F. T., and Guidry, M., 2006, MAGic: a Phanerozoic model for the geochemical cycling of major rock-forming components, *Am. J. Sci.* **306**:135–190.

Aya, I., Yamane, K., and Nariai, H., 1997, Solubility of CO_2 and density of CO_2 hydrate at 30 MPa, *Energy* **22**:263–271.

Bacastow, R. B., and Dewey, R. K., 1996, Effectiveness of CO_2 sequestration in the post-industrial ocean, *Energy Convers. Management* **37**:1079–1086.

Bacastow, R., and Keeling, C. D., 1973, Atmospheric carbon dioxide and radiocarbon in the natural carbon cycle: II. Changes from A. D. 1700 to 2070 as deduced from a geochemical model, in: *Carbon and the Biosphere* (G. M. Woodwell and E. V. Pecan, eds.), Nat. Tech. Inf. Serv., U. S. Dept. Commerce, Springfield, Va., pp. 86–135.

Baertschi, P., 1976, Absolute ^{18}O content of Standard Mean Ocean Water, *Earth Planet. Sci. Lett.* **31**:341–344.

Balzer, W., and Wefer, G., 1981, Dissolution of carbonate minerals in a subtropical shallow marine environment, *Mar. Chem.* **10**:545–558.

Barker, D. S., 1997, Carbonatites, *McGraw-Hill Encycl. Sci. Technol.*, 8th Ed., Volume 3, McGraw-Hill, New York, pp. 239–240.

Barnes, D. J., and Cuff, C., 2000, Solution of reef rock buffers seawater against rising atmospheric CO_2, in: *Proc. 9th Internat. Coral Reef Symp., Abstracts* (D. Hopley, M. Hopley, J. Tamelander, and T. Done, eds.), State Ministry for the Environment, Indonesia, p. 248.

Bassham, J. A., 1974, Photosynthesis, *Encycl. Brit., Macropaedia*, 1974 Ed., v. 14, Encyclopædia Britannica, Inc, Chicago, pp. 365–373.

Bates, N. R., and Brand, U., 1990, Secular variation of calcium carbonate mineralogy: an evaluation of ooid and micrite chemistries, *Geol. Rundschau* **79**:27–46.

Bates, N., and Samuels, L., 2001, Biogeochemical and physical factors influencing seawater $f CO_2$ and air-sea CO_2 exchange on the Bermuda coral reef, *Limnol. Oceanogr.* **46**:833–846.

Bathurst, R. G. C., 1974, Marine diagenesis of shallow water calcium carbonate sediments, *Ann. Rev. Earth Planet. Sci.* **2**:257–274.

Batjes, N. H., 1996, Total carbon and nitrogen in the soils of the world, *European J. Soil Sci.* **47**:151–163.

Bauer, J. E., Williams, P. M., and Druffel, E. R. M., 1992, Super ^{14}C activity of dissolved organic carbon fractions in the north-central Pacific and Sargasso Sea, *Nature* **357**:667–670.

Baumgartner, A., and Reichel, E., 1975, *The World Water Balance*, R. Oldenburg Verlag, München, 181 pp.

Ben-Yaakov, S., 1973, pH buffering of pore water of recent anoxic marine sediments, *Limnol. Oceanogr.* **18**:86–94.

Berger, W. H., 1982, Increase of carbon dioxide in the atmosphere during deglaciation: the coral reef hypothesis, *Naturwissenschaften* **69**:87–88.

Bernatowicz, T. J., Cowsik, R., Gibbons, P. C., Lodders, K., Fegley, B., Jr., Amari, S., and Lewis, R. S., 1996, Constraints on stellar grain formation from presolar graphite in the Murchison meteorite, *Astrophys. J.* **472**:760–782.

Berner, E. K., and Berner, R. A., 1996, *The Global Environment: Water, Air and Geochemical Cycle*, Prentice-Hall, Upper Saddle River, New Jersey, 376 pp.

Berner, R. A., 1994, Geocarb II: a revised model of atmospheric CO_2 over Phanerozoic time, *Am. J. Sci.* **294**:56–91.

Berner, R. A., 2004, *The Phanerozoic Carbon Cycle: CO_2 and O_2*, Oxford Univ. Press, New York, viii + 150 pp.

Berner, R. A., and Canfield, D. L., 1989, A new model for atmospheric oxygen over Phanerozoic time, *Am. J. Sci.* **289**:333–361.

Berner, R. A., and Kothavala, Z., 2001, GEOCARB III: a revised model of atmospheric CO_2 over Phanerozoic time, *Am. J. Sci.* **301**:182–204.

Berner, R. A., and Maasch, K. A., 1996, Chemical weathering and controls on atmospheric O_2 and CO_2: fundamental principles were enunciated by J. J. Ebelmen in 1845, *Geochim. Cosmochim. Acta* **60**:1633–1637.

Berner, R. A., Lasaga, A. C., and Garrels, R. M., 1983, The carbonate-silicate geochemical cycle and its effect on atmospheric carbon dioxide over the past 100 million years, *Am. J. Sci.* **283**:641–683.

Berry, J. P., and Wilkinson, B. H., 1994, Paleoclimatic control on the accumulation of North American cratonic sediment, *Geol. Soc. America Bull.* **106**:855–865.

Bessat, F., and Buiges, A. D., 2001, Two centuries of variation in coral growth in a massive *Porites* colony from Moorea (French Polynesia): a response of ocean-atmosphere variability from south central Pacific, *Paleogeogr. Paleoclimat. Paleoecol.* **175**:381–392.

Bijma, J., Spero, H. J., and Lea, D. W., 1999, Reassessing foraminiferal stable isotope geochemistry: impacts of the oceanic carbonate system (experimental results), in: *Use of Proxies in Paleoceanography: Example from the South Atlantic* (G. Fisher and G. Wefer, eds.), Springer-Verlag, New York, pp. 489–512.

Birch, F., 1966, Compressibility; elastic constants, in: *Handbook of Physical Constants* (S. P. Clark, Jr., ed.), *Geol. Soc. America Mem.* **97**:97–173.

Bird, M. I., Lloyd, J., and Farquhar, G. D., 1994, Terrestrial carbon storage at the LGM, *Nature* **371**:566.

Bischoff, W. D., Bishop, F. C., and Mackenzie, F. T., 1983, Biogenically produced magnesian calcite: inhomogeneities in chemical and physical properties: comparison with synthetic phases, *Am. Mineral.* **68**:1183–1188.

Bischoff, W. D., Paterson, V., and Mackenzie, F. T., 1984, Geochemical mass balance for sulfur- and nitrogen-bearing acid components, Eastern United States, in: *Geological Aspects of Acid Deposition* (O. P. Bricker, ed.), Amer. Chem. Soc. Acid Precip. Ser. v. 7, Butterworth Publishers, Boston, pp. 1–21.

Bischoff, W. D., Sharma, S. K., and Mackenzie, F. T., 1985, Carbonate ion disorder in synthetic and biogenic magnesian calcites: a Raman spectral study, *Am. Mineral.* **70**:581–589.

Bischoff, W. D., Mackenzie, F. T., and Bishop, F. C., 1987, Stabilities of synthetic magnesian calcites in aqueous solution: comparison with biogenic materials, *Geochim. Cosmochim. Acta* **51**:1413–1423.

Bischoff, W. D., Bertram, M. A., Mackenzie, F. T., and Bishop, F. C., 1993, Diagenetic stabilization pathways of magnesian calcites, *Carbonates and Evaporites* **8**:82–89.

Bluth, G. J., and Kump, L. R., 1991, Phanerozoic paleogeology, *Am. J. Sci.* **291**:284–308.

Boehme, S. E., Sabine, C. L., and Reimers, C. E., 1998, CO_2 fluxes from a coastal transect: a time series approach, *Mar. Chem.* **63**:49–67.

Borges, A., 2005, Do we have enough pieces of the jigsaw to integrate CO_2 fluxes in the coastal ocean?, *Estuaries* **28**:3–27.

Borges, A. V., and Frankignoulle, M., 2001, Short-term variations of the partial pressure of CO_2 in surface waters of the Galician upwelling system, *Progr. Oceanogr.* **51**:283–302.

Borges, A. V., and Frankignoulle, M., 2002a, Distribution and air-water exchange of carbon dioxide in the Scheldt plume off the Belgian coast, *Biogeochemistry* **59**:41–67.

Borges, A. V., and Frankignoulle, M., 2002b, Distribution of surface carbon dioxide and air-sea exchange in the upwelling system off the Galician coast, *Global Biogeochem. Cycles* **16**(2), 14 pp., doi:10.1029/2000GB001385.

Borges, A. V., Delille, B., Schiettecatte, L.-S., Gazeau, F., Abril, G., and Frankignoulle, M., 2004, Gas transfer velocities of CO_2 in three European estuaries (Randers Fjord, Scheldt, and Thames), *Limnol. Oceanogr.* **49**(5):1630–1641.

Borowitzka, M. A., 1981, Photosynthesis and calcification in the articulates coralline red algae *Amphiroa anceps* and *A. foliacea*, *Mar. Biol.* **62**:17–23.

Boss, S. K., and Wilkinson, B. H., 1991, Planktogenic-eustatic control on cratonic-oceanic carbonate accumulation, *J. Geol.* **99**:497–513.

Bottinga, Y., 1968, Calculation of fractionation factors for carbon and oxygen exchange in the system calcite-carbon dioxide-water, *J. Phys. Chem.* **72**:800–808.

Bottinga, Y., 1969a, Calculated fractionation factors for carbon and hydrogen isotope exchange in the system calcite-carbon dioxide-graphite-methane-hydrogen-water vapor, *Geochim. Cosmochim. Acta* **33**:49–64.

Bottinga, Y., 1969b, Carbon isotope fractionation between graphite, diamond and carbon dioxide, *Earth Planet. Sci. Lett.* **5**:301–307.

Boucher, G., Clavier, J., Hily, C., and Gattuso, J.-P., 1998, Contribution of soft-bottoms to the community metabolism (primary production and calcification) of a barrier reef flat (Moorea, French Polynesia), *J. Exp. Mar. Biol. Ecol.* **225**:269–283.

Boyle, E. A., 1988a, The role of vertical chemical fractionation in controlling Late Quaternary atmospheric carbon dioxide, *J. Geophys. Res.* **93**:15,701–15,714.

Boyle, E. A., 1988b, Vertical oceanic nutrient fractionation and glacial/interglacial CO_2 cycles, *Nature* **331**:55–56.

BP, 2004, *Statistical Review of World Energy*, BP p.l.c., London, http://www.bp.com/

Brady, P. V., 1991, The effect of silicate weathering on global temperature and atmospheric CO_2, *J. Geophys. Res.* **96**:18,101–18,106.

Brady, P. V., and Carroll, S. A., 1994, Direct effects of CO_2 and temperature on silicate weathering: possible implications for climate control, *Geochim. Cosmochim. Acta* **58**:1853–1856.

Bragg, W. L., 1924, The structure of aragonite, *Proc. Roy. Soc. London* **105**:16–39.

Brand, U., and Veizer, J., 1980, Chemical diagenesis of a multicomponent carbonate system—1: Trace elements, *J. Sedim. Petrology* **50**:1219–1236.

Brand, U., and Veizer, J., 1981, Chemical diagenesis of a multicomponent carbonate system—2: Stable isotopes, *J. Sedim. Petrology* **51**:987–997.

Bray, J. R., 1961, An estimate of minimum quantum yield of photosynthesis based on ecologic data, *Plant Physiol.* **36**(3):371–373.

Brewer, P. G., Friederich, G. E., Peltzer, E. T., and Orr, F. M., Jr., 1999, Direct experiments on the ocean disposal of fossil fuel CO_2, *Science* **284**:943–945.

Brewer, P. G., Peltzer, E. T., Friederich, G., Aya, I., and Yamane, K., 2000, Experiments on the ocean sequestration of fossil fuel CO_2: pH measurements and hydrate formation, *Mar. Chem.* **72**:83–93.

Brock, T. D., 1973, Lower pH limits for the existence of the blue-green algae: evolutionary and ecological implications, *Science* **179**:480–483.

Brody, M., and Emerson, R., 1959, The quantum yield of photosynthesis in *Porphyridium cruentum*, and the role of chlorophyll a in the photosynthesis of red algae, *J. General Physiol.* **43**: 251–264.

Broecker, W. S., 1974, *Chemical Oceanography*, Harcourt Brace Jovanovich, New York, x + 214 pp.

Broecker, W. S., 1982, Ocean chemistry during glacial time, *Geochim. Cosmochim. Acta* **46**:1689–1705.

Broecker, W. S., 1985, *How to Build a Habitable World*, Eldigio Press, Lamont-Doherty Earth Observatory, Columbia Univ., Palisades, New York.

Broecker, W. S., 1995, *The Glacial World According to Wally*, Lamont-Doherty Earth Observatory, Columbia Univ., Palisades, New York, 316 + A6 pp.

Broecker, W. S., 2002, *The Glacial World According to Wally*, Eldigio Press, Lamont-Doherty Earth Observatory, Columbia Univ., Palisades, New York.

Broecker, W. S., and Henderson, G. M., 1998, The sequence of events surrounding Termination II and their implication for the cause of glacial-interglacial CO_2 changes, *Paleoceanography* **13**:352–364.

Broecker, W. S., and Peng, T.-H., 1982, *Tracers in the Sea*, Lamont-Doherty Geological Observatory, Columbia Univ., Palisades, New York, 690 pp.

Brown, L. R., Renner, M., and Flavin, C., 1997, *The Environmental Trends that are Shaping Our Future: Vital Signs 1997*, W. W. Norton, New York, 165 pp.

Buddemeier, R. W., Kleypas, J. A., and Aronson, R. B., 2004, *Coral Reefs & Global Climate Change*, Pew Center on Global Climate Change, Arlington, Va., vi +44 pp.

Buffett, B. A., 2000, Clathrate hydrates, *Ann. Rev. Earth Planet. Sci.* **28**:477–507.

Buffett, B. A., and Archer, D., 2004, Global inventory of methane clathrate: sensitivity to changes in the deep ocean, *Earth Planet. Sci. Lett.* **227**:185–199.

Busenberg, E., and Plummer, L. N., 1982, The kinetics of dissolution of dolomite in CO_2-H_2O systems at 1.5 to 65°C and 0 to 1 atm P_{CO2}, *Am. J. Sci.* **282**:45–78.

Busenberg, E., and Plummer, L. N., 1986, A comparative study of the dissolution and crystal growth kinetics of calcite and aragonite, in: *Studies in Diagenesis* (F. A. Mumpton, ed.), *U. S. Geol. Surv. Bull.* **1578**:139–168.

Busenberg, E., and Plummer, L. N., 1989, Thermodynamics of magnesian calcite solid-solutions at 25°C and 1 atm total pressure, *Geochim. Cosmochim. Acta* **53**:1189–1208.

Cai, W.-J., Wang, Z. A., and Wang, Y., 2003, The role of marsh-dominated heterotrophic continental margins in transport of CO_2 between the atmosphere, the land-sea interface and the ocean, *Geophys. Res. Lett.* **30**(16):1849, doi:10.1029/2003GL017633.

Caldeira, K., and Kasting, J. F., 1992, The life span of the biosphere revisited, *Nature* **360**:721–723.

Carlson, W. D., 1980, The calcite-aragonite equilibrium: effects of Sr substitution and anion orientational disorder, *Am. Mineral.* **65**:1252–1262.

Carslaw, H. S., and Jaeger, J. C., 1959, *Conduction of Heat in Solids*, 2nd Ed., Oxford Univ. Press, Oxford, x + 510 pp.

Cartigny, P., 2005, Stable isotopes and the origin of diamond, *Elements* **1**:79–84.

Cerling, T. E., 1992, Use of carbon isotopes in paleosols as an indicator of $P(CO_2)$ of the paleoatmosphere, *Global Biogeochem. Cycles* **6**:307–314.

Cerling, T. E., 1999, Paleorecords of C_4 plants and ecosystems, in: C_4 *Plant Biology* (R. F. Sage and R. K. Monson, eds.), Academic Press, San Diego, Calif. pp. 445–469.

Chacko, T., Cole, D. R., and Horita, J., 2001, Equilibriuim oxygen, hydrogen and carbon isotope fractionation factors applicable to geologic systems, in: *Stable Isotope Geochemistry* (J. W. Valley and D. R. Cole, eds.), *Rev. Mineralogy and Geochemistry* **43**:1–81, Mineralogical Society of America, Washington, D.C.

Chapman, R., 2000, *A Sea Water Equation of State Calculator*, The Johns Hopkins Univ., Applied Physics Laboratory, Laurel, Md, http://ioc.unesco.org/oceanteacher/resourcekit/ M3/Converters/SeaWaterEquationOfState/Sea Water Equation of State Calculator.htm

Chave, K. E., 1962, Factors influencing the mineralogy of carbonate sediments, *Limnol. Oceanogr.* **7**:218–223.

Chavez, F. P., and Toggweiler, J. R., 1995, Physical estimates of global new production: the up-welling contribution, in: *Upwelling in the Ocean: Modern Processes and Ancient Records* (C. P. Summerhayes, K.-C. Emeis, M. V. Angel, R. L. Smith, and B. Zeitzschel, eds.), Wiley, New York, pp. 313–320.

Chen, C.-T. A., Rodman, M. R., Wei, C.-L., Olson, E. J., Feely, R. A., and Gendron, J. F., 1986, *Carbonate Chemistry of the North Pacific Ocean*, U. S. Department of Energy, Washington D.C., 176 pp.

Chester, R., 2000, *Marine Geochemistry*, Blackwell, Oxford, xiv + 506 pp.

Chilingar, G. V., 1956, Relationship between Ca/Mg ratio and geological age, *Am. Assoc. Petroleum Geologists Bull.* **20**:153–170.

Chou, L., Garrels, R. M., and Wollast, R., 1989, Comparative study of the kinetics and mechanisms of dissolution of carbonate minerals, *Chem. Geol.* **78**:269–282.

Christianson, G. E., 1999, *Greenhouse: The 200-Year Story of Global Warming*, Penguin Books, New York, 305 pp.

Chung, S.-N., Park, G.-H., Lee, K., Key, R. M., Millero, F. J., Feely, R. A., Sabine, C. L., and Falkowski, P. G., 2004, Postindustrial enhancement of aragonite undersaturation in the upper tropical and subtropical Atlantic Ocean: the role of anthropogenic CO_2, *Limnol. Oceanogr.* **49**:315–321.

Church, M., Ducklow, H. W., and Karl, D. M., 2002, Temporal variability in dissolved organic matter stocks in the Central North Pacific Gyre, *Limnol. Oceanogr.* **47**:1–10.

Chyba, C. F., 1987, The cometary contribution to the oceans of primitive Earth, *Nature* **330**: 632–635.

Cicero, A. D., and Lohmann, K. C., 2001, Sr/Mg variation during rock-water interaction: implications for secular changes in the elemental chemistry of ancient seawater, *Geochim. Cosmochim. Acta* **65**:741–761.

Clark, I. D., and Fritz, P., 1997, *Environmental Isotopes in Hydrology*, Lewis Publishers, New York, 328 pp.

Clausen, C. D., and Roth, A. A., 1975, Effect of temperature and temperature adaptation on calcification rate in the hermatypic coral *Pocillopora damicornis, Mar. Biol.* **33**:93–100.

Cohen, A. L., and McConnaughey, T. A., 2003, Geochemical perspectives on coral mineralization, in: *Biomineralization* (P. M. Dove, J. J. De Yoreo, and S. Weiner, eds.), *Rev. Mineralogy and Geochemistry* **54**:151–187, Mineralogical Society of America, Washington, D.C.

Cohen, J. E., Small, C., Mellinger, A., Gallup, J., and Sachs, J., 1997, Estimates of coastal populations, *Science* **278**:1211–1212.

Cole, D. R., and Chakraborty, S., 2001, Rates and mechanisms of isotopic exchange, in: *Stable Isotope Geochemistry* (J. W. Valley and D. R. Cole, eds.), *Rev. Mineralogy and Geochemistry* **43**:83–223, Mineralogical Society of America, Washington, D. C.

Cole, D. R., and Monger, H. C., 1994, Influence of atmospheric CO_2 on the decline of C_4 plants during the last deglaciation, *Nature* **368**:533–536.

Conkright, M. E., Locarnini, R. A., Garcia, H. E., O'Brien, T. D., Boyer, T. P., Stephens, C., and Antonov, J. I., 2002, *World Ocean Atlas 2001: Objective Analyses, Data Statistics, and Figures*, CD-ROM Documentation, National Oceanographic Data Center, Silver Spring, Md, 17 pp.

Coplen, T. B., Kendall, C., and Hopple, J., 1983, Comparison of isotope reference samples, *Nature* **302**:236.

Coplen, T. B., Hopple, J. A., Böhlke, J. K., Peiser, H. S., Rieder, S. E., Krouse, H. R., Rosman, K. J. R., Ding, T., Vocke, R. D., Jr., Révész, K. M., Lamberty, A., Taylor, P., and De Biévre, P., 2002, Compilation of minimum and maximum isotope ratios of selected elements in naturally occurring terrestrial materials and reagents, *U. S. Geol. Surv., Water Resources Investigation Report* **01–4222**, Reston, Va.

Craig, H., 1957, Isotopic standards for carbon and oxygen and correction factors for mass-spectrometric analysis of carbon dioxide, *Geochim. Cosmochim. Acta* **12**:133–149.

Craig, H., 1961, Standard for reporting concentrations of deuterium and oxygen-18 in natural water, *Science* **133**:1833–1834.

Crommelynk, D., 2002, *Solar Constant*, Roy. Meteor. Inst. Belgium, Electronic report at http://estirm2.oma.be/solarconstant/solar.html.

Crowley, T. J., 1995, Ice age terrestrial carbon changes revisited, *Global Biogeochem. Cycles* **9**:377–389.

Crowley, T. J., and Baum, S. K., 1997, Effect of vegetation on ice-age climate model simulation, *J. Geophys. Res.* **102**:16463–16480.

Crowley, T. J., and Berner, R. A., 2001, CO_2 and climate change, *Science* **292**:870–872.

Crutzen, P. J., 2002, Geology of mankind, *Nature* **415**:23.

Crutzen, P. J., and Stoermer, E. F., 2000, The "Anthropocene", *IGBP Newsletter* **41**:16–18, Royal Swedish Academy of Sciences, Stockholm.

Cubasch, U., and Cess, R. D., 1990, Processes and modelling, in: *Climate Change: The IPCC Scientific Assessment* (J. T. Houghton, G. J. Jenkins, and J. J. Ephraums, eds.), Cambridge Univ. Press, Cambridge, pp. 77–79.

Cubasch, U., Meehl, G. A., Boer, G. J., Stouffer, R. J., Dix, M., Noda, A., Senior, C. A., Raper, S., and Yap, K. S., 2001, Projections of future climate change, in: *Climate Change 2001: The*

Scientific Basis (J. T. Houghton, Y. Ding, D. J. Griggs, M. Noguer, P. J. van der Linden, X. Dai, K. Maskell, and C. A. Johnson, eds.), Cambridge Univ. Press, Cambridge, pp. 525–582.

Cullen, J. T., Rosenthal, Y., and Falkowski, P. G., 2001, The effect of anthropogenic CO_2 on the carbon isotope composition of marine phytoplankton, *Limnol. Oceanogr.* **46**:996–998.

Dacey, M. F., and Lerman, A., 1983, Sediment growth and aging as Markov chains, *J. Geol.* **91**:573–590.

Dalai, T. K., Krishnaswami, S., and Sarin, M. M., 2002, Major ion chemistry in the headwaters of the Yamuna river system: chemical weathering, its temperature dependence and CO_2 consumption in the Himalaya, *Geochim. Cosmochim. Acta* **19**:3397–3416.

Darling, D., 2005, *The Encyclopedia of Astrobiology, Astronomy, and Spaceflight*, http://www.daviddarling.info/encyclopedia/ETEmain.html.

Dawson, A. G., 1992, *Ice Age Earth: Late Quaternary Geology and Climate*, Routledge, London, 293 pp.

Deer, W. A., Howie, R. A., and Zussman, J., 1962a, *Rock-Forming Minerals*, v. 3, *Sheet-Silicates*, Wiley, New York, x + 270 pp.

Deer, W. A., Howie, R. A., and Zussman, J., 1962b, *Rock-Forming Minerals*, v. 5, *Non-Silicates*, Wiley, New York, x + 371 pp.

Deevey, E. S., Jr., 1973, Sulfur, nitrogen and carbon in the atmosphere, in: *Carbon and the Biosphere* (G. M. Woodwell and E. V. Peacan, eds.), U. S. Atomic Energy Commission, CONF-720510, Washington, D.C., pp. 182–190.

DeGrandpre, M. D., Olbu, G. D., Beatty, C. M., and Hammar, T. R., 2002, Air-sea CO_2 fluxes on the U. S. Middle Atlantic Bight, *Deep-Sea Res.* II, **49**:4355–4367.

Deines, P., 1980a, The isotopic composition of reduced organic carbon, in: *Handbook of Environmental Isotope Geochemistry*, vol. 1, *The Terrestrial Environment A* (P. Fritz and J. Ch. Fontes, eds.), Elsevier, Amsterdam, pp. 329–406.

Deines, P., 1980b, The carbon isotopic composition of diamonds: relationship to diamond shape, color, occurrence and vapor composition, *Geochim. Cosmochim. Acta* **44**:943–962.

Deines, P., Langmuir, D., and Harmon, R. S., 1974, Stable carbon isotope ratios and the existence of a gas phase in the evolution of carbonate ground waters, *Geochim. Cosmochim. Acta* **38**:1147–1164.

De Jonge, V. N., Boynton, W., D'Elia, C. F., Elmgren, R., and Welsh, B. L., 1994, Responses to developments in eutrophication in four different North Atlantic ecosystems, in: *Changes in Fluxes in Estuaries* (K. R. Dyer and R. J. Orth, eds.), Olsen and Olsen, Fredensborg, Denmark, pp. 179–196.

Delaney, M. L., and Boyle, E. A., 1986, Lithium in foraminiferal shells: implications for high-temperature hydrothermal circulation fluxes and oceanic generation rates, *Earth Planet. Sci. Lett.* **80**:91–105.

Delgado, E., Mitchell, R. A. C., and Lawlor, D. W., 1994, Interacting effects of CO_2 concentration, temperature and nitrogen supply on the photosynthesis and composition of winter wheat leaves, *Plant, Cell and Environment* **17**:1205–1213.

Delwiche, C. C., and Likens, G. E., 1977, Biological response to fossil fuel combustion products, in: *Global Chemical Cycles and Their Alterations by Man* (W. Stumm, ed.), Dahlem Konferenzen, Springer-Verlag, Berlin, pp. 73–88.

Des Marais, D. J., 1985, Carbon exchange between the mantle and the crust, and its effect upon the atmosphere: today compared to Archean time, in: *The Carbon Cycle and Atmospheric CO_2: Natural Variations Archean to Present* (E. T. Sundquist and W. S. Broecker, eds.), *Geophys. Mon.* **32**:602–611, Am. Geophys. Union, Washington, D. C.

Des Marais, D. J., 2001, Isotopic evolution of the biogeochemical carbon cycle during the Precambrian, in: *Stable Isotope Geochemistry* (J. W. Valley and D. R. Cole, eds.), *Rev. Mineralogy and Geochemistry* **43**:555–578, Mineralogical Society of America, Washington, D.C.

Dickson, A., and Millero, F. J., 1987, A comparison of the equilibrium constants for the dissociation of carbonic acid in seawater media, *Deep-Sea Res.* **38**:1733–1743.

Dignon, J., 1992, NO_x and SO_x emissions from fossil fuels: a global distribution, *Atmosph. Environ.* **26**:1157–1163.

Dignon, J., and Hameed, S., 1989, Global emissions of nitrogen and sulfur oxides from 1860 to 1980, *J. Air Pollution Control Assoc.* **39**:180–186.

Di Pierro, S., Gnos, E., Grobety, B. H., Armbuster, T., Bernasconi, S. M., and Ulmer, P., 2003, Rock-forming moissanite (natural α-silicon carbide), *Am. Mineral.* **88**:1817–1821.

Dixon, J. E., and Clague, D. A., 2001, Volatiles in basaltic glasses from Loihi Seamount, Hawaii: evidence for a relatively dry plume component, *J. Petrology* **42**:627–654.

Dixon, J. E., Clague, D. A., Wallace, P., and Poreda, R., 1997, Volatiles in alkaline basalts from the North Arch Volcanic Field, Hawaii: extensive degassing of deep submarine-erupted alkalic series lavas, *J. Petrology* **38**:911–939.

Donnadieu, D., Goddéris, Y., Ramstein, G., Nédélec, A., and Meert, J., 2004, A "snowball Earth" climate triggered by continental break-up through changes in runoff, *Nature* **428**:303–306.

Drake, B. G., Gonzàlez-Meler, M. A., and Long, S. P., 1997, More efficient plants: a consequence of rising CO_2, *Ann. Rev. Plant Physiol. Plant Mol. Biol.* **48**:609–639.

Drake, C. L., and Burk, C. A., 1974, Geological significance of continental margins, in: *The Geology of Continental Margins* (C. A. Burk and C. L. Drake, eds.), Springer-Verlag, New York, pp. 3–10.

Drever, J. I., 1988, *The Geochemistry of Natural Waters*, 2nd Ed., Prentice Hall, Englewood Cliffs, New Jersey, x + 437 pp.

Drever, J. I., 1997, *The Geochemistry of Natural Waters: Surface and Groundwater Environments*, 3rd Ed., Prentice Hall, Upper Saddle River, New Jersey, xii + 436 pp.

Drever, J. I., Li, Y.-H., and Maynard, J. B., 1988, Geochemical cycles: the continental crust and the oceans, in: *Chemical Cycles in the Evolution of the Earth* (C. B. Gregor, R. M. Garrels, F. T. Mackenzie, and J. B. Maynard, eds.), Wiley, New York, pp. 117–154.

Duan, Z., Møller, N., Greenberg, J., and Weare, J. N., 1992a, The prediction of methane solubility in natural waters to high ionic strength from 0 to 250°C and from 0 to 1600 bar, *Geochim. Cosmochim. Acta* **56**:1451–1460.

Duan, Z., Møller, N., and Weare, J. N., 1992b, An equation of state for the CH_4-CO_2-H_2O system: II. Mixtures from 50 to 1000°C and 0 to 1000 bar, *Geochim. Cosmochim. Acta* **56**:2619–2631.

Ducklow, H. W., and McAllister, S. L., 2004, The biogeochemistry of carbon dioxide in the coastal oceans, in: *The Sea: The Global Coastal Ocean—Multi-scale Interdisciplinary Processes*, v. 13 (A. R. Robinson and K. Brink, eds.), Harvard Univ. Press, Cambridge, Mass., pp. 193–225.

Dudziak, A., and Halas, S., 1996, Influence of freezing and thawing on the carbon isotope composition in soil CO_2, *Geoderma* **69**:209–216.

Dumas, J., 1842, *Essai de Statique Chimique des Êtres Organisés*, 2ème édit., Fortin, Masson, Paris, 4 + 88 pp.

Dutton, H. J., 1997, Carotenoid-sensitized photosynthesis: quantum efficiency, fluorescence and energy transfer, *Photosynthesis Res.* **52**:175–185.

Dutton, J. A., 1995, Scientific priorities in global change research, in: *The State of Earth Science from Space* (G. Asrar and D. J. Dokken, eds.), Am. Inst. Physics Press, Woodbury, New York, pp. 89–90.

Eddy, J. A., 1976, The Maunder minimum, *Science* **192**:1189–1192.

Edmond, J. M., 1970, High precision determination of titration alkalinity and total carbon dioxide content of sea water, *Deep-Sea Res.* **17**:737–750.

Edmond, J. M., Palmer, M. R., Measures, C. I., Brown, E. T., and Huh, Y., 1996, Fluvial geochemistry of the eastern slope of the northeastern Andes and its foredeep in the drainage of the Orinoco in Colombia and Venezuela, *Geochim. Cosmochim. Acta* **60**:2949–2974.

Emerson, S., and Bender, M. L., 1981, Carbon fluxes at the sediment-water interface of the deep sea: calcium carbonate preservation, *J. Mar. Res.* **39**:139–162.

Emery, K. O., and Uchupi, E., 1984, *The Geology of the Atlantic Ocean*, Springer-Verlag, New York, 1050 pp.

Emiliani, C., 1992, *Planet Earth*, Cambridge Univ. Press, New York, xiv + 719 pp.

EPICA Community Members, 2004, Eight glacial cycles from an Antarctic ice core, *Nature* **429**:623–628.

Eugster, H., 1966, Seawater: its history, in: *The Encyclopedia of Oceanography* (R. W. Fairbridge, ed.), Reinhold, New York, pp. 709–802.

Evans, H. T., Jr., 1994, Ionic radii in crystals, in: *Handbook of Chemistry and Physics* (D. R. Lide, ed.-in-chief), 75th Ed., CRC Press, Boca Raton, Fla., pp. **12**-8–**12**-9.

Fairbanks, R. G., 1989, A 17,000-year glacio-eustatic sea level record; influence of glacial melting rates on the Younger Dryas event and deep-ocean circulation, *Nature* **342**:637–642.

Falkowski, P. G., and Raven, J. A., 1997, *Aquatic Photosynthesis*, Blackwell, Malden, Mass., 375 pp.

Falkowski, P., Scholes, R. J., Boyle, E., Canadell, J., Canfield, D., Elser, J., Gruber, N., Hibbard, K., Hogberg, P., Linder, S., Mackenzie, F. T., Moore III, B., Pedersen, T., Rosenthal, Y., Seitzinger, S., Smetacek, V., and Steffen, W., 2000, The global carbon cycle: a test of our knowledge of Earth as a system, *Science* **290**:291–296.

Farquhar, G. D., Ehleringer, J. R., and Hubick, K. T., 1989, Carbon isotope discrimination and photosynthesis, *Ann. Rev. Plant Physiol. Plant Mol. Biol.* **40**:503–537.

Farrell, J. W., and Prell, W. L., 1989, Climatic change and $CaCO_3$ production: an 800,000 year bathymetric reconstruction from the Central Equatorial Pacific Ocean, *Paleoceanography* **4**:447–466.

Farrell, J. W., Pedersen, T. F., Calvert, S. E., and Nielsen, B., 1995, Glacial-interglacial changes in nutrient utilization in the equatorial Pacific Ocean, *Nature* **377**:514–517.

Faure, G., 1986, *Principles of Isotope Geology*, 2nd Ed., Wiley, New York, xv + 589 pp.

Faure, G., 1998, *Principles and Applications of Geochemistry*, 2nd Ed., Prentice-Hall, Upper Saddle River, New Jersey, xvi + 600 pp.

Faure, G., and Mensing, T. M., 2004, *Isotopes: Principles and Applications*, 3rd Ed., Wiley, New York, xxviii + 897 pp.

Feely, R. A., Sabine, C. L., Lee, K., Berelson, W., Kleypas, J., Fabry, V. J., and Millero, F. J., 2004, Impact of anthropogenic CO_2 on the $CaCO_3$ system in the oceans, *Science* **305**:362–366.

Fenchel, T., King, G. M., and Blackburn, T. H., 1998, *Bacterial Biogeochemistry: The Ecophysiology of Mineral Cycling*, 2nd Ed., Academic Press, New York, viii + 307 pp.

Fer, I., and Haugan, P. M., 2003, Dissolution from a liquid CO_2 lake disposed in the deep ocean, *Limnol. Oceanogr.* **48**:872–883.

Fichefet, T., Hovine, S., and Duplessy, J.-C., 1994, A model study of the Atlantic thermohaline circulation during the last glacial maximum, *Nature* **372**:252–255.

Fischer, A. G., 1984, The two Phanerozoic super cycles, in: *Catastrophes in Earth History* (W. A. Berggren and J. A. Vancouvering, eds.), Princeton Univ. Press, Princeton, New Jersey, pp. 129–148.

Frakes, L. A., Francis, J. E., and Syktus, J. I., 1992, *Climate Mode of the Phanerozoic: the History of the Earth's Climate Over the Past 600 Million Years*, Cambridge Univ. Press, New York, xi + 274 pp.

François, R., Altabet, M. A., Yu, E. F., Sigman, D. M., Bacon, M. P., Frank, M., Bohrmann, G., Bareille, G., and Labeyrie, L., 1997, Contribution of Southern Ocean surface-water stratification to low atmospheric CO_2 concentrations during the last glacial period, *Nature* **389**:929–935.

Frank, L. A., and Sigwarth, J. B., 1993, Atmospheric holes and small comets, *Rev. Geophys.* **31**:1–28.

Frankignoulle, M., and Borges, A. V., 2001, European continental shelf as a significant sink for atmospheric carbon dioxide, *Global Biogeochem. Cycles* **15**:569–576.

Frankignoulle, M., Canon, C., and Gattuso, J.-P., 1994, Marine calcification as a source of carbon dioxide: positive feedback of increasing atmospheric CO_2, *Limnol. Oceanogr.* **39**:458–462.

Frankignoulle, M., Abril, G., Borges, A., Bourge, I., Canon, C., DeLille, B., Libert, E., and Théate, J.-M., 1998, Carbon dioxide emission from European estuaries, *Science* **282**: 434–436.

Friedlingstein, P., Prentice, K. C., Fung, I., John, J., and Brasseur, G., 1993, Carbon-biosphere-climate interactions in the last glacial maximum climate, *J. Geophys. Res.* **100**:7203–7221.

Friedlingstein, P., Fung, I. Y., Holland, E., John, J. G., Brasseur, G. P., Erickson, D., and Schimel, D., 1995, On the contribution of CO_2 fertilization to the missing biospheric sink, *Global Biogeochem. Cycles* **9**:541–556.

Friedman, I., and O'Neil, J. R., 1977, Compilation of stable isotope fractionation factors of geochemical interest, in: *Data of Geochemistry*, 6th Ed. (M. Fleischer, ed.), *U. S. Geol. Survey Prof. Pap.* **440-KK**.

Fuex, A. N., and Baker, D. R., 1973, Stable carbon isotopes in selected granitic, mafic and ultramafic igneous rocks, *Geochim. Cosmochim. Acta* **37**:2509–2521.

Gaffron, H., 1964, Photosynthesis, *Encycl. Brit.*, 1964 Ed., v. 17, Encyclopædia Britannica, Inc., Chicago, pp. 855–856B.

Gaillardet, J., Dupré, B., Allègre, C. J., and Negrel, P., 1997, Chemical and physical denudation in the Amazon River Basin, *Chem. Geol.* **142**:141–173.

Gaillardet, J., Dupré, B., Louvat, P., and Allègre, C. J., 1999, Global silicate weathering and CO_2 consumption rates deduced from the chemistry of large rivers, *Chem. Geol.* **159**:3–30.

Gaillardet, J., Millot, R., and Dupré, B., 2003, Chemical denudation rates of the western Canadian orogenic belt: the Stikine terrane, *Chem. Geol.* **201**: 257–279.

Gaines, A. M., 1974, Protodolomite synthesis at 100°C and atmospheric pressure, *Science* **183**:178–182.

Gaines, A. M., 1980, Dolomitization kinetics, recent experimental studies, in: *Concepts and Models of Dolomitization* (D. H. Zenger, J. B. Dunham, and R. L. Ethington, eds.), Soc. Econ. Paleontologists Mineralogists Spec. Pub. No. 28, Tulsa, Oklahoma, pp. 81–86.

Galy, A., and France-Lanord, C., 1999, Weathering processes in the Ganges-Brahmaputra basin and the riverine alkalinity budget, *Chem. Geol.* **159**:31–60.

Ganersham, R. S., Pedersen, T. F., Calvert, S. E., and Murray, J. W., 1995, Large changes in oceanic nutrient inventories from glacial to interglacial periods, *Nature* **376**:755–757.

Gao, K., Aruga, Y., Asada, K., Ishihara, T., Akano, T., and Kiyohara, M., 1993, Calcification in the articulated coralline alga *Corallina pilulifera*, with special reference to the effect of elevated CO_2 concentration, *Mar. Biol.* **117**:129–132.

Garrels, R. M., and Lerman, A., 1981, Phanerozoic cycles of sedimentary carbon and sulfur, *Proc. Nat. Acad. Sci. USA* **78**:4652–4656.

Garrels, R. M., and Lerman, A., 1984, Coupling of the sedimentary sulfur and carbon cycles—an improved model, *Am. J. Sci.* **284**:989–1007.

Garrels, R. M., and Mackenzie, F. T., 1967, Origin of the chemical compositions of some springs and lakes, in: *Equilibrium Concepts in Natural Water Systems* (W. Stumm, ed.), *Adv. Chem. Ser.* **67**:222–242.

Garrels, R. M., and Mackenzie, F. T., 1971a, *Evolution of Sedimentary Rocks*, W. W. Norton, New York, xvi + 397 pp.

Garrels, R. M., and Mackenzie, F. T., 1971b, Gregor's denudation of the continents, *Nature* **231**:382–383.

Garrels, R. M., and Mackenzie, F. T., 1972, A quantitative model for the sedimentary rock cycle, *Mar. Chem.* **1**:27–41.

Garrels, R. M., and Perry, E. A., Jr., 1974, Cycling of carbon, sulfur, oxygen through geologic time, in: *The Sea*, vol. 5 (E. D. Goldberg, ed.), Wiley, New York, pp. 303–336.

Garrels, R. M., Lerman, A., and Mackenzie, F. T., 1976, Controls of atmospheric O_2 and CO_2: past, present and future, *Am. Scientist* **64**:306–315.

Gattuso, J.-P., and Buddemeier, R. W., 2000, Ocean biogeochemistry: calcification and CO_2, *Nature* **407**:311–312.

Gattuso, J.-P., Pichon, M., Delesalle, B., and Frankignoulle, M., 1993, Community metabolism and air-sea CO_2 fluxes in a coral reef ecosystem (Moorea, French Polynesia), *Mar. Ecol. Progr. Ser.* **96**:259–267.

Gattuso, J.-P., Pichon, M., and Frankignoulle, M., 1995, Atmospheric CO_2, marine production and calcification, *Mar. Ecol. Progr. Ser.* **129**:307–312.

Gattuso, J.-P., Frankignoulle, M., Bourge, I., Romaine, S., and Buddemeier, R. W., 1998, Effect of calcium carbonate saturation of seawater on coral calcification, *Global Planet. Change* **18**:37–46.

Gattuso, J.-P., Allemand, P. D., and Frankignoulle, M., 1999, Interactions between the carbon and carbonate cycles at organism and community levels on coral reefs: a review of processes and control by carbonate chemistry, *Am. Zoolog.* **39**:160–188.

Gautelier, M., Oelkers, E. H., and Schott, J., 1999, An experimental study of dolomite dissolution rates as a function of pH from -0.5 to 5 and temperature from 25 to 80°C, *Chem. Geol.* **157**:13–26.

Gavish, E., and Friedman, G. M., 1969, Progressive diagenesis in Quaternary to Late Tertiary carbonate sediments: sequence and time scale, *J. Sedim. Petrology* **39**:980–1006.

Gibbs, M. T., and Kump, L. R., 1994, Global chemical erosion during the last glacial maximum and the present: sensitivity to changes in lithology and hydrology, *Paleoceanography* **9**:529–543.

Gíslason, S. R., Arnórsson, S., and Ármannsson, H., 1996, Chemical weathering of basalt in Southwest Iceland: effects of runoff, age of rocks and vegetative/glacial cover, *Am. J. Sci.* **296**:837–907.

Given, R. K., and Wilkinson, B. H., 1987, Dolomite abundance and stratigraphic age: constraints on rates and mechanisms of Phanerozoic dolostone formation, *J. Sedim. Petrology* **57**:1068–1079.

Goddéris, Y., and Veizer, J., 2000, Tectonic control of chemical and isotopic composition of ancient oceans: the impact of continental growth, *Am. J. Sci.* **300**:434–461.

Gold, T., 1999, *The Deep Hot Biosphere*, Springer-Verlag, New York, xiv + 235 pp.

Goldich, S., 1938, A study in rock-weathering, *J. Geol.* **46**:17–58.

Goldschmidt, V. M., 1933, Grundlagen der quantitativen Geochemie, *Fortsch. Mineral. Kristallog. Petrogr.* **17**:1–112.

Goldsmith, J. R., and Graf, D. L., 1958, Structural and compositional variations in some natural dolomite, *J. Geol.* **66**:678–793.

Goldsmith, J. R., and Heard, H. C., 1961, Subsolidus phase relations in the system $CaCO_3$-$MgCO_3$, *J. Geol.* **69**:45–74.

Goldsmith, J. R., Graf, D. L., and Heard, H. C., 1961, Lattice constants of calcium-magnesium carbonates, *Am. Mineral.* **46**:456–457.

Goody, R., 1976, Atmospheric evaporation, in: Tipler, P. A., *Physics*, pp. 241–243, Worth Publishers, New York, xxvi + 1026 pp.

Goody, R., 1995, *Principles of Atmospheric Physics and Chemistry*, Oxford Univ. Press, New York, xii + 324 pp.

Gordon, A. L., 2000, *Oceanographic Data Profiles from World Ocean Atlas*, Lamont-Doherty Earth Observatory, Columbia Univ., Palisades, New York, http://ingrid.ldgo.columbia. edu/SOURCES/.LEVITUS94/oceanviews2.html.

Gough, D. O., 1981, Solar interior structure and luminosity variations, *Solar Phys.* **74**:21–34.

Gould, S. J., 1991, *Bully for Brontosaurus: Reflections in Natural History*, W. W. Norton, New York, 540 pp.

Goyet, C., Millero, F. J., O'Sullivan, D. W., Eischeid, G., McCue, S. J., and Bellerby, R. G. J., 1998, Temporal variations of pCO_2 in surface seawater of the Arabian Sea in 1995, *Deep-Sea Res.* I, **45**:609–623.

Gradstein, F. M., Ogg, J. G., and Smith, A. G., 2004, *A Geologic Time Scale 2004*, Cambridge Univ. Press, New York, xix + 589 pp.

Graf, D. L., 1961, Crystallographic tables for the rhombohedral carbonates, *Am. Mineral.* **46**:1283–1316.

Graf, D. L., and Goldsmith, J. R., 1955, Dolomite-magnesian relations at elevated temperatures and CO_2 pressures, *Geochim. Cosmochim. Acta* **7**:109–128.

Graf, D. L., and Goldsmith, J. R., 1956, Some hydrothermal syntheses of dolomite and protodolomite, *J. Geol.* **64**:173–186.

Graf, D. L., and Goldsmith, J. R., 1958, The solid solubility of $MgCO_3$ in $CaCO_3$: a revision. *Geochim. Cosmochim. Acta* **13**:218–219.

Graham, L., 1974, Heat, *Encycl. Brit., Macropaedia*, 1974 Ed., v. 8, pp. 700–706; *Micropaedia*, 1974 Ed., vol. 4, Encyclopædia Britannica, Inc., Chicago, p. 1007.

Gregor, C. B., 1968, The rate of denudation in post-Algonkian time, *Koninkl. Ned. Akad. Wetenschap. Proc.* **71**:22–30.

Gregor, C. B., 1970, Denudation of the continents, *Nature* **228**:273–275.

Gregor, C. B., 1980, Weathering rates of sedimentary and crystalline rocks, *Kon. Ned. Akad. Wet. Proc., ser. B, Phys. Sci.* **83**:173–181.

Gregor, C. B., 1985, The mass-age distribution of Phanerozoic sediments, in: *Geochronology and the Geologic Record* (N. J. Snelling, ed.), *Geol. Soc. London Mem.* No. 10, pp. 284–289.

Gregor, C. B., 1988, Prologue: cyclic processes in geology, a historical sketch, in: *Chemical Cycles in the Evolution of the Earth* (C. B. Gregor, R. M. Garrels, F. T. Mackenzie, and J. B. Maynard, eds.), Wiley, New York, pp. 5–16.

Gregor, C. B., 1992, Some ideas on the rock cycle: 1788–1988, *Geochim. Cosmochim. Acta* **56**:2993–3000.

Grigg, R. W., 1982, Coral reef development at high latitudes in Hawaii, in: *Proceedings of the Fourth International Coral Reef Symposium* (E. D. Gomez, C. E. Birkeland, R. W. Buddemeier, R. E. Johannes, J. A. Marsh, Jr., and R. T. Tsuda, eds.), v. 1, Marine Sciences Center, Univ. of the Philippines, Manila, Philippines, pp. 687–693.

Grigg, R. W., 1997, Paleoceanography of coral reefs in the Hawaiian-Emperor Chain—revisited, *Coral Reefs* **16**:S33–S38.

Grotzinger, J. P., and James, N. P., 2000, Precambrian carbonates, evolution of understanding, in: *Carbonate Sedimentation and Diagenesis in the Evolving Precambrian World* (J. P. Grotzinger and N. P. James, eds.), Soc. Econ. Paleontologists Mineralogists Spec. Pub. No. 67, Tulsa, Okla, pp. 1–20.

Grotzinger, J. P., and Rothman, D. H., 1996, An abiotic model for stromatolite morphogenesis, *Nature* **383**:423–425.

Gung, Y., Panning, M., and Romanowicz, B., 2003, Global anisotropy and thickness of continents, *Nature* **422**:707–711.

Gurvich, L. V., Iorish, V. S., Yungman, V. S., and Dorofeeva, O. V., 1994, Thermodynamic properties as a function of temperature, in: *Handbook of Chemistry and Physics* (D. R. Lide, ed.-in-chief), 75th Ed., CRC Press, Boca Raton, Fla, pp. 5-48–5-71.

Hales, B., Bandstra, L., Takahashi, T., Covert, P., and Jennings, J., 2003, The Oregon coastal ocean: a sink for atmospheric CO_2?, *Newsletter of Coastal Ocean Processes* **17**:4–5.

Hallam, A., 1984, Pre-Quaternary sea-level changes, *Ann. Rev. Earth Planet. Sci.* **12**:205–243.

Halley, R. B., and Yates, K. K., 2000, Will reef sediments buffer corals from increased global CO_2?, *Proc. 9th Internat. Coral Reef Symp., Abstracts* (D. Hopley, M. Hopley, J. Tamelander, and T. Done, eds.), State Ministry for the Environment, Indonesia, 248 pp.

Hameed, S., and Dignon, J., 1992, Global emissions of nitrogen and sulfur oxides in fossil fuel combustion 1970–1986, *J. Air Waste Management Assoc.* **42**:159–163.

Hansen, J., Fung, I., Lacis, A., Rind, D., Lebedeff, S., Ruedy, R., and Russell, G., 1988, Global climate changes as forecast by Goddard Institute of Space Studies three-dimensional model, *J. Geophys. Res.* **93**(D8):9341–9364.

Hardie, L. A., 1987, Perspectives on dolomitization: a critical review of some current views, *J. Sedim. Petrology* **57**:166–183.

Hardie, L. A., 1996, Secular variation in seawater chemistry: an explanation for the coupled secular variation in the mineralogies of marine limestones and potash evaporites over the past 600 m.y., *Geology* **24**:279–283.

Harker, R. I., and Tuttle, O. F., 1955a, Studies in the system CaO-MgO-CO_2, Part 1. The thermal dissociation of calcite, dolomite, and magnesite, *Am. J. Sci.* **255**:209–224.

Harker, R. I. and Tuttle, O. F., 1955b. Studies in the system CaO-MgO-CO_2, Part 2. Limits of solid solution along the join $CaCO_3$-$MgCO_3$, *Am. J. Sci.* **253**:274–282.

Harlan, J. R., 1971, Agricultural origins: center and noncenters, *Science* **174**:468–474.

Harland, W. B., Armstrong, R. L., Cox, A. V., Craig, L. E., Smith, A. G., and Smith, D. G., 1990, *A Geologic Time Scale 1989*, Cambridge Univ. Press, Cambridge, xvi + 263 pp.

Harris, R. C., 1969, Boron regulation in the oceans, *Nature* **223**:290–291.

Harrison, K. G., Post, W. M., and Richter, D. D., 1995, Soil carbon turnover in a recovering temperate forest, *Global Biogeochem. Cycles* **9**:449–454.

Harte, J., 1988, *Consider a Spherical Cow: A Course in Environmental Problem Solving*, Univ. Science Books, Mill Valley, California, xvi + 283 pp.

Hay, W. W., 1985, Potential errors in estimates of carbonate rock accumulating through geologic time, in: *The Carbon Cycle and Atmospheric CO_2: Natural Variations Archean to Present* (E. T. Sundquist and W. S. Broecker, eds.), *Geophys. Mon.* **32**:573–584, Am. Geophys. Union, Washington, D. C.

Hayes, J. M., 2001, Fractionation of stable carbon and hydrogen isotopes in biosynthetic processes, in: *Stable Isotope Geochemistry* (J. W. Valley and D. R. Cole, eds.), *Rev. Mineralogy and Geochemistry* **43**:225–277, Mineralogical Society of America, Washington, D. C.

Hayes, J. M., Des Marais, D. J., Lambert, J. B., Strauss, H., and Summons, R. E., 1992a, Proterozoic biogeochemistry, in: *The Proterozoic Biosphere, a Multidisciplinary Study* (J. W. Schopf and C. Klein, eds.), Cambridge Univ. Press, Cambridge, pp. 81–134.

Hayes, J. M., Lambert, I. B., and Strauss, H., 1992b, The sulfur-isotopic record, in: *The Proterozoic Biosphere, a Multidisciplinary Study* (J. W. Schopf and C. Klein, eds.), Cambridge Univ. Press, Cambridge, pp. 129–132.

Hayes, J. M., Strauss, H., and Kaufman, A. J., 1999, The abundance of ^{13}C in marine organic matter and isotopic fractionation in the global biogeochemical cycle of carbon during the past 800 Ma, *Chem. Geol.* **161**:103–125.

Henderson-Sellers, A., and Robinson, P. J., 1986, *Contemporary Climatology*, Wiley and Longman Scientific and Technical, New York, xvi + 439 pp.

Hickson, S. J., 1911, On *Ceratopora*, the type of a new family of *Alcyonaria*, *Proc. Roy. Soc. London*, ser. B, **84**:195–200, pl. 6.

Hickson, S. J., 1912, Change in the name of a genus of *Alcyonaria*, *Zool. Anzeiger* **40**: 351.

Hillel, D., 1998, *Environmental Soil Physics*, Academic Press, New York, xxviii + 771 pp.

Hobbie, J. E., 1980, Major findings, in: *Limnology of Tundra Ponds, Barrow, Alaska* (J. E. Hobbie, ed.), Dowden, Hutchinson and Ross, Stroudsburg, Pa., pp. 1–18.

Hoefs, J., 1987, *Stable Isotope Geochemistry*, 3rd Ed., Springer, Berlin, x + 241 pp.

Hoefs, J., 1997, *Stable Isotope Geochemistry*, 4th Ed., Springer, Berlin, vii + 201 pp.

Hoffman, P. F., Kaufman, A. J., Halverson, G. P., and Schrag, D. P., 1998, A Neoproterozoic Snowball Earth, *Science* **281**:1342–1346.

Holden, N. E., 2002, Table of the isotopes, in: *Handbook of Chemistry and Physics* (D. R. Lide, ed.-in-chief), 85th Ed., 2004–2005, CRC Press, Boca Raton, Florida, pp. **11**-50–**11**-201.

Holland, H. D., 1978, *The Chemistry of the Atmosphere and Oceans*, Wiley, New York, xvi + 351 pp.

Holland, H. D., 1984, *The Chemical Evolution of the Atmosphere and Oceans*, Princeton Univ. Press, Princeton, New Jersey, xii + 582 pp.

Holland, H. D., 1997, Enhanced: evidence for life on Earth more than 3850 million years ago, *Science* **275**:38–39.

Holland, H. D., 2004, The geologic history of seawater, in: *The Oceans and Marine Chemistry* (H. Elderfield, ed.), v. 6, *Treatise on Geochemistry* (H. D. Holland and K. K. Turekian, eds.), Elsevier, Amsterdam, pp. 583–625.

Holland, H. D., and Zimmermann, H., 2000, The dolomite problem revisited. *Internat. Geology Rev.* **42**:481–490.

Holmén, K., 1992, The global carbon cycle, in: *Global Biogeochemical Cycles* (S. S. Butcher, R. J. Charlson, G. H. Orians, and G. V. Wolfe, eds.), Academic Press, New York, pp. 239–262.

Holmes, M. E., Schneider, R. R., Müller, P. J., Segl, M., and Wefer, G., 1997, Reconstruction of past nutrient utilization in the eastern Angola Basin based on sedimentary $^{15}N/^{14}N$ ratios, *Paleoceanography* **12**:604–614.

Holser, W. T., 1963, Chemistry of brine inclusions in Permian salt from Hutchinson, Kansas, *1st Symp. on Salt, Northern Ohio Geol. Soc.*, Cleveland, pp. 86–95.

Holser, W. T., 1984, Gradual and abrupt shifts in ocean chemistry during Phanerozoic time, in: *Patterns of Change in Earth Evolution* (H. D. Holland and A. F. Trendall, eds.), Dahlem Workshop, Springer-Verlag, Berlin, pp. 123–144.

Holser, W. T., and Kaplan, I. R., 1966, Isotope geochemistry of sedimentary sulfates, *Chem. Geol.* **1**:93–135.

Holser, W. T., Schidlowski, M., Mackenzie, F. T., and Maynard, J. B., 1988, Geochemical cycles of carbon and sulfur, in: *Chemical Cycles in the Evolution of the Earth* (C. B. Gregor, R. M. Garrels, F. T. Mackenzie, and J. B. Maynard, eds.), Wiley, New York, pp. 105–173.

Horita, J., Friedman, T. J., Lazar, B., and Holland, H. D., 1991, The composition of Permian seawater, *Geochim. Cosmochim. Acta* **55**:417–432.

Horita, J., Zimmermann, H., and Holland, H. D., 2002, The chemical evolution of seawater during the Phanerozoic: implications from the record of marine evaporates, *Geochim. Cosmochim. Acta* **66**:3733–3756.

HOT, 2003, *Hawaiian Ocean Time Series*, http://hahana.soest.hawaii.edu/hot/hot_jgofs.html.

Houghton, J. T., 1991, *The Physics of Atmospheres*, 2nd Ed., Cambridge Univ. Press, Cambridge, xvi + 271 pp.

Houghton, J. T., Ding, Y., Griggs, D. J., Noguer, M., van der Linden, P. J., Dai, X., Maskell, K., and Johnson, C. A. (eds.), 2001, *Climate Change 2001: The Scientific Basis*, Cambridge Univ. Press, Cambridge, x + 881 pp.

Houghton, R. A., 1995, Effects of land-use change, surface temperature and CO_2 concentration on the terrestrial stores of carbon, in: *Biotic Feedbacks in the Global Climatic System: Will the Warming Feed the Warming?* (G. M. Woodwell and F. T. Mackenzie, eds.), Oxford University Press, Oxford, pp. 333–350.

Houghton, R. A., Schlesinger, W. H., Brown, S., and Richards, J. F., 1985, Carbon dioxide exchange between the atmosphere and terrestrial ecosystems, in: *Atmospheric Carbon Dioxide and the Global Carbon Cycle* (J. R. Trabalka, ed.), U. S. Department of Energy, Washington, D. C., pp. 113–140.

Houghton, R. A., Davidson, E. A., and Woodwell, G. M., 1998, Missing sinks, feedbacks, and understanding the role of terrestrial ecosystems in the global carbon balance, *Global Biogeochem. Cycles* **12**:25–34.

Hoyt, D. V., and Schatten, K. H., 1997, *The Role of the Sun in Climate Change*, Oxford Univ. Press, New York, 279 pp.

Huh, Y., and Edmond, J. M., 1999, The fluvial geochemistry of the rivers of eastern Siberia: III. Tributaries of the Lena and Anabar draining the basement terrain of the Siberian Craton and the Trans-Baikal Highlands, *Geochim. Cosmochim. Acta* **63**:967–987.

Huh, Y., Tsoi, M.-Y., Zaitsev, A., and Edmond, J. M., 1998a, The fluvial geochemistry of the rivers of Eastern Siberia: I. Tributaries of the Lena River draining the sedimentary platform of the Siberian Craton, *Geochim. Cosmochim. Acta* **62**:1657–1676.

Huh, Y., Panteleyev, G., Babich, D., Zaitsev, A., Edmond, J. M., 1998b, The fluvial geochemistry of the rivers of Eastern Siberia: II. Tributaries of the Lena, Omoloy, Yana, Indigirka, Kolyma, and Anadyr draining the collisional/accretionary zone of the Verkhoyansk and Cherskiy ranges, *Geochim. Cosmochim. Acta* **62**:2053–2075.

Iglesias-Rodriguez, M. D., Armstrong, R., Feely, R., Hood, R., Kleypas, J., Milliman, J. D., Sabine, C., and Sarmiento, J., 2002, Progress made in study of ocean's calcium carbonate budget, *Eos, Am. Geophys. Un. Trans.* **83**(34):365 and 374–375.

Imbrie, J., Hays, J. D., Martinson, D. G., McIntyre, A., Mix, A. C., Morley, J. J., Pisias, N. G., Prell, W. L., and Shackleton, N. J., 1984, The orbital theory of Pleistocene climate: support from a revised chronology of the marine $\delta^{18}O$ record, in: *Milankovitch and Climate*, Part I (A. Berger, J. Imbrie, J. D. Hays, and G. K. B. Saltzman, eds.), Reidel, Dordrecht, pp. 269–305.

Indermühle, A., Stocker, T. F., Joos, F., Fischer, H., Smith, H. J., Wahlen, M., Deck, B., Mastroianni, D., Tschumi, J., Blunier, T., Meyer, R., and Stauffer, B., 1999, Holocene carbon-cycle dynamics based on CO_2 trapped in ice at Taylor Dome, Antarctica, *Nature* **398**:121–126.

Jacob, D. J., 2003, The oxidizing power of the atmosphere, in: *Handbook of Weather, Climate, and Water: Atmospheric Chemistry, Hydrology, and Societal Impacts* (T. D. Potter and B. R. Colman, eds.), Wiley, New York, Chapter 2.

Jacobson, A. D., Blum, J. D., and Walter, L. M., 2002, Reconciling the elemental and Sr isotope composition of Himalayan weathering fluxes: insights from the carbonate geochemistry of stream waters, *Geochim. Cosmochim. Acta* **66**:3417–3429.

James, N. P., 1997, The cool-water carbonate depositional realm, in: *Cool-Water Carbonates* (N. P. James and J. A. D. Clarke, eds.), Soc. Sedimentary Geol. Spec. Pub. No. 56, Tulsa, Okla. pp. 1–20.

Joly, J., 1899, An estimate of the geological age of the Earth, *Royal Dublin Soc., Sci. Trans* [2], 7:23–66.

Kah, L. C., Lyons, T. W., and Frank, T. D., 2004, Low marine sulfate and protracted oxygenation of the Proterozoic biosphere, *Nature* **421**:834–838.

Kampschulte, A., 2001, Schwefelisotopenuntersuchungen an strukturell substituierten Sulfaten in marinen Karbonaten des Phanerozoikums: Implikationen für die geochemische Evolution des Meerwassers und die Korrelation verschiedener Stoffkreisläufe, Fakultät für Geowissenschaften, Ruhr-Universität Bochum, Doctoral Dissertation, xii + 152 pp.

Kampschulte, A., and Strauss, H., 1998, The isotopic composition of trace sulphates in Paleozoic biogenic carbonates: implications for coeval seawater and geochemical cycles, *Mineral. Mag.* **62**A:744–745.

Karhu, J. A., and Holland, H. D., 1996, Carbon isotopes and the rise of atmospheric oxygen, *Geology* **24**:867–870.

Karim, A., and Veizer, J., 2000, Weathering processes in the Indus River Basin: implications from riverine carbon, sulfur, oxygen, and strontium isotopes, *Chem. Geol.* **170**:153–177.

Kasting, J. F., 1987, Theoretical constraints on oxygen and carbon dioxide concentrations in the precambrian atmosphere, *Precambrian Res.* **34**:205–229.

Kasting, J. F., and Toon, O. B., 1988, How climate evolved on the terrestrial planets, *Sci. Am.* **258**(2):90–97.

Kasting, J. F., Whitmire, D. P., and Reynolds, R. T., 1993, Habitable zones around main sequence stars, *Icarus* **101**:108–128.

Kattenberg, A., Giorgi, F., Grassl, H., Meehl, G. A., Mitchell, J. F. B., Stouffer, R. J., Tokioka, T., Weaver, A. J., and Wigley, T. M. L., 1996, Climate models—projections of future climate, in: *Climate Change 1995: The Science of Climate Change* (J. T. Houghton, L. G. Meira Filho, B. A. Callander, N. Harris, A. Kettenberg, and K. Maskell, eds.), Cambridge Univ. Press, Cambridge, pp. 285–358.

Katz, A., and Matthews, A., 1977, The dolomitization of $CaCO_3$, an experimental study at 252–295°C, *Geochim. Cosmochim. Acta* **41**:297–308.

Keeling, C. D., and Whorf, T. P., 2003, Atmospheric CO_2 records from sites in the SIO air sampling network, in: *Trends: A Compendium of Data on Global Change*, Carbon Dioxide Information Analysis Center, Oak Ridge National Laboratory, U. S. Department of Energy, Oak Ridge, Tenn., http://cdiac.esd.ornl.gov/trends/co2/sio-mlo.htm.

Keeling, C. D., Mook, W. G., and Tans, P. P., 1979, Recent trends in the $^{13}C/^{12}C$ ratio of atmospheric carbon dioxide, *Nature* **277**:121–123.

Keeling, C. D., Carter, A. F., and Mook, W. G., 1984, Seasonal, latitudinal and secular variations in the abundance and isotopic ratios of atmospheric carbon dioxide. II. Results from oceanographic cruises in the tropical Pacific Ocean, *J. Geophys. Res.* **89**:4615–4628.

Keir, R. S., 1995, Is there a component of Pleistocene CO_2 change associated with carbonate dissolution cycles?, *Paleoceanography* **10**:871–880.

Keir, R. S., and Berger, W. H., 1985, Late Holocene carbonate dissolution in the equatorial Pacific: reef growth or Neoglaciation?, in: *The Carbon Cycle and Atmospheric CO_2: Natural Variations Archean to Present* (E. T. Sundquist and W. S. Broecker, eds.), *Geophys. Monogr.* **32**:208–219, Am. Geophys. Union, Washington, D. C.

Kempe, S., and Degens, E. T., 1985, An early soda ocean?, *Chem. Geol.* **53**:95–108.

Kempe, S., and Kaźmierczak, J., 1994, The role of alkalinity in the evolution of ocean chemistry, organization of living systems, and biocalcification processes, *Inst. Océanogr. Monaco Bull.*, num. spéc. **13**, pp. 61–117.

Kempe, S., and Pegler, K., 1991, Sinks and sources of CO_2 in coastal seas: the North Sea, *Tellus* **43**B:224–235.

Kempe, S., Kaźmierczak, J., and Degens, E. T., 1989, The soda ocean concept and its bearing on biotic evolution, in: *Origin, Evolution, and Modern Aspects of Biomineralization in Plants and Animals* (R. E. Crick, ed.), Plenum, New York, pp. 29–39.

Kern, R., and Weisbrod, A., 1967, *Thermodynamics for Geologists*, Freeman, Cooper, San Francisco, Calif., 304 pp.

Kerr, R. A., 2002, Deep life in the slow, slow lane, *Science* **296**:1056–1058.

Kerrick, D. M., 2001, Present and past nonanthropogenic CO_2 degassing from the solid Earth, *Rev. Geophys.* **39**:565–585.

Kerrick, D. M., and Connolly, J. A. D., 1998, Subduction of ophiocarbonates and recycling of CO_2 and H_2O, *Geology* **26**:375–378.

Kerrick, D. M., and Connolly, J. A. D., 2001a, Metamorphic devolatilization of subducted marine sediments and the transport of volatiles into the Earth's mantle, *Nature* **411**:293–296.

Kerrick, D. M., and Connolly, J. A. D., 2001b, Metamorphic devolatilization of subducted oceanic metabasalts: implications for seismicity, arc magmatism and volatile recycling, *Earth Planet. Sci. Lett.* **189**:19–29.

Ketchum, B. H., 1969, Productivity of marine communities, in: *Encyclopedia of Marine Resources*, (F. E. Firth, ed.), Van Nostrand Reinhold, New York, pp. 553–559.

Kheshgi, H. S., Jain, A. K., and Wuebbles, D. J., 1999, Model-based estimation of the global carbon budget and its uncertainty from carbon dioxide and carbon isotope records, *J. Geophys. Res.* **104**:31,127–31,143.

Kiessling, W., 2002, Secular variations in the Phanerozoic reef ecosystem, in: *Phanerozoic Reef Patterns* (W. Kiessling, E. Flügel, and J. Golonka, eds.), Soc. Econ. Paleontologists Mineralogists Spec. Pub. No. 72, Tulsa, Okla., pp. 625–690.

Kimball, J. W., 2004, The energy relationships in cellular respiration and photosynthesis: the balance sheet, *Kimball's Biology Pages*, http://biology-pages.info

Kleypas, J. A., Buddemeier, R. W., Archer, D., Gattuso, J.-P., Langdon, C., and Opdyke, B. N., 1999, Geochemical consequences of increased atmospheric carbon dioxide on coral reefs, *Science* **284**:118–120.

Kleypas, J. A., Buddemeier, R. W., and Gattuso, J.-P., 2001, The future of coral reefs in an age of global change, *Internat. J. Earth Sci. (Geol. Rundschau)* **90**:426–437.

Köhler, P., Fischer, H., Munhoven, G., and Zeebe, R. E., 2005, Quantitative interpretation of atmospheric carbon records over the last glacial termination, *Global Biogeochem. Cycles* **20**, GB00234, doi: 10.1029/2004, 33 pp.

Kohfeld, K. E., Le Quéré, C., Harrison, S. P., and Anderson, R. F., 2005, Role of marine biology in the glacial-interglacial cycles, *Science* **308**:74–78.

Krom, M. D., Groom, S., and Zohary, T., 2003, The Eastern Mediterranean, in: *Biogeochemistry of Marine Systems* (K. D. Black and G. B. Shimmield, eds.), Blackwell, Oxford, pp. 91–126.

Kroopnick, P., 1980, The distribution of ^{13}C in the Atlantic Ocean, *Earth Planet. Sci. Lett.* **49**:469–494.

Kroopnick, P., 1985, The distribution of ^{13}C of ΣCO_2 in the world oceans, *Deep-Sea Res.* **32**:57–84.

Kumar, N., Anderson, R. F., Mortlock, R. A., Froelich, P. N., Kubik, P., Dittrich-Hannen, B., and Suter, M., 1995, Increased biological productivity and export production in the glacial Southern Ocean, *Nature* **378**:675–680.

Kump, L. R., 1989, Alternative modeling approaches to the geochemical cycles of carbon, sulfur, and strontium isotopes, *Am. J. Sci.* **289**:390–410.

Kump, L. R., 1991, Interpreting carbon-isotope excursions: Strangelove oceans, *Geology* **19**:299–302.

Kump, L. R., and Alley, R. B., 1994, Global geochemical weathering on glacial time scales, in: *Material Fluxes on the Surface of the Earth* (National Research Council, ed.), National Academy Press, Washington, D.C., pp. 46–60.

Kump, L. R., Kasting, J. F., and Crane, R. G., 1999, *The Earth System*, Prentice Hall, Upper Saddle River, New Jersey, xii + 351 pp.

Kump, L. R., Kasting, J. F., and Crane, R. G., 2004, *The Earth System*, 2nd Ed., Prentice Hall, Upper Saddle River, New Jersey, xii + 419 pp.

Kump, L. R., Pavlov, A., and Arthur, M. A., 2005, Massive release of hydrogen sulfide to the surface ocean and atmosphere during intervals of oceanic anoxia, *Geology* **33**:397–400.

Kvenvolden, K. A., 1988, Methane—a major reservoir of carbon on a shallow geosphere?, *Chem. Geol.* **71**:41–51.

Kvenvolden, K. A., and Lorenson, T. D., 2001, The global occurrence of natural gas hydrates, in: *Natural Gas Hydrates: Occurrence, Distribution, and Detection* (C. K. Paull and W. P. Dillon, eds.), *Geophys. Mon.* **124**:3–18, Am. Geophys. Union, Washington, D. C.

Lagrula, J., 1966, Hypsographic curve, in: *Encyclopedia of Geomorphology* (R. W. Fairbridge, ed.), Reinhold, New York, pp. 364–366.

Land, L. S., 1985, The origin of massive dolomite, *J. Geological Education* **33**:112–125.

Land, L. S., 1989, The carbon and oxygen isotopic chemistry of surficial Holocene shallow marine carbonate sediment and Quaternary limestone and dolomite, in: *Handbook of Environmental Isotope Geochemistry*, v. 3, *The Marine Environment* (P. Fritz and J. Ch. Fontes, eds.), Elsevier, Amsterdam, pp. 191–218.

Land, L. S., 1995, Comment on "Oxygen and carbon isotopic composition of Ordovician brachiopods: implications for coeval seawater by Qing, H., Veizer, J.", *Geochim. Cosmochim. Acta* **59**:2843–2844.

Langdon, C., Takahashi, T., Sweeney, C., Chipman, D., Goddard, J., Marubini, F., Aceves, H., Barnett, H., and Atkinson, M., 2000, Effect of calcium carbonate saturation state on the calcification rate of an experimental coral reef, *Global Biogeochem. Cycles* **14**:639–654.

Langdon, C., Broecker, W. S., Hammond, D. E., Glenn, E., Fitzsimmons, K., Nelson, S. G., Peng, T.-H., Hajdas, I., and Bonani, G., 2003, Effect of elevated CO_2 on the community metabolism of an experimental coral reef, *Global Biogeochem. Cycles* **17**, doi:10.1029/2002GB001941.

Lasaga, A. C., 1998, *Kinetic Theory in the Earth Sciences*, Princeton Univ. Press, Princeton, New Jersey, x + 811 pp.

Lasaga, A. C., and Ohmoto, H., 2002, The oxygen geochemical cycle: dynamics and stability, *Geochim. Cosmochim. Acta* **66**:361–381.

Lasaga, A. C., Berner, R. A., and Garrels, R. M., 1985, An improved geochemical model of atmospheric CO_2 fluctuations over the past 100 million years, in: *The Carbon Cycle and Atmospheric CO₂: Natural Variations Archean to Present* (E. T. Sundquist and W. S. Broecker, eds.), *Geophys. Mon.* **32**:397–411, Am. Geophys. Union, Washington, D. C.

Laws, E. A., Popp, B. N., Bidigare, R. R., Kennicutt, M. C., and Macko, S. A., 1995, Dependence of phytoplankton carbon isotopic composition on growth rate and [CO₂]ₐq: theoretical considerations and experimental results, *Geochim. Cosmochim. Acta* **59**:1131–1138.

Laws, E. A., Bidigare, R. R., and Popp, B. N., 1997, Effect of growth rate and CO₂ concentration on carbon fractionation by the marine diatom *Phaeodactylum tricornutum, Limnol. Oceanogr.* **42**:1552–1560.

Lazar, B., and Holland, H. D., 1988, The analysis of fluid inclusions in halite, *Geochim. Cosmochim. Acta* **52**:485–490.

Lear, C. H., Elderfield, H., and Wilson, P. A., 2003, A Cenozoic seawater Sr/Ca record from benthic foraminiferal calcite and its application in determining global weathering fluxes, *Earth Planet. Sci. Lett.* **208**:69–84.

Leclercq, N., Gattuso, J.-P., and Jaubert, J., 2000, CO₂ partial pressure controls the calcification rate of a coral community, *Global Change Biol.* **6**:329–334.

Leclercq, N., Gattuso, J.-P., and Jaubert, J., 2002, Primary production, respiration, and calcification of a coral reef mesocosm under increased CO₂ partial pressure, *Limnol. Oceanogr.* **47**:558–564.

Lee, C., 1992, Controls on organic carbon preservation: the use of stratified water bodies to compare intrinsic rates of decomposition in oxic and anoxic systems, *Geochim. Cosmochim. Acta* **56**:3323–3335.

Lemmon, E. W., McLinden, M. O., and Friend, D. G., 2003, Thermophysical properties of fluid systems, in: *NIST Chemistry WebBook, NIST Standard Reference Database Number 69* (P. J. Linstrom and W. G. Mallard, eds.), National Institute of Standards and Technology, Gaithersburg, Md., http://webbook.nist.gov/.

Le Quéré, C., and Metzl, N., 2004, Natural processes regulating the ocean uptake of CO₂, in: *The Global Carbon Cycle* (C. B. Field and M. R. Raupach, eds.), SCOPE 62, Island Press, Washington, D. C., pp. 243–255.

Lerman, A., 1979, *Geochemical Processes—Water and Sediment Environments*, Wiley, New York, viii + 481 pp.; 1988, reprint ed., Kruger Press, Malabar, Fla.

Lerman, A., 1994, Surficial weathering fluxes and their geochemical controls, in: *Material Fluxes on the Surface of the Earth* (W. W. Hay and others, eds.), Studies in Geophysics, Nat. Res. Council, Nat. Acad. Press, Washington, D. C., pp. 28–45.

Lerman, A., and Mackenzie, F. T., 2004, CO₂ air-sea exchange due to calcium carbonate and organic matter storage: pre-industrial and Last Glacial Maximum estimates, *Biogeosciences Discussions* **1**:429–495.

Lerman, A., and Mackenzie, F. T., 2005, CO₂ air-sea exchange due to calcium carbonate and organic matter storage, and its implications for the global carbon cycle, *Aquatic Geochem.* **11**:345–390.

Lerman, A., and Wu, L., 2006, CO₂ and sulfuric acid controls of weathering and river water composition, *J. Geochem. Exploration* **88**:427–430.

Lerman, A., Mackenzie, F. T., and Garrels, R. M., 1975, Modeling of geochemical cycles: phosphorus as an example, *Geol. Soc. America Mem.* **142**:205–218.

Lerman, A., Weiser, N. M., and Plummer, L. N., 1996, Geochemical aspects of deep acid waste injection, in: *Deep Injection Disposal of Hazardous and Industrial Wastes* (J. A. Apps and C.-F. Tsang, eds.), Academic Press, New York, pp. 585–600.

Lerman, A., Mackenzie, F. T., and Ver, L. M., 2004, Coupling of the perturbed C-N-P cycles in industrial time, *Aquatic Geochem.* **10**:3–32.

Lewis, E., and Wallace, D. W. R., 1998, Program Developed for CO$_2$ System Calculations. ORNL/CDIAC-105. Carbon Dioxide Information Analysis Center, Oak Ridge National Laboratory, U. S. Department of Energy, Oak Ridge, Tenn.

Li, Y.-H., 2000, *A Compendium of Geochemistry*, Princeton Univ. Press, Princeton, New Jersey, xiv + 475 pp.

Lide, D. R. (ed.-in-chief), 1994, Bond lengths and angles in gas-phase molecules, in: *Handbook of Chemistry and Physics*, 75th Ed., CRC Press, Boca Raton, Fla., pp. **9**-15–**9**-21.

Lide, D. R., (ed.-in-chief), 1994, *Handbook of Chemistry and Physics*, 75th Ed., CRC Press, Boca Raton, Fla.

Lide, D. R., (ed.-in-chief), 2004, Vapor pressure of water from 0 to 370°C, pp. **6**-9–**6**-9; Chemical composition of the human body, p. **7**-11, in: *Handbook of Chemistry and Physics*, 85th Ed., CRC Press, Boca Raton, Fla.

Likens, G. E., Bormann, H. F., and Johnson, N. M., 1981, Interactions between major biogeochemical cycles in terrestrial ecosystems, in: *Some Perspectives of the Major Biogeochemical Cycles* (G. E. Likens, ed.), SCOPE 17, Wiley, New York, pp. 93–112.

Liou, K. N., 1992, *Radiation and Cloud Processes in the Atmosphere*, Oxford Univ. Press, New York, x + 487 pp.

Locklair, R. E., and Lerman, A., 2005, A model of Phanerozoic cycles of carbon and calcium in the global ocean: evaluation and constraints on ocean chemistry and input fluxes, *Chem. Geol.* **217**:113–126.

Lotka, A. J., 1925, *Elements of Physical Biology*, Williams & Wilkins, Baltimore, Md, xxx + 460 pp. Published as Lotka, A. J., 1956, *Elements of Mathematical Biology*, Dover, New York, xxx + 465 pp.

Lough, J. M., and Barnes, D. J., 2000, Environmental control on the massive coral *Porites, J. Exper. Mar. Biol. Ecol.* **245**:225–243.

Lovelock, J. E., 1988, *The Ages of Gaia*, Oxford Univ. Press, New York, 272 pp.

Lovelock, J. E., 1995, *Gaia, a New Look at Life on Earth*, Oxford Univ. Press, New York, xviii + 148 p.

Lowenstam, H. A., 1986, Mineralization processes in monerans and protoctists, in: *Biomineralization in Lower Plants and Animals* (B. S. C. Leadbeater and R. Riding, eds.), Oxford Univ. Press, Oxford, pp. 1–17.

Lowenstam, H. A., and Weiner, S., 1989, *On Biomineralization*, Oxford Univ. Press, Oxford, ix + 324 pp.

Lowenstein, T. K., Timofeeff, M. N., Brennan, S. T., Hardie, L. A., and Demicco, R. V., 2001, Oscillations in Phanerozoic seawater chemistry: evidence from fluid inclusions, *Science* **294**:1086–1088.

Lumsden, D. N., 1985, Secular variations in dolomite abundance in deep marine sediments, *Geology* **13**:766–769.

Lumsden, D. N., Snipe, L. G., and Lloyd, R. V., 1989, Mineralogy and Mn geochemistry of laboratory synthesized dolomite, *Geochim. Cosmochim. Acta* **53**:2325–2329.

Lüttge, A., Winkler, U., and Lasaga, A. C., 2003, Interferometric study of the dolomite dissolution; a new conceptual model for mineral dissolution, *Geochim. Cosmochim. Acta* **67**:1099–1116.

Lyle, M., Zahn, R., Prahl, F., Dymond, J., Collier, R., Pisias, N., and Suess, E., 1992, Paleoproductivity and carbon burial across the California Current: the Multitracers Transect, 42°N, *Paleoceanography* **7**:251–272.

Machel, H. G., and Mountjoy, E. W., 1986, Chemistry and environments of dolomitization—a reappraisal, *Earth Sci. Rev.* **23**:175–222.

Mackenzie, F. T., 1975, Sedimentary cycling and the evolution of seawater, in: *Chemical Oceanography*, v. 1 (J. P. Riley and G. Skirrow, eds.), 2nd Ed., Academic Press, New York, pp. 309–364.

Mackenzie, F. T., 1992, Chemical mass balance between rivers and oceans, in: *Encyclopedia of Earth System Science*, v. 1, Academic Press, New York, pp. 431–445.

Mackenzie, F. T., 2003, *Our Changing Planet: An Introduction to Earth System Science and Global Environmental Change*, 3rd Ed., Prentice Hall, Upper Saddle River, New Jersey, xii + 580 pp.

Mackenzie, F. T., and Agegian, C., 1989, Biomineralization and tentative links to plate tectonics, in: *Origin, Evolution and Modern Aspects of Biomineralization in Plants and Animals* (R. E. Crick, ed.), Plenum Press, New York, pp. 11–28.

Mackenzie, F. T., and Garrels, R. M., 1966a, Silica-bicarbonate balance in the ocean and early diagenesis, *J. Sedim. Res.* **36**:1075–1084.

Mackenzie, F. T., and Garrels, R. M., 1966b, Chemical mass balance between rivers and oceans, *Am. J. Sci.* **264**:507–525.

Mackenzie, F. T., and Morse, J. W., 1992, Sedimentary carbonates through Phanerozoic time, *Geochim. Cosmochim. Acta* **56**:3281–3295.

Mackenzie, F. T., and Pigott, J. P., 1981, Tectonic controls of Phanerozoic sedimentary rock cycling, *J. Geol. Soc. London* **138**:183–196.

Mackenzie, F. T., Bischoff, W. D., Bishop, F. C., Loijens, M., Schoonmaker, J., and Wollast, R., 1983, Magnesian calcites: low-temperature occurrence, solubility, and solid solution behavior, in: *Carbonates: Mineralogy and Chemistry* (R. J. Reeder, ed.), *Rev. Mineralogy* **11**:97–144, Mineralogical Society of America, Washington, D. C.

Mackenzie, F. T., Ver, L. M., Sabine, C., Lane, M., and Lerman, A., 1993, C, N, P, S global biogeochemical cycles and modeling of global change, in: *Interactions of C, N, P and S Biogeochemical Cycles and Global Change* (R. Wollast, F. T. Mackenzie, and L. Chou, eds.), Springer-Verlag, New York, pp. 1–62.

Mackenzie, F. T., Vink, S., Wollast, R., and Chou, L., 1995, Comparative geochemistry of marine saline lakes, in: *Physics and Chemistry of Lakes* (A. Lerman, D. Imboden, and J. Gat, eds.), 2nd Ed., Springer-Verlag, Berlin, pp. 265–278.

Mackenzie, F. T., Lerman, A., and Ver, L. M., 1998, Role of the continental margin in the global carbon balance during the past three centuries, *Geology* **26**:423–426.

Mackenzie, F. T., Lerman, A., and Ver, L. M. B., 2001, Recent past and future of the global carbon cycle, in: *Geological Perspectives of Global Climate Change* (L. C. Gerhard, W. E. Harrison, and B. M. Hanson, eds.), AAPG Studies in Geology No. 47, Am. Assoc. Petroleum Geologists, Tulsa, Okla, pp. 51–82.

Mackenzie, F. T., Ver, L. M., and Lerman, A., 2002, Century-scale nitrogen and phosphorus controls of the carbon cycle, *Chem. Geol.* **190**:13–32.

Mackenzie, F. T., Lerman, A., and Andersson, A. J., 2004, Past and present of sediment and carbon biogeochemical cycling models, *Biogeosciences* **1**:27–85, http://www. biogeosciences.net/bgd/1/27/.

Madigan, M. T., Takigiku, R., Lee, R. G., Gest, H., and Hayes, J. M., 1989, Carbon isotope fractionation by thermophilic phototrophic sulfur bacteria: evidence of autotrophic growth in natural populations, *Appl. Env. Microbiol.* **55**(3):639–644.

Maier-Reimer, E., 1993, Geochemical cycles in an ocean general circulation model: preindustrial tracer distributions, *Global Biogeochem. Cycles* **7**:645–677.

Maier-Reimer, E., and Hasselmann, K. F., 1987, Transport and storage of CO_2 in the ocean—an inorganic ocean-circulation carbon cycle model, *Climate Dyn.* **2**:63–90.

Manabe, S., and Stouffer, R. J., 1994, Multiple-century response of a coupled ocean-atmosphere model to increase of atmospheric carbon dioxide, *J. Climate* **7**:5–23.

Manabe, S., and Stouffer, R. J., 1999, The role of thermohaline circulation in climate, *Tellus*, **51A**(1):91–109.

Mantoura, R. F. C., Martin, J. M., Wollast, R., and Jickells, T. D. (eds.), 1991, *Ocean Margin Processes in Global Change: Report of the Dahlem Workshop on Ocean Margin Processes in Global Change, Berlin, 1990*, Wiley, New York, xvi + 469 pp.

Margulis, L., and Schwartz, K. V., 1998, *Five Kingdoms*, 3rd Ed., Freeman, New York, xx + 520 pp.

Marland, G., Boden, T. A., and Andres, R. J., 2003, Global, regional, and national CO_2 emissions, in: *Trends: A Compendium of Data on Global Change*, Carbon Dioxide Information Analysis Center, Oak Ridge National laboratory, Oak Ridge, Tenn., http://cdiac.esd.ornl.gov/trends/emis/tre_glob.htm.

Marshall, T. J., Holmes, J. W., and Rose, C. W., 1996, *Soil Physics*, 3rd Ed., Cambridge Univ. Press, Cambridge, xiv + 453 pp.

Marty, B., and Tolstikhin, I. N., 1998, CO_2 fluxes from mid-ocean ridges, arcs and plumes, *Chem. Geol.* **145**:233–248.

Marubini, F., and Atkinson, M. J., 1999, Effects of lowered pH and elevated nitrate on coral calcification, *Mar. Ecol. Prog. Ser.* **188**:117–121.

Marubini, F., and Thake, B., 1999, Bicarbonate addition promotes coral growth, *Limnol. Oceanogr.* **44**:716–720.

Marubini, F., Barnett, H., Langdon, C., and Atkinson, M. J., 2001, Dependence of calcification on light and carbonate ion concentration for the hermatypic coral *Porites compressa, Mar. Ecol. Prog. Ser.* **220**:153–162.

Marubini, F., Ferrier-Pagés, C., and Cuif, J.-P., 2003, Suppression of skeletal growth in scleractinian corals by decreasing ambient carbonate-ion concentration: a cross-family comparison, *Proc. Roy. Soc. London*, ser. B, **270**:179–184.

Mason, B., 1958, *Principles of Geochemistry*, 2nd Ed. Wiley, New York, vii + 310 pp.

McArthur, J. M., 1994, Recent trends in strontium isotope stratigraphy, *Terra Nova* **6**:331–358.

McConnaughey, T.A., and Whelan, J. F., 1997, Calcification generates protons for nutrient and bicarbonate uptake, *Earth Sci. Rev.* **42**:95–117.

McConnaughey, T. A., Adey, W. H., and Small, A. M., 2000, Community and environmental influences on reef coral calcification, *Limnol. Oceanogr.* **45**:1667–1671.

McCorkle, D. C., Veeh, H. H., and Heggie, D. T., 1994, Glacial–Holocene paleoproductivity off western Australia: a comparison of proxy records, in: *Carbon Cycling in the Glacial Ocean: Constraints on the Ocean's Role in Global Change* (R. Zahn, T. F. Pedersen, M. A. Kaminski, and L. Labeyrie, eds.), NATO ASI Ser. I, Springer-Verlag, Heidelberg, pp. 443–479.

McGuire, A. D., Melillo, J. M., and Joyce, L. A., 1995, The role of nitrogen in the response of forest net primary production to elevated atmospheric carbon dioxide, *Ann. Rev. Ecol. Systematics* **26**:473–503.

McKenzie, J. A., 1991, The dolomite problem: an outstanding controversy, in: *Controversies in Modern Geology: Evolution of Geological Theories in Sedimentology, Earth History and Tectonics* (D. W. Muller, J. A. McKenzie, and H. Weissert, eds.), Academic Press, London, pp. 37–54.

McNeil, B. I., Matear, R. J., and Barnes, D. J., 2004, Coral reef calcification and climate change: the effect of ocean warming, *Geophys. Res. Lett.* **31**:L22309, doi:10.1029/2004GL021541.

Meadows, D. H., Meadows, D. L., Randers, J., and Behrens III, W. W., 1972, *The Limits to Growth*, Universe Books, New York, 205 pp.

Mehrbach, C., Culberson, C. H., Hawley, J. E., and Pytkowicz, R. M., 1973, Measurement of the apparent dissociation constants of carbonic acid in seawater at atmospheric pressure, *Limnol. Oceanogr.* **18**:897–907.

Meybeck, M., 1979, Concentrations des eaux fluviales en éléments majeurs et apports en solution aux océans, *Rev. Géol. Dyn. Géogr. Phys.* **21**(3):217–246.

Meybeck, M., 1982, Carbon, nitrogen, and phosphorus transport by world rivers, *Am. J. Sci.* **282**:401–450.

Meybeck, M., 1984, Les fleuves et le cycle géochimique des éléments, Thèse de Doctorat d'Etat ès Sciences Naturelles, Nº 84–35, Univ. Pierre et Marie Curie, Paris.

Meybeck, M., 1987, Global chemical weathering of surficial rocks estimated from river dissolved loads, *Am. J. Sci.* **287**:401–428.

Meybeck, M., 1988, How to establish and use world budgets of riverine materials, in: *Physical and Chemical Weathering in Geochemical Cycles* (A. Lerman and M. Meybeck, eds.), Kluwer, Dordrecht, The Netherlands, pp. 247–272.

Meybeck, M., and Ragu, A., 1995, River Discharges to the Oceans: an Assessment of Suspended Solids, Major Ions and Nutrients, United Nations Environment Programme, ii + 245 pp.

Meyer, B. S., 1964, Plant Physiology, *Encycl. Brit.*, 1964 Ed., v. 18, Encyclopædia Britannica, Inc, Chicago, pp. 16–31.

Milankovitch, M., 1920, *Théorie Mathématique des Phénomènes Thermiques Produits par la Radiation Solaire*, Gauthier-Villars, Paris, xvi + 338 pp.

Milankovitch, M., 1930, *Mathematische Klimalehre und astronomische Theorie der Klimaschwankungen*, Gebrüder Borntraeger, Berlin, iv + 176 pp.

Milankovitch, M., 1941, *Kanon der Erdbestrahlung und seine Anwendung auf das Eiszeitenproblem*, Académie Royale Serbe, édit. spec. tome 132, Sect. Sci. Math. Nat. tome 33, Belgrade, xx + 663 pp. English translation: 1969, *Canon of Insolation and Ice-Age Problem*, xxiii + 484 pp., translated by Israel Program for Scientific Translations, Jerusalem, available from U. S. Dept. Commerce, Clearinghouse for Scientific and Technical Information, Springfield, Va.

Miller, S. L., 1953, A production of amino acids under possible primitive earth conditions, *Science* **117**:528–529.

Miller, S. L., and Orgel, L. E., 1974, *The Origins of Life on the Earth*, Prentice-Hall, Englewood Cliffs, New Jersey, x + 229 pp.

Millero, F. J., 1996, *Chemical Oceanography*, 2nd Ed., CRC Press, Boca Raton, Fla., 469 pp.

Millero, F. J., 2001, *The Physical Chemistry of Natural Waters*, Wiley, New York, xix + 654 pp.

Milliman, J. D., 1993, Production and accumulation of calcium carbonate in the ocean: budget of a nonsteady state, *Global Biogeochem. Cycles* **7**:927–957.

Milliman, J. D., and Syvitski, J. P. M., 1994, Geomorphic/tectonic control of sediment discharge to the ocean: the importance of small mountainous rivers, in: *Material Fluxes on the Surface of the Earth* (National Research Council, ed.), National Academy Press, Washington, D.C., pp. 74–85.

Milliman, J. D., Troy, P. J., Balch, W. M., Adams, A. K., Li, Y.-H., and Mackenzie, F. T., 1999, Biologically mediated dissolution of calcium carbonate above the chemical lysocline?, *Deep-Sea Res.* I, **46**:1653–1669.

Millot, R., Gaillardet, J., Dupré, B., and Allègre, C. J., 2003, Northern latitude chemical weathering rates: clues from the Mackenzie River Basin, Canada, *Geochim. Cosmochim. Acta* **67**:1305–1329.

Minschwaner, K., and Dessler, A. E., 2004, Water vapor feedback in the tropical upper troposphere: model results and observations, *J. Climate* **17**:1272–1282.

Moeller, T., 1952, *Inorganic Chemistry: An Advanced Textbook*, Wiley, New York, xii + 966 pp.

Mohr, F., 1875, *Geschichte der Erde*, 2. Aufl., Verlag Max Cohen & Sohn, Bonn, xx +554 pp.

Mojzsis, S. J., Arrhenius, G., McKeegan, K. D., Harrison, T. M., Nutman, A. P., and Friend, C. R. L., 1996, Evidence for life on Earth by 3800 Myr, *Nature* **384**:55–59.

Mook, W. G., and Tan, F. C., 1991, Stable carbon isotopes in rivers and estuaries, in: *Biogeochemistry of Major World Rivers* (E. T. Degens, S. Kempe, and J. E. Richey, eds.), SCOPE 42, Wiley, Chichester, U. K., Chapter 11.

Mook, W. G., Bommerson, J. C., and Staverman, W. H., 1974, Carbon isotope fractionation between dissolved bicarbonate and gaseous carbon dioxide, *Earth Planet. Sci. Lett.* **22**:169–176.

Mook, W. G., Koopmans, M., Carter, A. F., and Keeling, C. D., 1983, Seasonal, latitudinal and secular variations in the abundance and isotopic ratios of atmospheric carbon dioxide, *J. Geophys. Res.* **88**:10915–10933.

Morse, J. W., 2004, Formation and diagenesis of carbonate sediments, in: *Sediments, Diagenesis and Sedimentary Rocks*, v. 7 (F. T. Mackenzie, ed.), *Treatise on Geochemistry* (H. D. Holland and K. K. Turekian, eds.), Elsevier-Pergamon, Oxford, pp. 67–85.

Morse, J. W., and Arvidson, R. S., 2002, The dissolution kinetics of major sedimentary carbonate minerals, *Earth Sci. Rev.* **58**:51–84.

Morse, J. W., and Mackenzie, F. T., 1990, *Geochemistry of Sedimentary Carbonates*, Elsevier, New York, xvi + 707 pp.

Morse, J. W., and Mackenzie, F. T., 1998, Hadean ocean carbonate geochemistry, *Chem. Geol.* **4**:301–319.

Morse, J. W., De Kanel, J., and Harris, J., 1979, Dissolution kinetics of calcium carbonate in seawater. VII: The dissolution kinetics of synthetic aragonite and pteropod tests, *Am. J. Sci.* **279**:482–502.

Morse, J. W., Zullig, J. J., Bernstein, L. D., Millero, F. J., Milne, P., Mucci, A., and Choppin, G. R., 1985, Chemistry of calcium carbonate-rich shallow water sediments, *Am. J. Sci.* **285**:147–185.

Mortatti, J., and Probst, J.-L., 2003, Silicate rock weathering and atmospheric/soil CO_2 uptake in the Amazon basin estimated from river water geochemistry: seasonal and spatial variations, *Chem. Geol.* **197**:177–196.

Moser, H., and Stichler, W., 1980, Environmental isotopes in ice and snow, in: *Handbook of Environmental Isotope Geochemistry*, v. 1, *The Terrestrial Environment A* (P. Fritz and J. Ch. Fontes, eds.), Elsevier, Amsterdam, pp. 141–178.

Mottl, M. J., 2003, Partitioning of energy and mass fluxes between mid-ocean ridge axes and flanks at high and low temperature, in: *Energy and Mass Transfer in Marine Hydrothermal Systems* (P. E. Halbach, V. Tunnicliffe, and J. R. Hein, eds.), Dahlem Univ. Press, Berlin, pp. 271–286.

Moulin, E., Jordens, A., and Wollast, R., 1985, Influence of the aerobic bacterial respiration on the early dissolution of carbonates in coastal sediments, *Proc. Progr. Belgium Oceanographic Res.*, Brussels, pp. 196–208.

Mucci, A., 1983, The solubility of calcite and aragonite in seawater at various salinities, temperatures, and one atmosphere total pressure, *Am. J. Sci.* **283**:780–799.

Müller, P. J., and Suess, E., 1979, Productivity, sedimentation rate, and sedimentary organic matter in the oceans, I. Organic carbon preservation, *Deep-Sea Res.* **26**:1347–1362.

Munhoven, G., 2002, Glacial-interglacial changes of continental weathering: estimates of the related CO_2 and HCO_3^- flux variations and their uncertainties. *Global Planet. Change* **33**:155–176.

Munhoven, G., and François, L. M., 1996, Glacial-interglacial variability of atmospheric CO_2 due to changing continental silicate rock weathering, *J. Geophys. Res.* **101**(D16):21,423–21,437.

Munk, W. H., 1966, Abyssal recipes, *Deep-Sea Res.* **13**:707–730.

Murray, C. N., Visitini, L., Bidoglio, G., and Henry, B., 1996, Permanent storage of carbon dioxide in the marine environment: the solid CO_2 penetrator, *Energy Convers. Management* **37**:1067–1072.

Murray, C. N., Mangin, A., Bidoglio, G., 1999, Technologies for the permanent disposal of CO_2 in deep marine sediment formations, in: *Greenhouse Gas Control Technologies* (P. W. F. Reimer, A. Y. Smith, and K. V. Thambimuthu, eds.), Elsevier, Amsterdam, pp. 261–267.

Nakićenović, N., and others, 2000, *Emissions Scenarios, a Special Report of Working Group III of the Intergovernmental Panel on Climate Change (IPCC)*, Cambridge University Press, Cambridge, 599 pp.

Nebel, B. J., 1990, *Environmental Science, the Way the World Works*, 3rd Ed., Prentice-Hall, Englewood Cliffs, New Jersey, xx + 603 pp.

Nebel, B. J., and Kormondy, E. J., 1981, *Environmental Science, the Way the World Works*, Prentice-Hall, Englewood Cliffs, New Jersey, xvii + 715 pp.

Nelson, C. S., 1988, An introductory perspective on non-tropical shelf carbonates, *Sedim. Geol.* **60**:3–12.

Neumann, A. C., 1965, Processes of recent carbonate sedimentation in Harrington Sound, Bermuda, *Bull. Mar. Sci.* **15**:987–1035.

Nightingale, P. J., Malin, G., Law, C. S., Watson, A. J., Liss, P. S., Liddicoat, M. I., Boutin, J., and Upstill-Goddard, R. C., 2000, In situ evaluation of air-sea gas exchange parametrizations using novel conservative and volatile tracers, *Global Biogeochem. Cycles* **14**:373–387.

NIST, 1992, *Report of Investigation, Reference Materials 8543–8546*, National Institute of Standards and Technology, Gaithersburg, Md., 3 pp.

NIST, 2002, *Measurement and Definition Standards, Deuterium and Carbon-13*, National Institute of Standards and Technology, Gaithersburg, Md., 3 pp.

Nozaki, Y., and Oba, T., 1995, Dissolution of calcareous tests in the ocean and atmospheric carbon dioxide, in: *Biogeochemical Processes and Ocean Flux in the Western Pacific* (H. Sakai and Y. Nozaki, eds.), Terra Scientific Publishing Company, Tokyo, pp. 83–92.

Oaillard, D., and Labeyrie, L., 1994, Role of the thermohaline circulation in the abrupt warming after Heinrich events, *Nature* **372**:162–164.

Ohde, S., and van Woesik, R., 1999, Carbon dioxide flux and metabolic processes of a coral reef, Okinawa, *Bull. Mar. Sci.* **65**:559–576.

Ohgaki, K., Makihara, Y., and Takano, K., 1993, Formation of CO_2 hydrate in pure and sea waters, *J. Chem. Engineering Japan* **26**:558–564.

O'Leary, M. H., 1988, Carbon isotopes in photosynthesis, *BioScience* **38**:328–335.

O'Leary, M. H., 1993, Biochemical basis of carbon isotope fractionation, in: *Stable Isotopes and Plant Carbon-Water Relations* (J. R. Ehleringer, A. E. Hall, and G. D. Farquhar, eds.), Academic Press, New York, pp. 19–28.

Oliver, L., Harris, N., Bickle, M., Chapman, H., Dise, N., and Horstwood, M., 2003, Silicate weathering rates decoupled from the $^{87}Sr/^{86}Sr$ ratio of the dissolved load during Himalayan erosion, *Chem. Geol.* **201**:119–139.

Oliver, M. K., 2002, What is the chemical composition of Lake Malawi water? How does it compare with Lakes Victoria and Tanganyika?, http://malawicichlids.com/mw01011.htm.

O'Neil, J. R., 1986, Theoretical and experimental aspects of isotopic fractionation, in: *Stable Isotopes in High Temperature Geological Processes* (J. W. Valley, H. P. Taylor, Jr., and J. R. O'Neil, eds.), *Rev. Mineralogy* **16**:1–40, Mineralogical Society of America, Washington, D. C.

O'Neil, J. R., Clayton, R. N., and Mayeda, T. K., 1969, Oxygen isotope fractionation in divalent metal carbonates, *J. Chem. Phys.* **51**:5547–5548.

Opdyke, B. N., and Walker, J. C. G., 1992, Return of the coral reef hypothesis: basin to shelf partitioning of $CaCO_3$ and its effect on atmospheric CO_2, *Geology* **20**:733–736.

Ormerod, W. G., Freund, P., Smith, A., and Davison, J., 2002, Ocean Storage of CO_2, International Energy Agency, Greenhouse Gas R&D Programme, http://www.ieagreen.org.uk/oceanrep.pdf, 27 pp.

Orr, J. C., Fabry, V. J., Aumont, O., Bopp, L., Doney, S. C., Feely, R. M., Gnanadesikan, A., Gruber, N., Ishida, A., Joos, F., Key, R. M., Lindsay, K., Maier-Reimer, E., Matear, R., Monfray, P., Mouchet, A., Najjar, R. A., Plattner, G.-K., Rodgers, K. B., Sabine, C. L., Sarmiento, J. L., Schlitzer, R., Slater, R. D., Totterdel, I. J., Weirig, M-F., Yamanaka, Y., and Yool, A., 2005, Anthropogenic ocean acidification over the 21st century and its impact on marine calcifying organisms, *Nature* **437**:681–686.

Overton, W. S., Kanciruk, P., Hook, L. A., Eilers, J. M., Landers, D. H., Brakke, D. F., Blick, D. J., Jr., Linthurst, R. A., DeHaan, M. D., and Omernik, J. M., 1986, *Characteristics of Lakes in the Eastern United States*, Vol. II: *Lakes Sampled and Descriptive Statistics for Physical and Chemical Variables*, U. S. Environmental Protection Agency, Washington, D. C., EPA/600/4-86/007b, xxiv + 374 pp.

Paasche, E., and Brubak, S., 1994, Enhanced calcification in the coccolithophorid *Emiliania huxleyi* (Haptophyceae) under phosphorus limitation, *Phycologia* **33**:324–330.

Pace, N. R., 1997, A molecular view of microbial diversity and the biosphere, *Science* **276**:734–740.

Pace, N. R., 2001, The universal nature of biochemistry, *Proc. Nat. Acad. Sci. USA* **98**:805–808.

Palmer, M. R., and Pearson, M. N., 2003, A 23,000-year record of surface water pH and P_{CO_2} in the Western Equatorial Pacific Ocean, *Nature* **300**:480–482.

Paquette, J., and Reeder, R. J., 1990, Single crystal X-ray structure refinements of two biogenic magnesian calcite crystals, *Am. Mineral.* **75**:1151–1158.

Patterson, C. C., 1956, Age of meteorites and the Earth, *Geochim. Cosmochim. Acta* **10**:230–237.

Pavlov, A. A., Kasting, J. F., Brown, L. L., Rages, K. A., and Freeman, R., 2000, Greenhouse warming by CH_4 in the atmosphere of early Earth, *J. Geophys. Res.* **105**(E5):11,981–11, 990.

Pavlov, A. A., Kasting, J. F., Eigenbrode, J. L., and Freeman, K. H., 2001, Organic haze in Earth's early atmosphere: source of low-^{13}C Late Archean kerogens?, *Geology* **29**:1003–1006.

Paytan, M., Kastner, M., Campbell, D., and Thiemens, M. H., 1998, Sulfur isotopic composition of Cenozoic seawater sulfate, *Science* **282**:1459–1462.

Paytan, M., Kastner, M., Campbell, D., and Thiemens, M. H., 2004, Seawater sulfur isotope fluctuations in the Cretaceous, *Science* **304**:1663–1665.

Peacock, S. M., 1993, Large-scale dehydration of the lithosphere above subducting slabs, *Chem. Geol.* **108**:49–59.

Peacor, D. R., Essene, E. J., and Gaines, A. M., 1987, Petrologic and crystal-chemical implications of cation-disorder in kutnahorite [$CaMn(CO_3)_2$], *Am. Mineral.* **72**:319–328.

Pedersen, T. F., 1983, Increased productivity in the eastern equatorial Pacific during the last glacial maximum (19,000 to 14,000 B. P.), *Geology* **11**:16–19.

Peltzer, E. T., Brewer, P. G., Friederich, G., and Rehder, G., 2000, Direct observation of the fate of oceanic carbon dioxide release at 800 m, *Am. Chem. Soc., Div. Fuel Chem.* **45**:794–798.

Pennisi, E., 2002, Geobiologists: as diverse as the bugs they study, *Science* **296**:1058–1060.

Pernetta, J. C., and Milliman, J. D. (eds.), 1995, *Land-Ocean Interactions in the Coastal Zone, Implementation Plan*, IGBP Report No. 33, Stockholm, LOICZ Office, Netherlands Institute of Sea Research, Texel, The Netherlands, 215 pp.

Petit, J.-R., Jouzel, J., Raynaud, D., Barkov, N. I., Barnola, J.-M., Basile, I., Bender, M., Chappellaz, J., Devis, M., Delaygue, G., Delmotte, G. M., Kotlyakov, V. M., Legrand, M.,

Lipenkov, V. Y., Lorius, C., Pepin, L., Ritz, C., Saltzman, E., and Stievenard, M., 1999, Climate and atmospheric history of the past 420,000 years from the Vostok ice core, Antarctica, *Nature* **399**:429–436.

Petsch, S. T., 2004, The global oxygen cycle, in: *Biogeochemistry*, v. 6 (W. H. Schlesinger, ed.), *Treatise on Geochemistry* (H. D. Holland and K. K. Turekian, eds.), Elsevier-Pergamon, Amsterdam, pp. 515–555.

Pettijohn, F. J., 1957, *Sedimentary Rocks*, 2nd Ed., Harper, New York, pp. 502–508.

Pingitore, N. E., 1976, Vadose and phreatic diagenesis: processes, products and their recognition in corals, *J. Sedim. Petrology* **46**:985–1006.

Plummer, L. N., 1975, Mixing of sea water with calcium carbonate ground water, *Geol. Soc. America Mem.* **142**:219–236.

Plummer, L. N., and Busenberg, E., 1982, The solubilities of calcite, aragonite and vaterite in CO_2-H_2O solutions between 0 and 90°C, and an evaluation of the aqueous model for the system $CaCO_3$-CO_2-H_2O, *Geochim. Cosmochim. Acta* **46**:1011–1040.

Plummer, L. N., and Mackenzie, F. T., 1974, Predicting mineral solubility from rate data: application to the dissolution of magnesian calcites, *Am. J. Sci.* **274**:61–83.

Plummer, L. N., Wigley, T. M. L., and Parkhurst, D. L., 1978, The kinetics of calcite dissolution in CO_2-water systems at 5° to 60°C and 0.0 to 1.0 atm CO_2, *Am. J. Sci.* **278**:179–216.

Pokrovsky, O. S., Schott, J., and Thomas, F., 1999, Dolomite surface speciation and reactivity in aquatic systems, *Geochim. Cosmochim. Acta* **63**:3133–3143.

Poldervaart, A., 1955, Chemistry of the Earth's surface, in: *Crust of the Earth* (A. Poldervaart, ed.), *Geol. Soc. America Spec. Pap.* **62**, pp. 119–144.

Polley, H. W., Johnson, H. B., Marino, B. D., and Mayeux, H. S., 1993, Increase in C_3 plant water-use efficiency and biomass over glacial to present CO_2 concentrations, *Nature* **361**:61–64.

Popp, B. N., Takigiku, R., Hayes, J. M., Louda, J. W., and Baker, E. W., 1989, The post-Paleozoic chronology and mechanism of ^{13}C depletion in primary marine organic matter, *Am. J. Sci.* **289**:436–454.

Popp, B. N., Laws, E. A., Bidigare, R. R., Dore, J. E., Hanson, K. L., and Wakeham, S. G., 1998, Effect of phytoplankton cell geometry on carbon isotopic fractionation, *Geochim. Cosmochim. Acta* **62**:69–77.

Poulton, S. W., Fralick, P. W., and Canfield, D. E., 2004, The transition to a sulfidic ocean ~1.84 billion years ago, *Nature* **431**:173–177.

Prentice, I. C., and Sykes, M. T., 1995, Vegetation geography and global carbon storage changes, in: *Biotic Feedback in the Global Climate System: Will the Warming Feed the Warming?* (G. M. Woodwell and F. T. Mackenzie, eds.), Oxford Univ. Press, Oxford, pp. 304–312.

Prentice, K. C., and Fung, I. Y., 1990, The sensitivity of terrestrial carbon storage to climate change, *Nature* **346**:48–51.

Pruess, K., 2004, Numerical simulation of CO_2 leakage from a geologic disposal reservoir, including transitions between super- and sub-critical conditions, and boiling of liquid CO_2, *J. Soc. Petroleum Engineers*, June 2004, pp. 237–248.

Quay, P. D., Tilbrook, B., and Wong, C. S., 1992, Oceanic uptake of fossil fuel CO_2: carbon-13 evidence, *Science* **256**:74–79.

Rabouille, C., Mackenzie, F. T., and Ver, L. M., 2001, Influence of the human perturbation on carbon, nitrogen, and oxygen biogeochemical cycles in the global coastal ocean, *Geochim. Cosmochim. Acta* **65**:3615–3639.

Rahmstorf, S., Archer, D., Ebel, D. S., Eugster, O., Jouzel, J., Maraun, D., Neu, U., Schmidt, G. A., Severinghaus, J., Weaver, A. J., and Zachos, J., 2004, Cosmic rays, carbon dioxide, and climate, *Eos, Am. Geophys. Un. Trans.* **85**(4):38–41.

Ramanathan, V., and Coakley, J. A., Jr., 1978, Climate modeling through radiative-convective models, *Rev. Geophys. Space Phys.* **16**:465–489.

Ramaswamy, V., Boucher, O., Haigh, J., Hauglustaine, D., Haywood, J., Myhre, G., Nakajima, T., and Shi, G. Y., 2001, Radiative forcing of climate change, in: *Climate Change 2001: The Scientific Basis* (J. T. Houghton, Y. Ding, D. J. Griggs, M. Noguer, P. J. van der Linden, X. Dai, K. Maskell, and C. A. Johnson, eds.), Cambridge Univ. Press, Cambridge, pp. 349–416.

Randerson, J. T., van der Werf, G. R., Collatz, G. J., Giglio, L., Still, C. J., Kasibhatla, P., Miller, J. B., White, J. W. C., DeFries, R. S., and Kasischke, E. S., 2005, Fire emissions from C_3 and C_4 vegetation and their influence on interannual variability of atmospheric CO_2 and $\delta^{13}CO_2$, *Global Biogeochem. Cycles*, **19**, **GB2019**, 13 pp., doi:10.1029/2004GB002366.

Rankama, K., and Sahama, Th. G., 1950, *Geochemistry*, Univ. Chicago Press, Chicago, xvi + 912 pp.

Rau, G. H., Takahashi, T., Des Marais, D. J., Repeta, D. J., and Martin, J. H., 1992, The relationship between ^{13}C of organic matter and $[CO_2(aq)]$ in ocean surface water: data from a JGOFS site in the northeast Atlantic Ocean and a model, *Geochim. Cosmochim. Acta* **56**:1413–1419.

Raval, A., and Ramanathan, V., 1989, Observational determination of the greenhouse effect, *Nature* **342**:758–761.

Raven, J. A., 1993, Limits on growth rates, *Nature* **361**:209–210.

Raven, J. A., 1997, Inorganic carbon acquisition by marine autotrophs, *Adv. Botanical Res.* **27**:85–209.

Redfield, A. C., Ketchum, B. H., and Richard, F. A., 1963, The influence of organisms on the composition of seawater, in: *The Sea*, v. 2 (M. N. Hill, ed.), Wiley, New York, pp. 26–77.

Reid, G. C., and Solomon, S., 1986, On the existence of an extraterrestrial source of water in the middle atmosphere, *Geophys. Res. Lett.* **13**:1129–1131.

Renard, M., 1986, Pelagic carbonate chemostratigraphy (Sr, Mg, ^{18}O, ^{13}C), *Mar. Micropaleontol.* **10**:117–164.

Retellack, G. J., 2001, A 300-million-year record of atmospheric carbon dioxide from fossil plant cuticles, *Nature* **411**:287–290.

Revelle, R., and Munk, W., 1977, The carbon dioxide cycle and the biosphere, in: *Energy and Climate*, Studies in Geophysics, National Academy Press, Washington, D. C., pp. 140–158.

Reynaud, S., Leclercq, N., Romaine-Lioud, S., Ferrier-Pagés, C., Jaubert, J., and Gattuso, J.-P., 2003, Interacting effects of CO_2 partial pressure and temperature on photosynthesis and calcification in a scleractinian coral, *Global Change Biol.* **9**:1660–1668, doi: 10.1046/j.1529-8817.2003.00678.x

Rich, J., 1998, The role of regional bioproductivity in atmospheric CO_2 changes, Ph. D. Dissertation, Northwestern Univ., Evanston, Ill., viii + 98 pp.

Richardson, K., and Heilmann, J. P., 1995, Primary production in the Kattegat: past and present, *Ophelia* **41**:317–328.

Richerson, P. J., Boyd, R., and Bettinger, R. L., 2001, Was agriculture impossible during the Pleistocene but mandatory during the Holocene? A climate change hypothesis, *Am. Antiquity* **66**(3):387–412.

Richet, P., Bottinga, Y., and Javoy, M., 1977, A review of hydrogen, carbon, nitrogen, oxygen, sulphur, and chlorine stable isotope fractionation among gaseous molecules, *Ann. Rev. Earth Planet. Sci.* **5**:65–110.

Riebesell, U., Zondervan, I., Rost, B., Tortell, P. D., Zeebe, R. E., and Morel, F. M. M., 2000, Reduced calcification of marine plankton in response to increased atmospheric CO_2, *Nature* **407**:364–367.

Robie, R. A., Hemingway, B. S., and Fisher, J. R., 1978, Thermodynamic properties of minerals and related substances at 298.15 K and 1 bar (10^5 pascals) pressure and at higher temperatures, *U. S. Geol. Surv. Bull.* **1452**:1–456.

Ronov, A. B., 1964, Common tendencies in the chemical evolution of the Earth's crust, ocean and atmosphere, *Geochem. Internat.* **1**:713–737.

Ronov, A. B., 1980, *Osadochnaya Obolochka Zemli (Kolichestvennyye Zakonomernosti Stroyeniya, Sostava i Evolyutsii)*, Nauka, Moscow, 80 pp.

Rosenberg, P. E., and Holland, H. D., 1964, Calcite-dolomite-magnesite stability relations in solutions at elevated temperatures, *Science* **145**:700–701.

Rosenberg, P. E., Burt, D. M., and Holland, H. D., 1967, Calcite-dolomite-magnesite stability relations in solutions: the effect of ionic strength, *Geochim. Cosmochim. Acta* **31**:391–396.

Rosing, M. T., 1999, ^{13}C-depleted carbon microparticles in >3700-Ma sea-floor sedimentary rocks from West Greenland, *Science* **283**:674–676.

Rowley, D. B., 2002, Rate of plate creation and destruction, 180 Ma to present, *Geol. Soc. America Bull.* **114**:927–933.

Rubey, W. W., 1951, Geologic history of seawater, an attempt to state the problem, *Geol. Soc. America Bull.* **62**:1111–1147.

Rubey, W. W., 1955, Development of the hydrosphere and atmosphere, with special reference to probable composition of the early atmosphere, in: *Crust of the Earth* (A. Poldervaart, ed.), *Geol. Soc. America Spec. Pap.* **62**, pp. 631–650.

Rubinson, M., and Clayton, R. N., 1969, Carbon-13 fractionation between aragonite and calcite, *Geochim. Cosmochim. Acta* **33**:997–1002.

Rudnick, R. L., and Gao, S., 2003, Composition of the continental crust, in: *The Crust*, v. 3 (R. Rudnick, ed.), *Treatise on Geochemistry* (H. D. Holland and K. K. Turekian, eds.), Elsevier-Pergamon, Amsterdam, pp. 1–64.

Runnegar, B., and Bengtson, S., 1990, Origin of hard parts—early skeletal fossils, in: *Paleobiology: A Synthesis* (D. E. G. Briggs and P. R. Crowther, eds.), Blackwell, Oxford, pp. 24–29.

Rye, R., and Holland, H. D., 1998, Paleosols and the evolution of atmospheric oxygen: a critical review, *Am. J. Sci.* **298**:621–672.

Sabine, C. L., Feeley, R. A., Gruber, N., Key, R. M., Lee, K., Bullister, J. L., Wanninkhof, R., Wong, C. S., Wallace, D. W. R., Tilbrook, B., Millero, F. J., Peng, T.-H., Kozyr, A., Ono, T., and Rios, A. F., 2004, The oceanic sink for anthropogenic CO_2, *Science* **305**:367–371.

Sackett, W. M., 1989, Stable carbon isotope studies on organic matter in the marine environment, in: *Handbook of Environmental Isotope Geochemistry*, v. 3, *The Marine Environment* (P. Fritz and J. Ch. Fontes, eds.), Elsevier, Amsterdam, pp. 139–170.

Sagan, C., and Chyba, C., 1997, The early faint Sun paradox: organic shielding of ultraviolet-labile greenhouse gases, *Science* **276**:1217–1221.

Salomons, W., and Mook, W. G., 1986, Isotope geochemistry of carbonates in the weathering zone, in: *Handbook of Environmental Isotope Geochemistry*, v. 2, *The Terrestrial Environment B* (P. Fritz and J. Ch. Fontes, eds.), Elsevier, Amsterdam, pp. 239–269.

Sandberg, P. A., 1975, New interpretation of great salt lake ooids and of ancient nonskeletal carbonate mineralogy, *Sedimentology* **22**:497–538.

Sandberg, P. A., 1983, An oscillating trend in non-skeletal carbonate mineralogy, *Nature* **305**:19–22.

Sandberg, P. A., 1985, Nonskeletal aragonite and pCO_2 in the Phanerozoic and Proterozoic, in: *The Carbon Cycle and Atmospheric CO_2: Natural Variations Archean to Present* (E. T. Sundquist and W. S. Broecker, eds.), *Geophys. Mon.* **32**:585–594, Am. Geophys. Union, Washington, D. C.

Sanger, R. G., 1964, Ellipse, *Encycl. Brit.*, 1964 Ed., v. 8, Encyclopædia Britannica, Inc, Chicago, pp. 295–296.

Sanyal, A., Hemming, G., Hansen, G., and Broecker, W. S., 1995, Evidence for a higher pH in the glacial ocean from boron isotopes in foraminifera, *Nature* 373:234–237.

Sarmiento, J. L., and Gruber, N., 2002, Sinks for anthropogenic carbon, *Phys. Today* 55(8):30–36.

Sarmiento, J. L., Orr, J. C., and Siegenthaler, U., 1992, A perturbation simulation of CO_2 uptake in an ocean general circulation model, *J. Geophys. Res.* 97:3621–3645.

Sarntheim, M., Winn, K., Duplessy, J.-C., and Fontugne, M. R., 1988, Global variations in surface ocean productivity in low and mid latitudes: influence on CO_2 reservoirs of the deep ocean and atmosphere during the last 21,000 years, *Paleoceanography* 3:361–399.

Savin, S. M., 1980, Oxygen and hydrogen isotope effects in low-temperature mineral-water interactions, in: *Handbook of Environmental Isotope Geochemistry*, v. 1, *The Terrestrial Environment A* (P. Fritz and J. Ch. Fontes, eds.), Elsevier, Amsterdam, pp. 283–328.

Schidlowski, M., 1988, A 3,800-million-year isotopic record of life from carbon in sedimentary rocks, *Nature* 333:313–318.

Schidlowski, M., Hayes, J. M., and Kaplan, I. R., 1983, Isotopic inferences of ancient bio-chemistries: carbon, sulfur, hydrogen, and nitrogen, in: *Earth's Earliest Biosphere* (J. W. Schopf, ed.), Princeton Univ. Press, Princeton, New Jersey, pp. 149–186.

Schlesinger, W. H., 1997, *Biogeochemistry: An Analysis of Global Change*, Academic Press, San Diego, Calif, 588 pp.

Schmalz, R. F., and Chave, K. E., 1963, Calcium carbonate: affecting saturation in ocean waters of Bermuda, *Science* 139:1206–1207.

Schmitz, W. J., Jr., 1995, On the interbasin-scale thermohaline circulation. *Rev. Geophys.* 33:151–174.

Schneider, S. H., and Kellogg, W. W., 1973, The chemical basis for climate change, in: *Chemistry of the Lower Atmosphere* (S. I. Rasool, ed.), Plenum, New York, pp. 203–249.

Schoonmaker, J. E., 1981, Magnesian calcite–seawater reactions: solubility and recrystallization behavior, Ph. D. Dissertation, Northwestern Univ., Evanston, Ill, xiv + 264.

Schopf, T. J. M., 1980, *Paleoceanography*, Harvard Univ. Press, Cambridge, Mass., xii + 341 pp.

Schultz-Guttler, R., 1986, The influence of disordered, non-equilibrium dolomites on the Mg-solubility in calcite in the system $CaCO_3$-$MgCO_3$, *Contrib. Miner. Petrol.* 93:395–398.

Sciandra, A., Harlay, J., Lefèvre, D., Lemée, R., Rimmelin, R., Denis, M., and Gattuso, J.-P., 2003, Response of coccolithophorid *Emiliania huxleyi* to elevated partial pressure of CO_2 under nitrogen limitation, *Mar. Ecol. Prog. Ser.* 261:111–122.

Sclater, J. G., Jaupart, C., and Galson, D., 1980, The heat flow through oceanic and continental crust and the heat loss of the Earth, *Rev. Geophys. Space Phys.* 18:269–311.

Scoffin, T. P., Tudhope, A. W., Brown, B. E., Chansang, H., and Cheeney, R. F., 1992, Patterns and possible environmental controls of skeletogenesis of *Porites lutea*, South Thailand, *Coral Reefs* 11:1–11.

Seinfeld, J. H., and Pandis, S. N., 1998, *Atmospheric Chemistry and Physics*, Wiley, New York, xxviii + 1326 pp.

Shannon, R. D., 1976, Revised effective ionic radii and systematic studies of interatomic distances in halides and chalcogenides, *Acta Crystallogr.*, ser. A, 32:751–767.

Sharma, A., Scott, J. H., Cody, G. D., Fogel, M. L., Hazen, R. M., Hemley, R. J., and Huntress, W. T., 2002, Microbial activity at gigapascal pressures, *Science* 295:1514–1516.

Shaviv, N. J., and Veizer, J., 2003, Celestrial driver of climate?, *GSA Today* 13:4–10.

Shields, G., and Veizer, J., 2002, The Precambrian marine carbonate isotope database: version 1, *Geochem. Geophys. Geosyst.* 3(6), June 6, 2002, http://g-cubed.org/gc2002/2001GC000266.

Shiklomanov, I. A., 1993, World fresh water resources, in: *Water in Crisis; a Guide to the World's Fresh Water Resources* (P. H. Gleick, ed.), Oxford Univ. Press, New York, pp. 13–24.

Shimada, S., Takahashi, H., Haraguchi, A., and Kaneko, M., 2001, The carbon content characteristics of tropical peats in Central Kalimantan, Indonesia: estimating their spatial variability and density, *Biogeochemistry* **53**:249–267.

Shine, K. P., Derwent, R. G., Wuebbles, D. J., and Morcrette, J.-J., 1990, Radiative forcing of climate, in: *Climate Change: The IPCC Scientific Assessment* (J. T. Houghton, G. J. Jenkins, and J. J. Ephraums, eds.), Cambridge Univ. Press, Cambridge, pp. 51–55.

Shine, K. P., Fouquart, Y., Ramaswamy, V., Solomon, S., and Srinivasan, J., 1995, Radiation forcing, in: *Climate Change 1994: Radiative Forcing of Climate Change and an Evaluation of the IPCC IS92 Emission Scenarios* (J. T. Houghton, L. G. Meira Filho, J. Bruce, H. Lee, B. A. Callander, E. Haites, N. Harris, and K. Maskell, eds.), Cambridge Univ. Press, Cambridge, pp. 163–203.

Sibley, D. F., 1990, Unstable to stable transformations during dolomitization, *J. Geol.* **98**:739–748.

Siegenthaler, U., and Oeschger, H., 1987, Biospheric CO_2 emissions during the past 200 years reconstructed by deconvolution of ice core data, *Tellus* **39B**:140–154.

Siegenthaler, U., Stocker, T. F., Monnin, E., Lüthi, D., Schwander, J., Stauffer, B., Raynaud, D., Barnola, J.-M., Fischer, H., Masson-Delmotte, V., and Jouzel, J., 2005, Stable carbon cycle-climate relationship during the Late Pleistocene, *Science* **310**:1313–1317.

Sienell, S., 2003, *Versiegelte Schreiben* No. 437 (März 1904), No. 527 (April 1907); *Allgemeine Akten* No. 666 (1913), Archiv, Österreichische Akademie der Wissenschaften, Wien, *personal communication*, March 2003.

Sigman, D. M., and Boyle, E. A., 2000, Glacial/interglacial variations in atmospheric carbon dioxide, *Nature* **407**:859–869.

Sigman, D. M., McCorkle, D. C., and Martin, W. R., 1998, The calcite lysocline as a constraint on glacial/interglacial low latitude production changes, *Global Biogeochem. Cycles* **12**:409–428.

Singsaas, E. L., Orr, D. R., and DeLucia, E. H., 2001, Variation in measured values of photosynthetic quantum yield in ecophysiological studies, *Oecologia* **128**:15–23.

Sjöberg, E. L., 1976, A fundamental equation for calcite dissolution kinetics, *Geochim. Cosmochim. Acta* **40**:441–447.

Sjöberg, E. L., and Rickard, D. T., 1984, Temperature-dependence of calcite dissolution kinetics between 1°C and 62°C at pH 2.7 to 8.4 in aqueous solutions, *Geochim. Cosmochim. Acta* **48**:485–493.

Skinner, B. J., 1966, Thermal expansion, in: *Handbook of Physical Constants* (S. P. Clark, Jr., ed.), *Geol. Soc. America Mem.* **97**:75–96.

Sleep, N. H., 1979, Thermal history and degassing of the Earth: some simple calculations, *J. Geol.* **87**:671–686.

Sloan, E. D., Jr., 1998, *Clathrate Hydrates of Natural Gases*, 2nd Ed., Marcel Dekker, New York, 705 pp.

Smith, A. D., and Roth, A. A., 1979, Effect of carbon dioxide concentration on calcification in the red coralline alga *Bossiella orbigniana*, *Mar. Biology* **52**:217–225.

Smith, D. E., 1964, Numerals, *Encycl. Brit.*, 1964 Ed., v. 16, Encyclopædia Britannica, Inc, Chicago, pp. 610–614.

Smith, H. J., Fisher, H., Wahlen, M., Mastroianni, D., and Deck, B., 1999, Dual modes of the carbon cycle since the Last Glacial Minimum, *Nature* **400**:248–250.

Smith, P. E., Brand, U., and Farquhar, R. M., 1994, U-Pb systematics and alteration trends of Pennsylvanian-aged aragonite and calcite, *Geochim. Cosmochim. Acta* **58**:313–322.

Smith, S. V., 1985, Physical, chemical and biological characteristics of CO_2 gas flux across the air-water interface, *Plant, Cell and Environment* **8**:387–398.

Smith, S. V., and Hollibaugh, J. T., 1993, Coastal metabolism and the oceanic organic carbon balance, *Rev. Geophys.* **31**:75–89.

Smith, S. V., and Mackenzie, F. T., 1987, The ocean as a net heterotrophic system: Implications from the carbon biogeochemical cycle, *Global Biogeochem. Cycles* **1**:187–198.

Smith, S. V., Buddemeier, R. W., Redaije, R. D., and Houck, J. E., 1979, Strontium-calcium thermometry in coral skeletons, *Science* **204**:404–407.

Sorby, H. C., 1879, The structure and orgin of limestones, *Proc. Geol. Soc. London* **35**:56–59.

Span, R., and Wagner, W., 1996, A new equation of state for carbon dioxide covering the fluid region from the triple-point temperature to 1100 K at pressure up to 800 MPa, *J. Phys. Chem. Ref. Data* **25**(6):1509–1596.

Speer, J. A., 1983, Crystal chemistry and phase relations of orthorhombic carbonates, in: *Carbonates: Mineralogy and Chemistry* (R. J. Reeder, ed.), *Rev. Mineralogy* **11**:145–189, Mineralogical Society of America, Washington, D. C.

Sperber, C. M., Wilkinson, B. H., and Peacor, D. R., 1984, Rock composition, dolomite stoichiometry and rock water reactions in dolomitic carbonate rocks, *J. Geol.* **92**:609–622.

Spero, H. J., Bijma, J., Lea, D. W., and Bemis, B. E., 1997, Effect of seawater carbonate concentration on foraminiferal carbon and oxygen isotopes, *Nature* **390**:497–500.

Stacey, F. D.,1992, *Physics of the Earth*, 3rd Ed., Brookfield Press, Brisbane, Australia, xii + 513 pp.

Stallard, R. F., 1988, Weathering and erosion in the humid tropics, in: *Physical and Chemical Weathering in Geochemical Cycles* (A. Lerman and M. Meybeck, eds.), Kluwer, Dordrecht, The Netherlands, pp. 225–246.

Stanley, S. M., and Hardie, L. A., 1998, Secular oscillations in the carbonate mineralogy of reef-building and sediment-producing organisms driven by tectonically forced shifts in seawater chemistry, *Palaeogeogr. Palaeoclimat. Palaeoecol.* **144**:3–19.

Stehli, F. G., and Hower, J., 1961, Mineralogy and early diagenesis of carbonate sediments, *J. Sedim. Petrology* **31**:358–371.

Steuber, T., and Veizer, J., 2002, A Phanerozoic record of plate tectonic control of seawater chemistry and carbonate sedimentation, *Geology* **30**:1123–1126.

Stumm, W., and Morgan, J. J., 1981, *Aquatic Chemistry: An Introduction Emphasizing Chemical Equilibria in Natural Waters*, 2nd Ed., Wiley, New York, xiv + 780 pp.

Stumm, W., and Morgan, J. J., 1996, *Aquatic Chemistry: Chemical Equilibria and Rates in Natural Waters*, 3rd Ed., Wiley, New York, xvi + 1022 pp.

Suess, E., Balzer, W., Heese, K.-F., Müller, P. J., Ungerer, C. A., and Wefer, G., 1982, Calcium carbonate hexahydrate from organic-rich sediments of the Antarctic shelf: precursors to glendonites, *Science* **216**:1128–1131.

Suess, H. E., 1955, Radiocarbon concentration in modern wood, *Science* **122**:415–417.

Sumner, D. Y., 1997, Carbonate precipitation and oxygen stratification in late Archean seawater as deduced from facies and stratigraphy of the Gamohaan and Frisco Formations, Transvaal Supergroup, South Africa, *Am. J. Sci.* **297**:455–487.

Sundquist, E. T., and Broecker, W. S. (eds.), 1985, *The Carbon Cycle and Atmospheric CO2: Natural Variations Archean to Present*, Am. Geophys. Union, Washington, D. C., 627 pp.

Sundquist, E. T., and Visser, K., 2004, The geologic history of the carbon cycle, in: *Biogeochemistry*, v. 8 (W. H. Schlesinger, ed.), *Treatise on Geochemistry* (H. D. Holland and K. K Turekian, eds.), Elsevier-Pergamon, Amsterdam, pp. 425–472.

Suzuki, A., 1998, Combined effects of photosynthesis and calcification on the partial pressure of carbon dioxide in seawater, *J. Oceanogr.* **54**:1–7.

Svensmark, H., 1998, Influence of cosmic rays on Earth's climate, *Phys. Rev. Lett.* **81**(22):5027–5030.

Takahashi, T., 1989, The carbon dioxide puzzle, *Oceanus* **32**(2):22–29.

Takahashi, T., Broecker, W. S., Brainbridge, A. E., and Weiss, R. F., 1980, Carbonate chemistry of the Atlantic, Pacific and Indian Oceans: The results of the GEOSECS expeditions, 1972–1978, Technical Report No. 1., Lamont-Doherty Geological Observatory, Palisades, New York.

Takahashi, T., Sutherland, S. C., Sweeney, C., Poisson, A., Metzl, N., Tilbrook, B., Bates, N., Wanninkhof, R., Feely, R. A., Sabine, C., Olafsson, J., and Nojiri, Y., 2002, Global sea-air CO_2 flux based on climatological surface ocean pCO_2, and seasonal biological and temperature effects, *Deep-Sea Res.* II, **49**:1601–1622.

Tan, F. C., 1988, Stable carbon isotopes in dissolved inorganic carbon in marine and estuarine environments, in: *Handbook of Environmental Isotope Geochemistry*, v. 3, *The Marine Environment* (P. Fritz and J. Ch. Fontes, eds.), Elsevier, Amsterdam, pp. 171–190.

Tans, P. P., Fung, I. Y., and Takahashi, T., 1990, Observational constraints on the global atmospheric CO_2 budget, *Science* **247**:1431–1438.

Tardy, Y., 1986, *Le Cycle de l'Eau: Climats, Paléoclimats et Géochimie Globale*, Masson, Paris, France, 338 pp.

Tardy, Y., Bustillo, V., and Boeglin, J.-L., 2004, Geochemistry applied to the watershed survey: hydrograph separation, erosion and soil dynamics. A case study: the basin of the Niger River, Africa, *Appl. Geochem.* **19**:469–518.

Teng, H., Yamasaki, A., and Shindo, Y., 1999, The fate of CO_2 hydrate released in the ocean, *Internat. J. Energy Res.* **23**:295–302.

Thode, H. G., Shima, M., Rees, C. E., and Krishnamurty, K. V., 1965, Carbon-13 isotope effects in systems containing carbon dioxide, bicarbonate, carbonate, and metal ions, *Can. J. Chem.* **43**:582–595.

Thomas, H., and Schneider, B., 1999, The seasonal cycle of carbon dioxide in Baltic Sea surface waters, *J. Mar. Systems* **22**:53–67.

Thomas, H., Bozec, Y., Elkalay, J., de Baar, H. J. W., 2004, Enhanced open ocean storage of CO_2 from shelf sea pumping, *Science* **304**:1005–1008.

Thompson, P. A., and Calvert, S. E., 1994, Carbon isotope fractionation by a marine diatom: The influence of irradiance, daylength, pH, and nitrogen source, *Limnol. Oceanogr.* **39**:1835–1844.

Thorstenson, D. C., and Plummer, L. N., 1977, Equilibrium criteria for two component solids reacting with fixed composition in an aqueous phase—example: the magnesian calcites, *Am. J. Sci.* **277**:1203–1223.

Tian, F., Toon, O. B., Pavlov, A. A., and De Sterck, H., 2005, A hydrogen-rich early Earth atmosphere, *Science* **308**:1014–1017.

Tice, M. M., and Lowe, D. R., 2004, Photosynthetic microbial mats in the 3,416-Myr-old ocean, *Nature* **431**:549–552.

Tidwell, W. D., and Nambudiri, E. M. V., 1989, *Tomlinsonia thomassonii,* gen. et sp. nov., a permineralized grass from the upper Miocene Ricardo Formation, California (USA), *Rev. Paleobot. Palynol.* **60**:165–178.

Tiessen, H. (ed.), 1995, *Phosphorus in the Global Environment: Transfers, Cycles and Management*, SCOPE 54, Wiley, Chichester, U.K. 462 pp.

Ting, I. P., 1994, CO_2 and crassulacean acid metabolism plants, in: *Regulation of Atmospheric CO₂ and O₂ by Photosynthetic Carbon Metabolism* (N. E. Tolbert and J. Preiss, eds.), Oxford Univ. Press, New York, pp. 176–198.

Tribble, J. S., and Mackenzie, F. T., 1998, Recrystallization of magnesian calcite overgrowths on calcite seeds suspended in seawater, *Aquatic Geochem.* **4**:337–360.

Tribble, J. S., Arvidson, R. S., Lane III, M., and Mackenzie, F. T., 1995, Crystal chemistry, and thermodynamic and kinetic properties of calcite, dolomite, apatite, and biogenic silica: applications to petrologic problems, *Sedim. Geol.* **95**:11–37.

Tsunogai, S., Watanabe, S., and Sato, T., 1999, Is there a "continental shelf pump" for the absorption of atmospheric CO_2?, *Tellus* **51**B:701–712.

Turekian, K. K., 1996, *Global-Environmental Change—Past, Present, and Future*, Prentice Hall, Upper Saddle River, New Jersey, viii + 200.

Turner, R. K., and Adger, W. N., 1996, *Coastal Zone Resources Assessment Guidelines*, Netherlands Institute of Sea Research, Texel, The Netherlands, Land-Ocean Interactions in the Coastal Zone Core Project of the IGBP, LOICZ/R&S/96-4, 101 p.

Tyndall, J., 1863, On the relation of radiant heat to aqueous vapour, *Phil. Mag.*, 4th ser., **26**:30–44.

Tyrrell, T., 1999, The relative influences of nitrogen and phosphorus on oceanic primary productivity, *Nature* **400**:525–531.

Tyrrell, T., and Zeebe, R. E., 2004, History of carbonate ion concentration over the last 100 million years, *Geochim. Cosmochim. Acta* **68**:3521–3530.

Udachin, K. A., Ratcliffe, C. I., and Ripmeester, J. A., 2001, Structure, composition, and thermal expansion of CO_2 hydrate from single crystal X-ray diffraction measurements, *J. Phys. Chem.* B **105**:4200–4204.

Udachin, K. A., Ratcliffe, C. I., and Ripmeester, J. A., 2002, Single crystal diffraction studies of Structure I, II and H hydrates: structure, cage occupancy and composition, *J. Supramol. Chem.* **2**:405–408.

United Nations, 1999, *The World at Six Billion*, Table 1, World Population From Year 0 to Stabilization, http://www.un.org/esa/population/publications/sixbillion/sixbilpart1.pdf.

United Nations Food and Agricultural Organization, 2001, *State of the World's Forests*, Food and Agricultural Organization, United Nations, Rome, Italy, 181 pp.

Urey, H. C., 1947, The thermodynamic properties of isotopic substances, *J. Chem. Soc. (London)*, pp. 562–581.

Urey, H. C., 1952, *The Planets: Their Origin and Development*, Yale Univ. Press, New Haven, Connecticut, xvii + 245 pp.

U. S. Census Bureau (USCB), 2005, *Total Midyear Population for the World: 1950–2050*, Data updated 4–26–2004, http://www.census.gov/ipc/www/worldpop.html.

Vail, P. R., Mitchum, R. W., and Thompson, S., 1977, Seismic stratigraphy and global changes of sea level, 4: Global cycles of relative changes of sea level, *Am. Assoc. Petroleum Geologists Mem.* **26**:83–97.

Valley, J. W., Peck, W. H., King, E. M. and Wilde, S. A., 2002, A cool early earth, *Geology* **30**:351–354.

Van Campo, E., Guiot, J., and Peng, C., 1993, A data-based re-appraisal of the terrestrial carbon budget at the last glacial maximum, *Global Planet. Change* **8**:189–201.

Van Cappellen, P., and Ingall, E. D., 1996, Redox stabilization of the atmosphere and oceans by phosphorus-limited marine productivity, *Science* **271**:493–496.

Van der Meer, L. G. H., 1993, The conditions limiting CO_2 storage in aquifers, *Energy Convers. Manag.* **34**(9–11):959–966.

Van Houten, F. B., and Bhattacharyya, D. P., 1982, Phanerozoic oolitic ironstones — geologic record and facies model, *Ann. Rev. Earth Planet. Sci.* **10**:441–457.

Veizer, J., 1973, Sedimentation in geologic history: recycling vs. evolution or recycling with evolution, *Contrib. Mineral. Petrol.* **38**:261–278.

Veizer, J., 1988, The evolving exogenic cycle, in: *Chemical Cycles in the Evolution of the Earth* (C. B. Gregor, R. M. Garrels, F. T. Mackenzie, and J. B. Maynard, eds.), Wiley, New York, pp. 175–261.

Veizer, J., 1995, Reply to comment by L. S. Land on "Oxygen and carbon isotopic composition of Ordovician brachiopods: implications for coeval seawater: discussion", *Geochim. Cosmochim. Acta* **59**:2845–2856.

Veizer, J., 2005, Celestial climate driver: a perspective from four billion years of the carbon cycle, *Geosci. Canada* **32**:13–29.

Veizer, J., and Mackenzie, F. T., 2004, Evolution of sedimentary rocks, in: *Sediments, Diagenesis and Sedimentary Rocks*, vol. 7 (F. T. Mackenzie, ed.), *Treatise on Geochemistry* (H. D. Holland and K. K. Turekian, eds.), Elsevier-Pergamon, Oxford, pp. 369–407.

Veizer, J., Ala, D., Azmy, K., Bruckschen, P., Buhl, D., Bruhn, F., Carden, G. A. F., Diener, A., Ebneth, S., Godderis, Y., Jasper, T., Korte, C., Pawellek, F., Podlaha, O. G., and Strauss, H., 1999, $^{87}Sr/^{86}Sr$, $\delta^{13}C$ and $\delta^{18}O$ evolution of Phanerozoic seawater, *Chem. Geol.* **161**: 59–88.

Veizer, J., Goddéris, Y., and François, L. M., 2000, Evidence for decoupling of atmospheric CO_2 and global climate during the Phanerozoic Eon, *Nature* **408**:698–701.

Ver, L. M. B., 1998, Global kinetic models of the coupled C, N, P, and S biogeochemical cycles: implications for global environmental change, Ph. D. Dissertation, Univ. of Hawaii, Honolulu, xxii + 681 pp.

Ver, L. M. B., Mackenzie, F. T., and Lerman, A., 1999, Biogeochemical responses of the carbon cycle to natural and human perturbation: past, present, and future, *Am. J. Sci.* **299**: 762–801.

Verardo, D. J., and McIntyre, A., 1994, Production and destruction: control of biogenous sedimentation in the tropical Atlantic, 0–300,000 years B. P., *Paleoceanography* **9**:63–86.

Vinogradov, A. P., and Ronov, A. B., 1956a, Composition of the sedimentary rocks of the Russian platform in relation to the history of its tectonic movements, *Geochemistry* **6**:533–559.

Vinogradov, A. P., and Ronov, A. B., 1956b, Evolution of the chemical composition of clays in the Russian platform, *Geochemistry* **2**:123–129.

Vogel, J. C., 1980, *Fractionation of the carbon isotopes during photosynthesis*, Springer-Verlag, Berlin, 29 pp. (Series *Sitzungsberichte der Heidelberger Akad. Wiss., Math.-Naturwiss. Kl., Jahrgang* 1980, 3. *Abh.*).

Vogel, J. C., 1993, Variability of carbon isotope fractionation during photosynthesis, in: *Stable Isotopes and Plant Carbon-Water Relations* (J. R. Ehleringer, A. E. Hall, and G. D. Farquhar, eds.), Academic Press, New York, pp. 29–46.

Vogel, J. C., Grootes, P. M., and Mook, W. G., 1970, Isotopic fractionation between gaseous and dissolved carbon dioxide, *Zeitschr. Physik* **230**:225–238.

Votintsev, K. K., 1993, On the natural conditions of Lake Baikal in connection with the development of its water quality standard, *Water Res.* **20**:595–604.

Wahlen, M., 2002, *Carbon-Isotopic Composition of Atmospheric CO_2 Since the Last Glacial Maximum*, National Snow and Ice Data Center, digital media, Boulder, Colo., http://nsidc.org/data/docs/agdc/nsidc0108_wahlen/index.html.

Walker, J. C. G., 1977, *Evolution of the Atmosphere*, Macmillan, New York, xiv + 318 pp.

Walker, J. C. G., 1986, Carbon dioxide on the early Earth, *Origins of Life* **16**:117–127.

Walker, J. C. G., and Opdyke, B. N., 1995, The influence of variable rates of shelf carbonate deposition on atmospheric carbon dioxide and pelagic sediments, *Paleoceanography* **10**:415–427.

Walker, L. J., Wilkinson, B. H., and Ivany, L. C., 2002, Continental drift and Phanerozoic carbonate accumulation in shallow-shelf and deep-marine settings, *J. Geol.* **110**:75–87.

Wallace, P. J., 1998, Water and partial melting in mantle plumes: inferences from the dissolved H_2O concentrations of Hawaiian basaltic magmas, *Geophys. Res. Lett.* **25**:3639–3642.

Wallmann, K., 2001, The geological water cycle and the evolution of marine $\delta^{18}O$, *Geochim. Cosmochim. Acta* **65**:2469–2485.

Wallmann, K., 2004, Impact of atmospheric CO_2 and galactic cosmic radiation on Phanerozoic climate change and the marine $\delta^{18}O$ record, *Geochem. Geophys. Geosyst.* **5**(6):Q06004, http://www.agu.org/journals/gc/.

Walter, L. M., and Morse, J. W., 1984, Reactive surface area of skeletal carbonate during dissolution: effect of grain size, *J. Sedim. Petrology* **54**:1081–1090.

Walter, L. M., and Morse, J. W., 1985, The dissolution kinetics of shallow marine carbonates in seawater: a laboratory study, *Geochim. Cosmochim. Acta* **49**:1503–1513.

Wang, S.-L., Chen, C.-T. A., Hong, G.-H., and Chung, C.-S., 2000, Carbon dioxide and related parameters in the East China Sea, *Continental Shelf Res.* **20**:525–544.

Wanninkhof, R., 1992, Relationships between wind speed and gas exchange over the ocean, *J. Geophys. Res.* **97**:7373–7382.

Ward, W. R., 1974, Climatic variations on Mars. I. Astronomical theory of insolation, *J. Geophys. Res.* **79**:3375–3395.

Ware, J. R., Smith, S. V., and Reaka-Kudla, M. L., 1992, Coral reefs: Sources or sinks of atmospheric CO_2, *Coral Reefs* **11**:127–130.

Warthmann, R., van Lith, Y., Vasconcelos, C., McKenzie, J. A., and Karpoff, A. M., 2000, Bacterially induced dolomite precipitation in anoxic culture experiments, *Geology* **28**:1091–1094.

Wayne, R. P., 2000, *Chemistry of Atmospheres*, 3rd ed., Oxford Univ. Press, New York, xxx + 775 pp.

WEC (World Energy Council), 2001, Peat, in: *19th Survey of Energy Resources*, London, http://www.worldenergy.org/wec-geis/publications/reports/ser/peat/peat.asp.

Wedepohl, H. K., 1995, The composition of the continental crust, *Geochim. Cosmochim. Acta* **59**:1217–1232.

Weiler, R. R., and Chawla, V. K., 1969, Dissolved mineral quality of Great Lakes waters, in: *Proceedings 12th Conf. Great Lakes Res.*, Internat. Assoc. Great Lakes. Res., Ann Arbor, Michigan, pp. 801–818.

Weiss, R. F., 1974, Carbon dioxide in water and seawater: the solubility of a non-ideal gas, *Mar. Chem.* **2**:203–215.

Weisstein, E. W., 2004, Stephan-Boltzmann law, *Eric Weisstein's World of Physics*—A Wolfram Web Resource, http://scienceworld.wolfram.com/physics/.

Weisstein, E. W., 2005, Sphere Packing, From *MathWorld*—A Wolfram Web Resource, http://mathworld.wolfram.com/SpherePacking.html.

Wenk, H. R., Barber, D. J., and Reeder, R. J., 1983, Microstructures in carbonates, in: *Carbonates: Mineralogy and Chemistry* (R. J. Reeder, ed.), *Rev. Mineralogy* **11**: 301–367, Mineralogical Society of America, Washington, D. C.

Wenk, H. R., Meisheng, H., and Frisia, S., 1993, Partially disordered dolomite: microstructural characteristics of Abu Dhabi carbonates, *Am. Mineral.* **78**:769–774.

White, A. F., 1995, Chemical weathering rates in soils, in: *Chemical Weathering Rates of Silicate Minerals* (A. F. White and S. L. Brantley, eds.), *Rev. Mineralogy* **31**:407–458, Mineralogical Society of America, Washington, D. C.

White, A. F., and Brantley, S. L., 2003, The effect of time on the weathering of silicate minerals: why do weathering rates differ in the laboratory and field?, *Chem. Geol.* **202**:479–506.

Whitmarsh, J., and Govindjee, 1995, The photosynthetic process, in: *Concepts in Photobiology: Photosynthesis and Photomorphogenesis* (G. S. Singhal, G. Renger, S. K. Soppory, K.-D. Irrgang, and Govindjee, eds.), Kluwer, Dordrecht, The Netherlands, pp. 11–51.

Whittaker, E. J. W., and Muntus, R., 1970, Ionic radii for use in geochemistry, *Geochim. Cosmochim. Acta* **34**:945–956.

Wigley, T. M. L., 1994, The contribution from emissions of different gases to the enhanced greenhouse effect, in: *Climate Change and the Agenda for Research* (T. Hanisch, ed.), Westview Press, Boulder, Colo., pp. 193–222.

Wilkinson, B. H., and Algeo, T. J., 1989, Sedimentary carbonate record of calcium and magnesium cycling, *Am. J. Sci.* **289**:1158–1194.

Wilkinson, B. H., and Walker, J. C. G., 1989, Phanerozoic cycling of sedimentary carbonate, *Am. J. Sci.* **289**:525–548.

Wilkinson, B. H., Owen, R. M., and Carroll, A. R., 1985, Submarine hydrothermal weathering, global eustasy and carbonate polymorphism in Phanerozoic marine oolites, *J. Sedim. Petrology* **55**: 171–183.

Willett, H. C., 1964, Meteorology, *Encycl. Brit.*, 1964 Ed., v. 15, Encyclopædia Britannica, Inc, Chicago, pp. 341–357.

Williams, R. S., and Hall, D. K., 1993, Glaciers, in: *Atlas of Earth Observations Related to Global Change* (R. J. Gurney, J. L. Foster, and C. L. Parkinson, eds.), Cambridge Univ. Press, Cambridge, pp. 401–422.

Willson, R. C., 1997, Total solar irradiance trend during solar cycles 21 and 22, *Science* **277**: 1963–1965.

Winn, C. D., Li, Y.-H., Mackenzie, F. T., and Karl, D. M., 1998, Rising surface ocean dissolved inorganic carbon at the Hawaii Ocean Time Series site, *Mar. Chem.* **60**: 33–47.

Winter, M. J., 2003, *Web-Elements Periodic Table*, Copyright 1993-2003, Mark J. Winter, University of Sheffield and WebElements, Ltd., U. K., http://www.webelements.com/.

Woese, C. R., 2000, Interpreting the universal phylogenetic tree, *Proc. Nat. Acad. Sci. USA* **97**:8392–8396.

Wolery, T. J., and Sleep, N. H., 1998, Interactions of geochemical cycles with the mantle, in: *Chemical Cycles in the Evolution of the Earth* (C. B. Gregor, R. M. Garrels, F. T. Mackenzie, and J. B. Maynard, eds.), Wiley, New York, pp. 77–104.

Wollast, R., 1994, The relative importance of bioremineralization and dissolution of $CaCO_3$ in the global carbon cycle, in: *Past and Present Biomineralization Processes: Considerations about the Carbonate Cycle* (F. Doumenge, D. Allemand, and A. Toulemont, eds.), *Inst. Océanogr. Monaco Bull., num. spéc.* **13**, pp. 13–34.

Wollast, R., 1998, Evaluation and comparison of the global carbon cycle in the coastal zone and in the open ocean, in: *The Sea: The Global Coastal Ocean*, v. 10 (K. H. Brink and A. R. Robinson, eds.), Wiley, New York, pp. 213–252.

Wollast, R., and Mackenzie, F. T., 1989, Global biogeochemical cycles and climate, in: *Climate and Geo-Sciences* (A. Berger, S. Schneider, and J.-C. Duplessy, eds.), Kluwer, Dordrecht, The Netherlands, pp. 453–473.

Wollast, R., Garrels, R. M., and Mackenzie, F. T., 1980, Calcite-seawater reactions in ocean surface waters, *Am. J. Sci.* **280**: 831–848.

Wood, B. J., Pawley, A., and Frost, D. R., 1996, Water and carbon in the Earth's mantle, *Phil. Trans. Roy. Soc. London*, ser. A, **354**:1495–1511.

Wood, J., and Long, G., 2000, *Long-Term World Oil Supply (A Resource Base/Production Path Analysis)*, U. S. Dept. of Energy, Energy Information Administration, Washington, D. C., 21 pp., http://www.eia.doe.gov/pub/oil_gas/petroleum/presentations/2000/long_term_supply/index.htm.

Woodhouse, S. C., 1910, *English-Greek Dictionary—A Vocabulary of the Attic Language*, George Routledge, London, viii + 1029 pp.

Woodwell, G. M., 1995, Biotic feedbacks from the warming of the Earth, in: *Biotic Feedbacks in the Global Climatic System: Will the Warming Feed the Warming?* (G. M. Woodwell and F. T. Mackenzie, eds.), Oxford Univ. Press, New York, pp. 3–21.

Woodwell, G. M., Mackenzie, F. T., Houghton, R. A., Apps, M., Gorham, E., and Davidson, E., 1998, Biotic feedbacks in the warming of the earth, *Climatic Change* **40**:495–518.

World Resources Institute, 2000, Marine jurisdictions: continental shelf areas, calculated from L. Pruett and J. Cimino, unpublished data, Global Maritime Boundaries Database (GMBD), Veridian—MRJ Technology Solutions, Fairfax, Virginia, http://earthtrends.wri.org/searchable_db/index.cfm.

Yapp, C. J., and Poths, H., 1992, Ancient atmospheric CO_2 pressures inferred from natural goethites, *Nature* **355**:342–344.

Yentsch, C. S., 1966, Planktonic photosynthesis, in: *Encyclopedia of Oceanography* (R. W. Fairbridge, ed.), Reinhold, New York, pp. 716–718.

Yoo, C. S., Cynn, H., Gygi, F., Galli, G., Nicol, M., Häusermann, D., Carlson, S., and Mailhiot, C., 1999, Crystal structures of carbon dioxide at high pressures: "superhard" polymeric carbon dioxide, *Phys. Rev. Lett.* **83**(26):5527–5530.

Yoo, C. S., Kohlmann, H., Cynn, H., Nicol, M. F., Iota, V., and LeBihan, T., 2002, Crystal structure of pseudo-six-fold carbon dioxide phase II at high pressures and temperatures, *Phys. Rev. B.* **65**:104103.

Yu, E. F., François, R., and Bacon, M. P., 1996, Similar rates of modern and last-glacial ocean thermohaline circulation inferred from radiochemical data, *Nature* **379**:689–694.

Zeebe, R. E., 2001, Seawater pH and isotopic paleotemperatures of Cretaceous oceans, *Paleogeogr., Paleoclimat., Paleoecol.* **170**:49–57.

Zeebe, R. E., and Wolf-Gladrow, D., 2001, *CO₂ in Seawater: Equilibrium, Kinetics, Isotopes*, Elsevier, Amsterdam, xiii + 346 pp.

Zenger, D. H., 1989, Dolomite abundance and stratigraphic age: constraints on rates and mechanisms of Phanerozoic dolostone formation, *J. Sedim. Petrol.* **59**:162–164.

Zhong, S., and Mucci, A., 1989, Calcite and aragonite precipitation from seawater solutions of various salinities: precipitation rates and overgrowth compositions, *Chem. Geol.* **78**:283–299.

Zondervan, I., Zeebe, R. E., Rost, B., and Riebesell, U., 2001, Decreasing marine biogenic calcification: a negative feedback on rising atmospheric pCO_2, *Global Biogeochem. Cycles* **15**:507–516.

Index